Kangaroos in Outback Australia

Kangaroos in Outback Australia

Comparative Ecology and Behavior of Three Coexisting Species

Dale R. McCullough
Yvette McCullough

Columbia University Press
New York

Columbia University Press
Publishers Since 1893
New York—Chichester, West Sussex

Library of Congress Cataloging-in-Publication Data

McCullough, Dale R., 1933–
    Kangaroos in outback Australia: comparative ecology and behavior of three
coexisting species / Dale R. McCullough, Yvette McCullough.
        p.   cm.
    Includes bibliographical references (p. 273)

    ISBN 978-0-231-11917-7 (pbk : alk. paper)

    1. Kangaroos—Australia—New South Wales.   I. McCullough,
Yvette.   II. Title.

QL737.M35 M34 2000
599.2′22′099449—dc21   00–24129

Printed in the United States of America

This book is dedicated to the late Graeme Caughley (1937–1994),

friend, colleague, and sponsor of this field project.

He was an exceptional scientist whose intellect

has had an indelible impact on wildlife ecology and management.

CONTENTS

PREFACE   ix

ACKNOWLEDGMENTS   xv

1. Introduction   1

2. Description of Yathong Nature Reserve   17

3. Methods   47

4. Population Size and Composition   80

5. Home Range and Movements   96

6. Activity Patterns   121

7. Social Grouping   152

8. Behavior   168

9. Feeding Ecology   200

10. Habitat and Niche Relationships   221

11. Predators and Competitors   233

12. Kangaroos, Ungulates, and Ecological
    and Evolutionary Models   249

LITERATURE CITED   273

INDEX   301

PREFACE

This book reports the results of research we conducted on kangaroos at Yathong Nature Reserve in western New South Wales, Australia. The study was conducted during the senior author's sabbatical leave from the University of California, Berkeley. Our goal was to study the behavior and ecology of kangaroos in the Australian outback. We were particularly interested in the eastern and western grey kangaroos, the closest ecological counterparts of the two North American deer species, the white-tailed and the mule deer, on which we have done considerable research. In addition to fulfilling our general interest in unusual life forms, we wished to compare the large Australian herbivores to their North American ecological equivalents in order to examine patterns in similarities and differences. The late Graeme Caughley of the Wildlife and Rangelands Research Division of the Commonwealth Scientific and Industrial Research Organization (CSIRO) in Canberra generously sponsored the project.

In May 1985, we arrived in Canberra with our two young children, Brent, age four, and Cheryl, one. Shortly afterward, anxious to begin, we were shepherded by Robin Barker of Graeme Caughley's lab into the outback to look at three potential study sites in western New South Wales: Kinchega National Park near Menindee, the site of a major kangaroo study reported by Caughley et al. (1987a), Willandra National Park near Hillston, and Yathong Nature Reserve, between Hillston and Cobar. Park or reserve status assured that the study would not be subsequently disrupted by a sheep station manager, and that domestic stock and station operations would not interfere with natural processes. Of course, this last desideratum could not be entirely met because all present National Parks and Reserves in western New South Wales are former sheep stations, and have an earlier history of sheep grazing and development for sheep management and production. Indeed, the present wide distribution and abundance of kangaroos is the result of water development, enhanced by conversion of woodland to grassland to favor sheep grazing and by control of dingoes. Additionally, introduced exotics, such as the European rabbit, fox, and feral goats and pigs, continue to be an unnatural influence on the landscape.

Kinchega and Willandra National Parks would have been the most favorable study sites for logistical reasons. They were close to human communities with materials, supplies, and social interactions, and they had accommodations with modern conveniences such as electricity, running water, and flush toilets. On the other hand, they had few eastern grey kangaroos, and those present were confined to heavily wooded riverbottoms. The landscape had virtually no habitat gradients because the river floodplain ended abruptly at a low bluff, demarcating the edge of the upland plain. There was relatively little diversity in the vegetation types or the topography.

Yathong Nature Reserve was far from civilization, being about 120 km by rough dirt road from the nearest community where one could obtain food, fuel, and supplies. The only modern accommodations were occupied, and the only remaining unclaimed structure was a dilapidated old cook shed at the former Shearers' Quarters that had no electricity, running water, or indoor plumbing. Shearers' Quarters houses the temporary crews that stay on a sheep station each year for a week or so and shear the flocks. Like most buildings in the outback, it was constructed from corrugated sheet metal over a cypress-pine frame set on wooden pilings.

What Yathong lacked in amenities, it made up for in biological opportunities. It had diverse habitats with sufficient gradients to allow choice by the kangaroos. There were good populations of eastern grey and western grey kangaroos and red kangaroos were abundant as well. Reds occur over much of the overlap zone of eastern and western grey kangaroos and provided a third species for community comparisons. A fourth species, the euro, was present on Yathong but in numbers too low to include in the study.

Yathong Nature Reserve proved to be a propitious choice for other, initially unrecognized reasons. It lies in a transitional area of particular ecological richness, where two great environmental forces—climate and topography—intersect. On a north-south axis, it falls near the center of overlap between northern summer and southern winter rainfall systems. On an east-west axis, it falls on the transition of the eastern mountains to the nearly level western shrublands. In this region, these elements converge to create a milieu of environmental variables found nowhere else. Yathong resembles a hybrid of four unrelated parental lines of climate and topography. It expresses some traits of each line, resembling one parent at one time and place and another at a different season or location. It is the site of a rich natural experiment where ecological and evolutionary processes, whose centers and origins are far away, converge to produce unique combinations. Here, three species of kangaroos, each with its exclusive domain lying far off in a different direction, come together. How they coexist promises rich biological lessons that will enlighten us as to why each species is successful in its exclusive range.

So, after gathering supplies and equipment in Canberra, we returned to Yathong to begin work at the end of May. We conducted continuous field work from June 1985 to July 1986. Stephen Thornton, our research assistant, continued selected studies through August and September 1986, and he did other follow-up work every six months through June 1988. We intertwined field work with renovation of the cook shed to render it habitable. The shed was cleared of animals (to the extent possible) and its windows and doors were replaced. Next followed cleaning and painting. Once linoleum was put down to stop the wind from whistling through the floor cracks, it wafted into the air like a parachute, illustrating the need for adhesive. A wood-burning stove was installed to ward off the freezing winter temperatures. Jon Hamblin, an American college student on summer break, arrived as part-time research assistant, part-time nanny. Shortly afterward, University Research Expedition Program volunteers arrived from the United States. Study methods were still in development, equipment had not arrived, and life was chaotic. Renovation halted so kangaroo work could proceed except when rain made travel impossible.

Gradually, the loose ends came together. Our major study methods were established, and the last of the volunteers and Jon Hamblin returned to the United States. Judy Graham, an Australian from Sydney, joined us as Jon's replacement. Our son Brent was enrolled in kindergarten with School of the Air, and the two-way radio and necessary school materials arrived. His teacher was 330 direct-line kilometers away in Broken Hill, a distance most schoolboys would envy. Thanks to his two-way radio and various school events, we came to know isolated families from all over western New South Wales and eastern South Australia. A propane system was installed to run the kitchen stove and refrigerator. Running water was piped to the house. On October 18, 1985, a memorable day, we turned the faucet on for the first automatic hot water shower. Electricity was installed soon after, although one had to start the generator before the switches would respond. The luxury of video movies suddenly was ours.

The kangaroo-catching team of CSIRO biologists and university students arrived from Canberra in December to place radio collars on kangaroos. One of the best catchers, Stephen Thornton, returned in January as a research assistant to collect most of the radio telemetry data.

Life and work at Yathong fell into a routine. With the arrival of summer came the heat and flies. Now a hot shower was the only kind available. "Swamp coolers" replaced the wood stove as the family gathering place, and the drone (or rattle, depending upon its state of repair) of the generator to keep the coolers going could be heard day and night. For relief from the heat, we swam in earthen water tanks, which, true to the Australian outback propensity for extremes, proved to be bone-chillingly cold.

We soon came to appreciate why many Australian researchers prefer commuting from Sydney or Canberra. The temperate zones of the world comply more or less with the Protestant ethic of everything in moderation. In the Australian outback, nothing happens in moderation—it is either too hot or too cold, too wet or too dry, and so on. One must endure stickers and spines that penetrate shoes and puncture tires, flies and bugs that crawl into eyes, nose, and mouth and become smeared on data sheets, poisonous snakes in the dunny (outhouse), appallingly large spiders everywhere, children ill with life-threatening dehydration, and other trials, large and small. Essential, or merely desirable, items are hundreds of miles away if they can be obtained at all. Yathong exacted its tribute, but it also paid due reward: wonderful kangaroos, fabulous showy birds, spectacular wildflower displays, horizon-to-horizon double rainbows, crystal-clear mornings, and salt-of-the-earth people. Yathong became home, a special place, far in distance and character from anything we had known, where the children could run free and life was lived with an intensity and immediacy usually buffered by the comforts and accoutrements of civilization.

This is an account of our studies of the kangaroos, and how their lives are shaped by a harsh and capricious climate in an immense and unforgiving landscape. For the most part—the exception being capture to fit collars for radio telemetry—our methods were passive. We intruded as little as possible in the lives of the animals. We assumed the role of students, and let the kangaroos instruct us through their selection of what to do, and when, where, and with whom to do it.

We hope this book will appeal to a wide range of readers with a common interest in kangaroos and the Australian outback. However, it is written from the North American perspective we brought to the study, and we expect that North Americans will constitute the major audience. We have used some colloquialisms for the interest of readers who may share our fascination with things Australian.

During this study, people in the outback frequently asked our opinions of the commercial shooting programs on kangaroos, an often contentious issue, both in Australia and the international conservation community. We do not address that issue here. When we first arrived in the outback, station owners assumed we must be agents for some environmental group. Why else would an American family, apparently normal in other respects, choose to live in the outback, ostensibly to study kangaroos? To any station manager who might chance to read this book, we say, "Yeah, mate, somebody is crazy enough to live in the outback for no reason other than to watch bloody kangaroos."

ACKNOWLEDGMENTS

Special thanks are due to the late Graeme Caughley of CSIRO Wildlife and Rangeland Research for the numerous ways he made it possible for us to conduct this study. He served as project sponsor, obtained funding and logistic support from CSIRO, and was instrumental in our obtaining myriad required visas, licenses, and permits. Besides funding, he furnished a four-wheel drive vehicle and a motorbike to the project. And, in Canberra, he and his family showed us gracious hospitality. Most importantly, he shared with us his extensive research experience on kangaroos.

What Graeme didn't do, his research group at Gungahlin (CSIRO headquarters near Canberra) did. Robin Barker was our "minder" (i.e., babysitter), and he took us on our first foray into the outback to select a study area. Thereafter, he facilitated the obtaining of supplies, the receiving and returning of equipment from the United States, the fulfillment of bureaucratic requirements, and he served as communication central for messages to and from Yathong and the rest of the world. Jeff Short, Bevan Brown, and David Grice shared their knowledge and experience, and they were instrumental in the great kangaroo-catching exercises. Other 'roo catchers included Grant Anderson, Mark Carson, Ross Carter, Jon Hamblin, Robert Palmer, Darren Smyth, and Stephen Thornton.

Others at Gungahlin who were helpful in sharing their knowledge, loaning equipment, or providing other favors included Brian Walker, Alan Newsome, John Calaby, Peter Catlin, Sue Briggs, and Vickie

Hodge. David Tongway and Ruth Windridge, of CSIRO Deniliquin, did the fecal nitrogen determinations, and David Spratt, of CSIRO Gungahlin, and Ian Beveridge from the Institute of Medical and Veterinary Science in Adelaide, identified parasites in fecal droppings.

New South Wales National Parks and Wildlife Service, Sydney, issued a license to do research on kangaroos and gave us permission to work at Yathong Nature Reserve. We are especially thankful to Neil Shepherd for his continuing support and help with the completion of legal requirements. He, David Priddel, and Graham Robertson shared their extensive experience studying kangaroos and working in the outback. Special thanks go to David Priddel for loaning us a radio telemetry receiver for several months while our equipment took a world tour, courtesy of air freight, and then disappeared despite Robin Barker's frantic efforts to find it. We suppose we should also thank the bloke at the warehouse in Sydney, who finally called Robin and asked, "Who is going to pay the storage on these boxes?" Others with National Parks and Wildlife who assisted us were John Giles, Geoff Ross, and Chris Dahlenberg of the Cobar office, and Judy Caughley, Janet Cohen, and Ross Bradstock, other researchers doing studies at Yathong. They generously shared their familiarity with the area and its plants and animals, and that was an enormous help; their company and friendship were also much appreciated.

At Berkeley, Lori Merkle, Jerry Morse, and William Carmen helped with logistics. University Research Expedition Program volunteers who helped with the field work included Sarah Armstrong, Caroline Brawner, Jan Davies, Kathleen DeVaney, Tim Foley, Mary Freiburg, Richard Freiburg, Helena Linde Gunderson, Constance Hefferan, Cynthia Hess, Benjamin Rinzler, Jean Simmons, Virginia Sims, Marcia Teitgen, Joan Young, and Patti Young.

Special appreciation goes to our field research assistants Jon Hamblin, Judy Graham, and Stephen Thornton. They became like family. Harry Warren, the manager at Yathong, assisted in innumerable ways, from construction to mechanical repair to kangaroo catching. He and his wife, Bev, were unending sources of information on ways to make do in the outback. They also taught us the nuances of outback etiquette. They and their children, Vanessa and Megan, completed our tight-knit resident community at Yathong. We will always remember and appreciate their generosity and kindness. John Warren, twin brother of Harry, worked on Yathong part of the time during our research, and he gave much assistance to the project. Neighbors on adjacent sheep stations, Rod and Marie Forsyth, and John and Janet Houghton, made us feel welcome, as did many families throughout western New South Wales and eastern South Australia we met through School of the Air. Ted Fryor, former owner of Yathong, gave us historical perspective on Yathong as a sheep station before it was acquired by National Parks and Wildlife. Ron Gibbins piloted the aircraft to locate the errant Scarlett O'Hara. We thank Paul Beier, Lyn Branch, Letitia Grenier, Steve Gu, Marni Koopman, Pam Muick, and Ben Sacks for their help with data analysis.

Lori Merkle, the senior author's secretary at Berkeley, was indispensable throughout the project—she did whatever needed to be done. She also did much of the word processing, often editing along the way. Margaret Jaeger also contributed greatly to editing and preparation of the manuscript, critically reading numerous drafts and locating, organizing, and verifying references. We are indebted to Elena Talamantez, Kathleen Jennings, Annie Pava, David Bise, Terry Bowyer, John Bissonette, Bruce Coblentz, Ian Hume, Peter Jarman, Graeme Coulson, Jeff Short, and Alan Newsome for reading various drafts. We give special thanks to Graeme Coulson, Peter Jarman, and Jeff Short, whose reviews were especially insightful thanks to their long experience with research on kangaroos. In addition, Jeff Short was our champion kangaroo catcher.

We owe special appreciation to Holly Hodder of Columbia University Press for her support and encouragement in bringing this book to completion. Her interest in kangaroos and confidence in the worth of this book were instrumental to its completion. We further thank Ron Harris of Columbia University Press for shepherding the book through the production process, and Maureen Jablinske, Lois Rankin, and Mark Smith of G&S Editors for improvements in the presentation.

We met all of the requirements for doing wildlife research in Australia. We had an Australian sponsor (CSIRO), a license from National Parks and Wildlife Service that authorized the use of Yathong Nature Reserve (with specific limitations), a humane treatment of animals review and approval by a committee in CSIRO, and a license to operate in our radio frequency range by the Australian Commonwealth Radio Commission.

We appreciate the encouragement and assistance of the many people who made this research and book possible. Any errors or misrepresentations, of course, are our responsibility alone.

Kangaroos in Outback Australia

1

INTRODUCTION

To compare it to any European animal would be impossible as it has not the least resemblance of any one I have seen. Its forelegs are extremely short and of no use to it in walking, its hind legs again as disproportionately long; with these it hops 7 or 8 feet at each hop in the same manner as the Gerbua, to which animal indeed it bears much resemblance except in size, this being in weight 38 lbs. and the Gerbua no larger than a common rat.   —Joseph Banks 1770 (Banks 1962)

A Most Unusual Animal

The kangaroo is indeed an odd animal. If it did not exist, who would imagine such a creature? It sits on its tail, hops on its hind legs, fights with its forepaws, and carries its young in a pocket. Kangaroos are among the first animals we learn as children, along with elephants, giraffes, rhinoceroses, and other oddities of the animal kingdom.

Even within the scientific community, the kangaroo was long regarded as a primitive aberration (Dawson 1977). Many biologists considered them dull and lacking fitness, as if surely the rules of natural selection had been suspended. Their reproductive biology is marsupialian. The fact that they give birth to minuscule, embryonic offspring, with most of their development occurring in an external pouch, inevitably labeled them as primitive. It was as if a mammalian fossil group continued to exist unchanged from paleontological times in the present. Surely this blueprint could work only in some remote and impoverished environment where the beast was isolated from competition with placental, herbivorous mammals with hooves.

Once the animals' uniqueness wore off, the first European settlers in Australia were not impressed by kangaroos. They were shy, retiring animals of no threat to anyone, despite their reported proclivity to disembowel a person with their hind feet. They were easily stalked and killed, did not attack when wounded, and carried no handsome trophies on their heads. Even today, there is no sport hunting of kangaroos beyond occasionally shooting them out of boredom. Initially, their meat was important to human settlers, but the abundance of introduced domestic livestock soon created an aversion to eating it. Today people living in the outback think there is something wrong in the head with anyone who would eat kangaroo. They say the meat is full of worms. In commercial hunting operations, kangaroo meat has been sold mainly as pet food, although in recent years there have been some marginally successful attempts to market it as a gourmet food for humans. The fact that the market for "gourmet kangaroo" is composed mainly of foreigners only confirms another firmly held opinion among outback folk.

This disparaging view of kangaroos is a distortion. Indeed, they do not have the sporting values or trophy structures of many ungulates. Few would deny that a history of competition with eutherian mammals would almost certainly have led to a different outcome. However, kangaroos have been evolving for a long time in an environment more extreme and demanding than that of most of the world; Australia is the driest continent on earth (Phillips 1977). Hardly primitive, their distinctive morphology represents extreme specialization. Their unusual pogo-stick–like hopping locomotion has been shown to be more energetically efficient than the four-footed locomotion of ungulates (Dawson and Taylor 1973; Dawson 1977; Baudinette 1994). Their temperature regulation in the face of heat is superior to that of most placental mammals (Dawson 1983). Marsupials have a lower body temperature and lower metabolic rate than similar-sized placental mammals, which is more efficient in a hot climate (Martin 1902; Dawson and Hulbert 1970; Dawson 1995), and also their needs for water are less (Denny and Dawson 1975). For example, Dawson (1995) estimated that kangaroos require only 10 percent of the water needed by domestic sheep. Also, their dietary nitrogen requirements are less than those of placental mammals—a distinct advantage during dry feed conditions (Brown 1968; Dawson 1995).

Their supposedly primitive reproductive biology allows them to produce young in a fraction of the time required by large eutherian mammals with long gestation periods, and to do so with lower prepartum cost to the female (Newsome 1975; Low 1978). This is no small advantage to an animal whose population is periodically decimated by drought and for which rapid population recovery is a priority. Furthermore, the same female can simultaneously produce different amounts and concentrations of milk to two young of different ages (Green et al. 1980; Green 1984; Merchant 1989).

Evolution of Kangaroos

Across most of the world, ungulates constitute the large-mammal herbivorous fauna, with two families, Cervidae and Bovidae, dominating in number of species and population sizes. The ascendancy of these two families has been relatively recent, for in earlier times horses, camels, mammoths, mastodons, ground sloths, and others held sway. Over evolutionary time, multiple experiments with successive models for converting plant tissues to animal flesh were tried and discarded, as species evolved, specialized, thrived, and became extinct. The large-mammal fauna we see today has a long history of developing diverse ways to garner a living from plants, in the face of counterselection on plants to avoid being eaten. And, of course, large and tasty herbivores fostered the evolution of an associated set of predators that the herbivores had to evade through alertness, guile, speed, weapons, or size.

Australia, an island continent with approximately the same land mass as the contiguous United States, was long isolated from these ungulate experiments by its remote location and oceanic barriers. It is an old continent, far removed from the uplifting forces of the junction of tectonic plates and worn nearly flat over geologic time. In this refugium, a separate set of experiments was under way (Keast 1981). Starting at the time of tectonic separation from Gondwanaland—about 55 million years ago, during the late Paleocene—evolution of a primitive marsupial stock in Australia's isolation produced a unique large herbivorous mammal solution—the kangaroo.

The family of mammals that includes kangaroos, the Macropodidae (big foot), is a diverse group of about 40 species that has radiated into a wide variety of niches and habitats (Flannery 1984, 1989). They range from rat-sized animals living in burrows, through the medium-sized wallabies, up to the large species, four of which are commonly called kangaroos: the red kangaroo (*Macropus rufus*); two sibling species, the eastern grey (*M. giganteus*) and western grey (*M. fuliginosus*) kangaroos; and the antilopine kangaroo (*M. antilopinus*). Two other large species (*M. robustus* and *M. bernardus*) are commonly called euros or wallaroos. Kangaroos, along with the Phalangeridae (brushtail possums and cuscuses; medium-sized, tree-dwelling marsupials), arose from a common ancestor in the mid-Miocene (Ride 1978; Flannery 1984, 1989). This ancestor is thought to have been a tree-dwelling form that fed on leaves and stems in the canopies of the extensive forests that covered most of Australia at that time, when the climate was much wetter. From the late Miocene through the Pliocene and into the Pleistocene, there was a drying of the climate, with expansion of grasslands and the corresponding decline of forests (Galloway and Kemp 1981). Aridity reached its extreme about 17,000 years ago (Archer 1984b).

Increasing aridity and expansion of grasslands led to a major radiation of macropodids through the late Miocene and Pleistocene (Bartholomai 1972; Archer 1984a; Archer and Clayton 1984). This radiation was characterized by enlarging body size and adaptation to the low-quality grass diet by development of foregut fermentation, analogous to ruminant digestion among eutherian mammals (Freudenberger et al. 1989). In many ways, the evolutionary experiment in Australia paralleled that in other parts of the world. At its peak, this diversity in Australia included many genera of huge mammals, including rhinoceros-like and lionlike marsupials, as well as *Palorchestes,* a bull-sized quadruped with a trunklike snout (Archer 1984a; Clemens et al. 1989). This fauna was equivalent to the megafauna extant at the same time in North America and Europe. Bartholomai (1975) proposed that the current eastern grey kangaroo was derived from a much larger ancestor, *Macropus titan,* recalling the downsizing elsewhere in modern times of a few Pleistocene survivors. Indeed, modern kangaroos arose late in evolutionary history (Dawson 1995).

The Pleistocene, with its alternating cycles of cold and warm, ice cap accumulation and melt, and shifting sea levels, was a time of major upheaval in Australia as well as in the Northern Hemisphere. Sea levels on the continent rose and fell repeatedly during the Pleistocene (Galloway and Kemp 1981). During the last glacial period, there was a cool, wet phase—about 45,000 to 25,000 years ago—and a dry, arid phase—about 25,000 to 13,000 years ago (Bowler 1978). During recurring dry, hot periods, the central area of Australia became extremely arid and unvegetated. The high-pressure systems associated with the hot climate resulted in strong winds and consequent dust storms (which occur today during periods of drought) that deposited wind-blown sand dunes over extensive areas of the nearly level landscape (Bowler 1980, 1983).

Extinction of much of the Australian megafauna occurred during these drastic oscillations in climate and sea level (Murray 1984, 1991). Burney (1993) noted that 19 total genera went to extinction, including 13 genera of large mammals. Only one genus of large mammals, *Macropus,* has persisted into recent times.

The genus *Macropus* includes the subgenus *Macropus,* which contains the closely related eastern and western grey kangaroos, and the subgenus *Osphranter,* which contains the more distantly related red kangaroo, which is aligned with the euro (*M. robustus*), antilopine kangaroo (*M. antilopinus*), and black wallaroo (*M. bernardus*). The euro is a heavy-bodied macropodid that lives in rocky, hilly areas distributed sporadically over eastern and central Australia (Poole 1983b). The antilopine kangaroo replaces the eastern grey kangaroo in the tropical woodlands of extreme northern Australia (Calaby 1983a), and the black wallaroo inhabits steep, rocky escarpments in one small area of central northern Australia (Calaby 1983b).

The eastern and western grey kangaroos are recently evolved sibling species (Kirsch and Poole 1967, 1972; Coulson 1997b). As aridity moved south across the continent, climatic extremes and fluctuating sea levels in the Great Australian Bight split the ancestral stock of eastern grey–like kangaroos, with part of the population going into the mesic eastern mountains and part into a refugium in southwestern Australia (Maynes 1989). While separated, these subpopulations diverged sufficiently that they did not interbreed upon secondary contact in more recent times. The eastern grey kangaroo was the stock already adapted to the mesic conditions of the eastern mountains. Most of the evolutionary change occurred in the southwestern refugium, a region that produced the western grey kangaroo. This southwestern refugium was important for the evolution of many other taxa as well (Merrilees 1984; Heatwole 1987; Main 1987; Clemens et al. 1989).

Clearly, the kangaroo blueprint has proved robust and adaptable to a suite of herbivore niches. It not only survived the climatic upheavals that caused the major extinctions of large mammals during the Pleistocene, but also withstood the predation pressure of a number of now extinct native marsupial predators, including the marsupial lion (*Thylacoleo*), a monitor lizard (Varanidae), and the thylacine wolf (*Thylacinus*). More recently, it survived the invasion of humans and the introduction of the eutherian dingo (*Canis familiaris dingo*) (Wells et al. 1982; Jarman and Coulson 1989; Robertshaw and Harden 1989). Some authors have attributed the extinction of the North American megafauna to overkill by invading humans (Martin and Klein 1984). Caughley (1987a) and Miller et al. (1999) attributed the Australian extinction of megafauna to the same cause. Whatever the cause—climate, overkill, or something else—the kangaroo, as an evolutionary model, survived the circumstances that proved fatal to a number of other evolutionary lines.

Since the arrival of Europeans in Australia, a number of the small and medium-sized macropods have become extinct or have been decimated (Poole 1979). However, kangaroos in Australia number about 20 million (Caughley et al. 1983). In fact, kangaroos have increased in the face of competition from millions of sheep and cattle, despite the help to live-stock and hindrance to the kangaroos by the graziers (Calaby 1971; Caughley et al. 1987a; Edwards 1989; Calaby and Grigg 1989; Dawson 1995). Controlled pasture grazing studies with domestic sheep and red kangaroos have shown that kangaroos are successful competitors (Edwards 1990; Dawson and Ellis 1994; Edwards et al. 1996). As with the white-tailed deer in North American parks (McShea et al. 1997), the eastern grey kangaroo in eastern Australian parks and reserves has become overabundant, with attendant management problems (Coulson 1998). Considering this success, one wonders how the idea of kangaroos' supposed inferior competitive ability with cervids and bovids got started in the first place.

Overview of Kangaroo Studies

Logistics are difficult in the harsh and lonely outback, characterized by climatic extremes and immense distances. Most kangaroo studies are done by commuting to the field from Canberra, Sydney, or other population centers. Early studies of kangaroos concentrated on reproduction (Kirkpatrick 1965b; Clark and Poole 1967; Kirsch and Poole 1972; Poole and Catling 1974), age determination (Kirkpatrick 1965a, 1965b; Dudzinski et al. 1977; Newsome et al. 1977), food habits (Griffiths and Barker 1966; Bell 1973; Griffiths et al. 1974; Taylor 1983), habitat selection (Bell 1973; Kaufmann 1974a; Caughley et al. 1977; Taylor 1980; Hill 1981, 1982; Johnson and Bayliss 1981), group size and composition (Caughley 1964; Kirkpatrick 1966; Grant 1973; Kaufmann 1974a, 1975; Taylor 1982; Johnson 1983a), and population size and structure (Wilson 1975; Caughley et al. 1977; Caughley and Grigg 1982; Short et al. 1983).

The book by Graeme Caughley and his colleagues (1987a) greatly advanced our knowledge of kangaroos in the wild. Extensive field studies by Peter Jarman, David Croft, Terry Dawson, Graeme Coulson, and their students and colleagues have substantially increased our knowledge of the natural ecology and behavior of kangaroos. Massive books (Strahan 1983; Dyne and Walton 1987; Walton and Richardson 1989; Grigg et al. 1989; Dawson 1995) have extended, summarized, and synthesized our understanding of kangaroos and their relatives. A particularly compact and cogent coverage that gives a rapid overview of what kangaroos are all about is Hume et al. (1989). In concert, these studies should help to dispel the reputation of kangaroos as primitive oddities and allow these animals to be recognized for what they are: highly specialized herbivores that have adapted to a harsh and unpredictable environment. Their ecology and behavior are of interest equal to that for any ungulate. Studies of the behavioral ecology of kangaroos also have increased. Ganslosser (1980) reviewed the behavioral literature on kangaroos up to 1979, and field studies were conducted by Coulson (1983), Croft (1983), Jaremovic (1983), Jarman (1983a), Johnson (1983b), and Priddel (1983). Descriptions of stereotyped behaviors and the behavioral repertoire were provided by Sharman and Calaby (1964), Russell (1970a, 1970b), Coulson (1983, 1997b), Croft (1983), and Ganslosser (1995).

Early work inferred little social organization in kangaroos because of their small group sizes and inconsistent associations. Similar conclusions might have been reached about deer from the same kind of evidence, but study of individuals showed that deer social organization was intricate, including dominance hierarchies in males and a matrilineal structure in females (Hawkins and Klimstra 1970; Hirth 1977; Brown and Hirth 1979; McCullough 1979; Ozoga et al. 1982; Nelson and Mech 1981). Recent studies pointed to a more complex social organization for kangaroos. Dominance hierarchies were reported by Russell (1970a) and Grant (1973). Kirkpatrick (1966) and Kaufmann (1974a, 1974b, 1975) reported organized social units of 20 to 25 individuals ("mobs") in eastern grey kangaroos with little interchange between mobs; Jaremovic (1983, 1984) and Jarman and Taylor (1983) supported this conclusion. Dawson (1995) suggested that female associations may be based on matrilines. Segregation of age and sex classes by social group and habitat was reported for eastern grey kangaroos by Kirkpatrick (1966), Grant (1973), and Jaremovic (1983), for western greys by Johnson (1983a), and for red kangaroos by Johnson and Bayliss (1981) and Johnson (1983a). Coulson (1983, 1990) reported habitat separation in the field between eastern and western grey kangaroo groups.

In earlier times, kangaroos were considered largely nomadic, a belief based partly on folklore and partly on scattered natural history observations. This view seemed consistent with the occurrence of aggregations (called "mobs" as well) of kangaroos at favorable food patches during poor seasons or droughts (Newsome 1971; Low et al. 1973), and long-distance movements of some marked individuals (Bailey 1971).

There are two possible explanations for formation of the large "mob" aggregations of macropods. First, they may be a socially organized grouping of related individuals of one or both sexes. Descriptions of mobs as nonoverlapping, noninteracting entities, with little exchange between mobs in eastern grey kangaroos (Kaufmann 1975; Jaremovic 1983, 1984; Jarman and Southwell 1986) would be consistent with this hypothesis. Johnson (1986) found that kinship played a role in female groups in the red-necked wallaby (*M. rufogriseus*), but thus far, kinship has only been inferred in kangaroos (Dawson 1995). The second explanation is that mobs are temporary aggregations of unrelated individuals around favorable resource patches during unfavorable resource periods (Caughley 1964; Newsome 1971; many others). The appearance of non-overlap of mobs and little exchange between mobs might simply be a function of the patchy distribution of favorable resources and the likelihood that a given individual or mother-offspring social unit would use only one or a few patches in close proximity.

Dispersal patterns obviously influence the relatedness of mobs. Because of the polygynous breeding system, one would expect dispersal to be predominantly by young males (Greenwood 1980; Dobson 1982), with females showing much higher philopatry, and this appears to be true (Johnson 1989; Stuart-Dick and Higginbottom 1989). Although Jarman and Taylor (1983) reported long-distance movements of adult female eastern grey kangaroos, and Bailey (1971) found no clear pattern of long-distance movements in red kangaroos, more recent work supports male bias in dispersal (Norbury et al. 1994; Edwards et al. 1994).

Group size is correlated with population density (Taylor 1982; Johnson 1983a; Southwell 1984). However, Jarman and Taylor (1983) reported relatively small home ranges of eastern grey kangaroos and wallaroos. Studies of eastern grey (Jaremovic 1983) and red kangaroos (Priddel 1987) showed that individual home ranges for these species were less than 10 km^2 in size, and aggregations were the function of redistribution of local individuals within their normal home ranges. Overall, these results indicated that the behavioral ecology of kangaroos was considerably more complex than had been thought.

The Nature of this Study

We had one advantage over Australian researchers—we were seeing kangaroos and their environment with new eyes. We had little in the way of preconceived notions, and nothing in the way of prior experience, this being our first trip to Australia. Our different perspective allowed us to note some things that Australians take for granted. Our perspective also differed in that it had been shaped by years of study of North American ungulates.

We employed a comparative approach to examine the community structuring of the three species of kangaroo and contrast their behavior and ecology on Yathong Nature Reserve (fig. 1.1). We hypothesized that the three species were sorted out by niche relationships among habitat and physical environmental gradients over space and time. Consequently, what would have been simply descriptive work if done on a

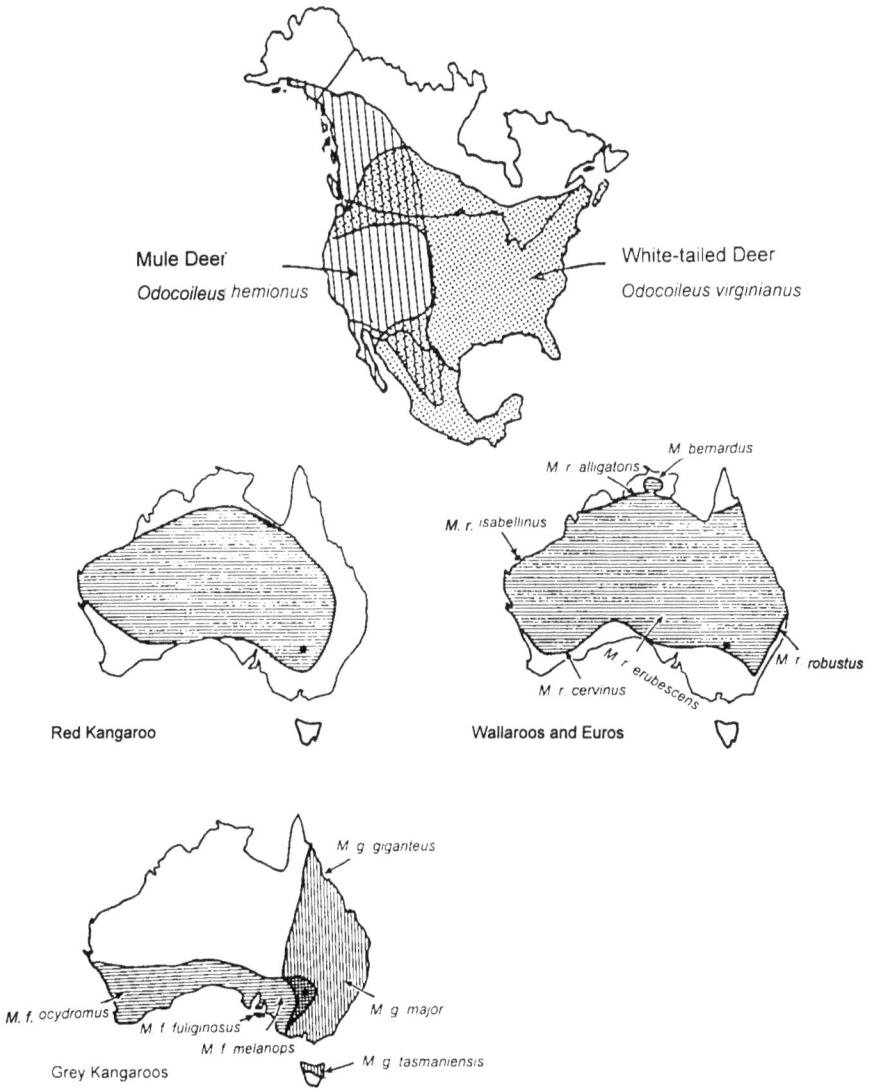

Figure 1.1. Distribution of odocoileids in North America (above, after Taylor 1956; Wallmo 1981; Baker 1984) and kangaroos (including the euro) in Australia (below, after Frith and Calaby 1969; Caughley 1987a). The dot indicates the location of Yathong Nature Reserve.

single species became analytical in the comparison of similarity and contrast between species in habitat use, spatial distribution, home range characteristics and movements, activity patterns, responses to weather, and other variables. Each of these aspects can be considered as a single dimension of the niche, according to the concept of the niche being an *n*-dimensional hyperspace (Hutchinson 1957).

Comparison of Kangaroos with Deer

Workers such as Kaufmann (1974a) and Jarman (1983a) have utilized the comparative approach broadly across the macropodid family; this can be extended to a comparison between kangaroos and their ecological equivalents, ungulates, which occur in other biological communities across the world (Eisenberg and Golani 1977; Jarman 1983b). Indeed, the original motivation for this study was to compare and contrast the behavioral ecology of kangaroos with their ecological counterparts in North America—deer of the genus *Odocoileus*.

There are two species of this genus in North America, the white-tailed deer, *O. virginianus*, primarily eastern in distribution (but with range extensions into the west) and the mule deer, *O. hemionus*, primarily western in distribution (fig. 1.1). There is an enormous literature on these deer. A number of hypotheses to account for the evolution of social organization, breeding systems, and habitat selection have been advanced for deer (McCullough 1979) and other ungulates (for example, Jarman 1974; Geist 1974), and for ungulates and kangaroos (Jarman 1983b). These hypotheses propose that environmental characteristics shape the behavior and ecology of large herbivores along predictable lines. Evidence in support of these hypotheses rests primarily on correlation. Cause and effect relationships can only be inferred. Experimental manipulation, the usual means by which cause-effect relationships are confirmed, is usually not feasible in studies of large, highly mobile mammals.

As an alternative, the comparative approach, which has proven useful in evolutionary studies, offers considerable potential for ecological studies to further test these hypotheses, because the same environmental variables can be examined for animals with independent evolutionary histories. Kangaroos are functional equivalents to deer. They occupy habitats grossly similar to those of North American deer. The two are similar in body size, foregut fermentation capability, and use of rapid locomotion to escape danger. On the other hand, kangaroos contrast with deer in phylogenetic origin, morphological form, gait, and reproductive system.

If the paradigms used to explain behavioral and ecological systems of North American deer (and other ungulates) are robust, and if the eco-logical variables as selective forces producing them are as strong as be-lieved, then similar conditions of environment (e.g., habitat structure, predation pressure, intra- and interspecific competition) should produce systems of behavior and ecology in kangaroos similar to those of deer (or other ungulates with comparable environments), despite their differ-ent phylogenetic origin and morphology. Conformance of kangaroos to hypotheses based on the ungulate model will strengthen the hypotheses, whereas partial support may suggest ways in which the hypotheses can be altered to be more robust. Conversely, failure of kangaroos to con-form to hypotheses based on ungulates must cast doubt on the validity, or at least the generalizability, of these hypotheses.

To maximize the power of the comparison, the closest counterparts to white-tailed and mule deer among the Australian macropods were selected for study. These counterparts are the eastern grey kangaroo, which is eastern in distribution, and the western grey kangaroo, which is primarily southern in distribution (fig. 1.1). In both deer and grey kangaroos, the eastern species lives predominantly in the more mesic habitats, characterized by deciduous forest, in which the prime habitat is secondary successional patches where the mature forest has been dis-turbed. The western species occurs in more xeric habitats with less forest canopy, more brush, and generally more open environments. In both deer and grey kangaroos, the eastern and western species have large areas of nonoverlapping distribution, but are sympatric over parts of their range (Caughley et al. 1984b). In areas of sympatry, the species are expected to sort out by habitat and other niche variables.

Coexisting with the eastern and western grey kangaroos is the red kangaroo, which has an extremely wide range over the Australia's arid openlands (fig. 1.1). There is no close analog to the red kangaroo among North American ungulates. The nearest counterpart is probably the elk, *Cervus elaphus*, a habitat generalist whose original distribution over-lapped much of the range of both white-tailed and mule deer, and cur-rently coexists in areas of sympatry with deer of both species in the northern Rocky Mountains (Thomas and Toweill 1982). Also coexisting with the red kangaroo, with virtually the same range, is the euro or wal-laroo, which has adapted to rocky hills (fig. 1.1). Unfortunately, their numbers were too sparse at Yathong, and their confinement to rocky areas made comparative work unfeasible by the methods used.

The principal original predators of North American deer were the wolf (*Canis lupus*) in the east, and the wolf, mountain lion (*Puma con-color*), and (mainly on young and infirm animals) the coyote (*Canis la-trans*) in the west. The principal predators of kangaroos were the Tasmanian wolf, marsupial lion, and more recently, the dingo (Corbett and Newsome 1987; Jarman and Coulson 1989; Robertshaw and Harden 1989). Thus, the ecological counterparts of the coursing (marsupial wolf, dingo) and ambush (marsupial lion) predators were present in Australia. Aboriginal hunters were important predators on both kangaroos and deer (Young, 1956; Tunbridge 1988, 1991; McCabe and McCabe 1984), as are modern humans (Spencer 1956; Stransky 1984; Shepherd and Caughley 1987; Dawson 1995).

It has long been recognized that macropod species are not randomly distributed over habitats, and on a gross scale these separations are understood. The eastern grey kangaroo (fig. 1.2) occurs in mesic deciduous forests in the more uniform rainfall area of eastern Australia (Taylor 1980; Hill 1981, 1982). It reaches highest densities in seral stages, where disturbed forest sites produce favorable combinations of food (primarily understory grasses) and concealment cover (primarily woody shrubs in densities low enough to allow movement) (Kaufmann 1974a, Taylor 1980; Hill 1982). Surrounding forests provide shade for thermal cover, and the canopy ameliorates the microclimate (Griffiths and Barker 1966; Bell 1973). Free water is relatively available, and eastern grey kangaroos associate with water and drink relatively frequently (Caughley 1964). These habitat characteristics are virtually identical to those of the white-tailed deer, with the notable exception of the high prevalence of grasses in the eastern grey kangaroo diet. White-tailed deer are highly selective of diet for quality (McCullough and Ullrey 1985) and consume grasses primarily in green, succulent growth (McCullough 1985; Beier 1987).

Toward the western end of its distribution, the eastern grey kangaroo follows river courses where thick forests and water are available (Kirsch and Poole 1972; Caughley et al. 1977, 1984b), as does the white-tailed deer in the northwestern extension of its distribution in the northern Rocky Mountains (Y. McCullough 1980) and Oregon (Gavin 1978; Gavin et al. 1984).

The western grey kangaroo (fig. 1.3) has a more southern distribution, inhabiting the Mediterranean climate zone (Short et al. 1983; Caughley et al. 1984b, 1987b) where rainfall is seasonal (occurring primarily in the winter) and more variable in amount. This species occupies open woodland and shrub habitats that are typically more mixed (Short et al. 1983; Arnold et al. 1994), generally more xeric, and with less free water available than those of the eastern grey kangaroo. The diet is predominantly grasses, but includes a greater mixture of forbs and browse. Available concealment cover is relatively heavy.

The red kangaroo (fig. 1.4) is widely distributed over most of Australia except for the forested eastern mountains (Frith and Calaby 1969).

A.   B.

Figure 1.2.   Eastern grey kangaroos. Large adult male (A) and adult female (B).

A.   B.

Figure 1.3.   Western grey kangaroos. Large adult male (A) and adult female with large pouch young (B). Note the dried seed stalks protruding from the mouth of the male, and the split right ear of the female.

It occupies arid habitats over which it moves extensively in search of high-quality, succulent grasses (Bailey 1971; Newsome 1971). The euro is also widely distributed over Australia but is adapted to use rocky cliffs for thermal cover; in northwestern Australia, it can also survive away from such areas if other resources are available (Ealey 1967).

The three species of macropodids coexisting as a community at Yathong feed largely on grasses (Griffiths and Barker 1966; Griffiths et al. 1974; Taylor 1983), but there appears to be a difference in the quality of grasses consumed. Eastern grey kangaroos consume coarse grasses and

Figure 1.4. Red kangaroos. Large adult male (A) and adult female with large pouch young (B). Note bimble box vegetation type behind the adult male.

have a well-developed foregut fermentation system (Taylor 1980, 1983). The red kangaroo is selective for high-quality grasses (Newsome 1971, 1977a), in contrast to the usual relationship of diet quality and body size. Newsome (1980) reported diet differences between the sexes in the red kangaroo. The western grey kangaroo feeds on coarser grasses and has a more developed fermentation system than the red kangaroo (Griffiths and Barker 1966), but the difference in diet and fermentation system between the eastern and western grey kangaroos is relatively poorly known.

Compared to odocoileids, and ungulates in general, macropods show extreme sexual dimorphism in body size and form (Jarman 1989; Weckerly 1998). For the macropods, females are about 56 percent of male mass (Weckerly 1998). Both sexes continue to grow throughout their lives, but females grow slowly after about five years of age (Poole et al. 1982a, 1982b; Jarman 1989; Edwards 1990; Moss and Croft 1999). Females are only 48 percent (Poole et al. 1982a) or 53 percent (Jarman 1989) of dominant male weight in eastern greys, 43 percent (Jarman 1989) or 51 percent (Poole et al. 1982b) in western greys of the subspecies (*M. f. melanops*) sympatric with the eastern grey, 40 percent (Jarman 1989) or 50 percent (Frith and Calaby 1969; Newsome 1977b) in red kangaroos. At Yathong, males are so much larger than females that their head height while standing on all fours is as high or higher than the head height of large, adult females standing on their hind legs and tail (fig. 1.5). The Kangaroo Island subspecies of the western grey kangaroo (*M. f. fuliginosus*) has extreme dimorphism, with females being only

Figure 1.5. Adult red kangaroos showing great dimorphism between large adult male (rear) and two adult females. The larger males are as tall standing on all four legs as females sitting up.

33 percent the weight of dominant males (Poole et al. 1982b). By comparison, females are about 65 percent of adult male weight in mule deer (subspecies *O. h. hemionus*) (Anderson et al. 1974), 62 percent in black-tailed (subspecies *O. h. columbianus*), and 77 percent in white-tailed deer (D. R. McCullough, unpublished data).

Predation on macropods over the last several thousand years has been primarily by a coursing predator, the dingo, and by aboriginal humans. Currently, humans are the predominant kangaroo predator, particularly through commercial harvest (Shaw 1979; Anderson 1980; Shepherd and Caughley 1987; Dawson 1995). Despite severe reductions in their numbers through control programs and a dog-proof fence around New South Wales, the dingo remains a significant predator of kangaroos in many areas (Caughley et al. 1980; Shepherd 1981; Corbett and Newsome 1987; Robertshaw and Harden 1989; Jarman and Coulson 1989). Eastern grey and red kangaroos flee as groups when disturbed (Caughley 1964; Kaufmann 1974b). Kangaroos in large groups spend much less time alert than those in small groups (Southwell 1976, 1981; Jarman 1987; Colagross and Cockburn 1993; Coulson 1999), as has also been reported for mule deer (Bowyer 1984).

The broad goal of the study, therefore, was to explore community relations among kangaroos. More specifically, our objectives may be summarized as follows:

1. To determine if the coexisting species of kangaroos segregate along one or more dimensions of ecological niche.
2. To compare behavior and ecology between the three species of kangaroos, as well as generally with ungulates, and more specifically, with deer of the genus *Odocoileus,* in order to examine convergence or divergence that may be instructive about the structure and function of large herbivore communities.
3. To determine if greater sexual dimorphism in kangaroos is the product of sexual selection, with large dominant males accounting for a major share of paternity in a highly polygynous breeding system.
4. To determine if group formation in macropods has been ultimately selected by predation pressure, with food quality and patchiness as proximal factors.

2

DESCRIPTION OF YATHONG NATURE RESERVE

Location

If one flew in a small plane from Sydney, starting at sea level on the east coast, on an azimuth slightly north of west (285°) through Yathong Nature Reserve, one would experience the essential topography of the Australian continent in a few hours (fig. 2.1). In a few minutes, one would traverse the narrow (50 km wide) coastal plain, and in 80 km, cross the forested crest of the Blue Mountains at about 1000 m above sea level. The Blue Mountains form part of the Great Dividing Range, the continental divide; this lies within 100 km of the eastern seaboard, and beyond here, the rivers drain westward. In another 30 km, one would descend the west slope of the mountains, and pass over the rolling hills and small villages of the croplands zone, the breadbasket of Australia, which produces most of the wheat and sheep. The rolling hills gradually dampen out in the next 130 km. At this point, some 320 km from Sydney, the flat plains extend as far as the eye can see. Indeed, give or take a low hilly range here and there, the plains continue along this transect some 3780 km to the Indian Ocean on the northwest coast.

Continuing west over the plains, the landscape becomes progressively drier, the fields more scattered, and the sheep fewer. Near the small town of Lake Cargelligo, some 450 km from Sydney, the fields disappear for the most part, but the sheep continue. At Lake Cargelligo, one crosses the Lachlan River; this and Lake Cargelligo are the last natural permanent water sources until the Darling River some 400 km beyond. Lake Cargelligo also marks the end of the paved road.

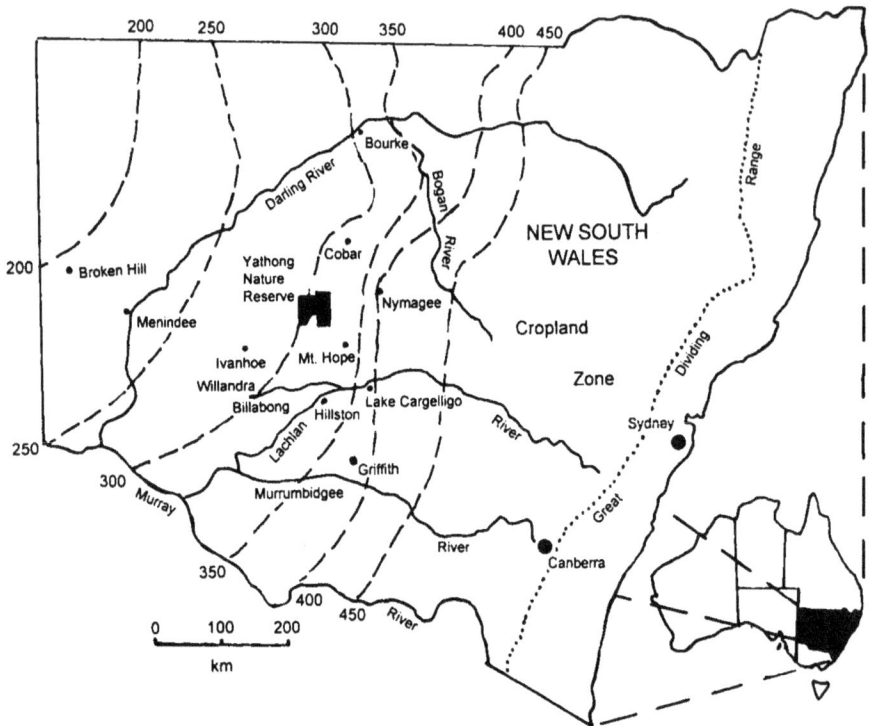

Figure 2.1.   Map of New South Wales showing the location of Yathong Nature Reserve, Great Dividing Range, major river courses, and towns mentioned in the text. Rainfall isohytes at 50 mm intervals are indicated for the western part of the state.

A few minutes past Lake Cargelligo, on the distant horizon, some low ridges rise about 200 m above the plain. Situated on the near ridge is Mt. Hope, previously a booming gold mining town of 3000 miners and their families, but now reduced to a few scattered buildings, the principal of which is the Mt. Hope Pub, watering hole for truck drivers on the dusty, rutted dirt road between Hillston in the south and Cobar, a copper mining town, to the north. The second ridge, 130 km from Lake Cargelligo and 580 airline km from Sydney, marks the eastern edge of Yathong Nature Reserve.

The vast hot and dry interior of Australia is variously called "the bush," "beyond the black stump," "back o' Bourke," but most commonly, "the outback." It is not unknown that children born on isolated sheep stations in the outback reach school age before having experienced rain. A folk ballad, quoted by Caughley et al. (1987a), sums up the outback well:

Where the girls and grass are scanty;
Where the creeks run dry or ten feet high,
And it's either drought or plenty.

Yathong Nature Reserve is a 1190 km² — half the size of Rhode Island — natural area in the Western District of New South Wales. Yathong is located nearly equidistant between Cobar to the northeast, Hillston to the south, Lake Cargelligo to the southeast, and Ivanhoe to the southwest (fig. 2.1). It is 120 km by dirt road from the nearest small town. Yathong is protected by the New South Wales National Parks and Wildlife Service. Other lands in the Western Division (approximately that area west of the 400 mm rainfall isohyte shown in fig. 2.1) are state-owned but held by individuals under 99-year leases. Whereas these lands are legally public property, in practice they are occupied, passed down in the family, and sold as if they were private property. The graziers on the sheep stations certainly consider themselves owners of the land, and trespass without permission is a serious breach of outback etiquette.

History

Aboriginal peoples used the Yathong area prior to European settlement, and signs of their previous occupation are still present. A small cave adjacent to a spring in the main ridge shows faint signs of paintings, and several other caves on surrounding sheep stations are said to have well-preserved paintings. Large old bimble box (*Eucalyptus populnea*) trees in the Yathong Homestead area bear scars where large pieces of bark were removed to make carrying pouches, although some scars may be due to early European graziers using the bark to construct sheep-watering troughs. Most people believe that Yathong was not permanently occupied by Aborigines, but rather, was probably visited during favorable seasons by small hunting and gathering parties.

Europeans moved into the area via two routes: one by crossing the Blue Mountains and moving westward to settle the croplands, and the other by graziers moving their sheep up the Darling River to the west and north of Yathong. There was no permanent water. Yathong lies at the center of one of the last regions of New South Wales to be settled by Europeans. Other than a small spring in Yathong's main ridge that dries to a trickle in dry seasons, the nearest permanent water is Willandra Billabong, a branch of the Lachlan River, 65 km airline distance to the south. The Bogan River, a branch of the Darling River, is 160 km to the northeast, and the Darling River is 240 km to the northwest.

The first sheep station that included Yathong was established around 1860. It was a vast area extending to Ivanhoe, some 125 airline km to the west. Sheep grazing on Yathong would have occurred during "greenup" periods following rains, when high moisture content of the vegetation removed the animals' need for free water. Because the original station was so large, there usually was green forage from localized rain storms available somewhere on the property, and sheep could be shifted about to reduce the impact of unpredictable weather. Only during more extended droughts would the entire station dry up.

Permanent occupation of Yathong around the turn of the century followed the development of a technique to construct earthen water tanks. These were created by dredging out a large square hole, like an inverted pyramid, in a depression and then cutting radiating ditches to funnel rainwater into the tank. Occasionally, shallow drainages were utilized, and the tank was also partly a dam. Larger drainages were not used because they carried too much water, and earthen dams were easily washed out because the typical red soil could not be consolidated. Drainages were usually not present in any case. On sheep stations, one or more earthen tanks was constructed for each separately fenced paddock. Presently, there are 31 earthen tanks on Yathong Nature Reserve.

The vast sheep stations of the late 1800s were divided into progressively smaller parcels following World War I (around 1920) and World War II (late 1940s). These subdivided station properties were distributed by lottery among returning veterans. Most of the stations have since changed hands, but at the time of this study, Rod Forsyth of Stanniford Sheep Station, adjacent to the northeast boundary of Yathong, still lived on the property he acquired following World War II.

From the postwar period through the 1970s, most sheep stations did well. The international market for sheep and wool was strong and prices were high. The predominant problem facing the graziers was highly unpredictable rainfall. Rains were necessary not only to rejuvenate the forage, but to replenish the earthen water tanks. It takes a substantial rainfall to create sufficient runoff to put water into tanks; they are replenished infrequently and irregularly, and commonly go dry before a sufficient rainstorm refills them. It is local storms, not regional weather patterns, that usually recharge tanks, and these sweep across the land, filling a few tanks in their path without influencing the rest. During extended droughts, all of the tanks dry up.

Some stations have bore holes (wells) that serve livestock as emergency water supplies. However, even if sufficient water can be delivered from bore holes, the forage supply is quickly exhausted. Sheep must be moved, artificially fed, or sold, or they will starve to death. Because extended droughts tend to be regional, bringing the sheep to feed, or feed to the sheep, is usually prohibitively expensive. Furthermore, because large numbers of graziers have sheep to dispose of during a drought, the market price is invariably depressed.

Economic survival on a sheep station is difficult. Often, graziers in the Yathong area either have too few sheep to eat all of the grass when the rains come, or they have too many to carry through a drought. After a drought, by the time they have bred up the numbers of sheep (purchasing sufficient stock from elsewhere is usually beyond their means), the next drought is liable to hit. And graziers are once again faced either with disposing of stock at unfavorable prices or investing in carrying it over until the rains come. Gambling on rainfall in outback Australia is a loser's game.

As if natural processes were not difficult enough to deal with, inflation in fuel and equipment costs, high labor costs, and depressed lamb and wool prices on the world market due to overproduction have combined to put additional pressure on graziers. Most stations can no longer afford to hire even an unskilled jackaroo (workman). More commonly, a nanny is hired to care for the children, and the wife goes to work in the paddocks. Many graziers are being forced to sell out and leave the land.

The New South Wales National Parks and Wildlife Service acquired Yathong Nature Reserve in 1978, and since then has expanded their holdings in the surrounding area, with the intent to create a Mallee National Park. These acquisitions have caused considerable resentment among local people for whom the National Parks and Wildlife Service makes a convenient scapegoat for the much greater forces that conspire against the independent grazier.

Yathong Nature Reserve (fig. 2.2) consists of three former sheep stations—Yathong, Irymple, and Glenlea. It is bounded on the south by a New South Wales State Forest and on the other three sides by private sheep stations. It is surrounded by a four-strand, smooth-wire sheep fence that prevents movement by sheep but is a trivial barrier to kangaroos, which are adept at crawling under, through, or jumping over wires. Dilapidated fences in the interior of Yathong are gradually being removed, and the posts burned as firewood.

Nature Reserves in Australia are dedicated to conservation of, and research on, the natural environment and biota. Unlike National Parks, they are not open to public access. However, two shire dirt roads (much like U.S. county roads) on Yathong, Gilgunnia on the eastern boundary and Belford in the center, pass through the reserve from north to south. Throughout this study, most use of the reserve roads was by the resident manager or researchers working in the area; through traffic on Belford

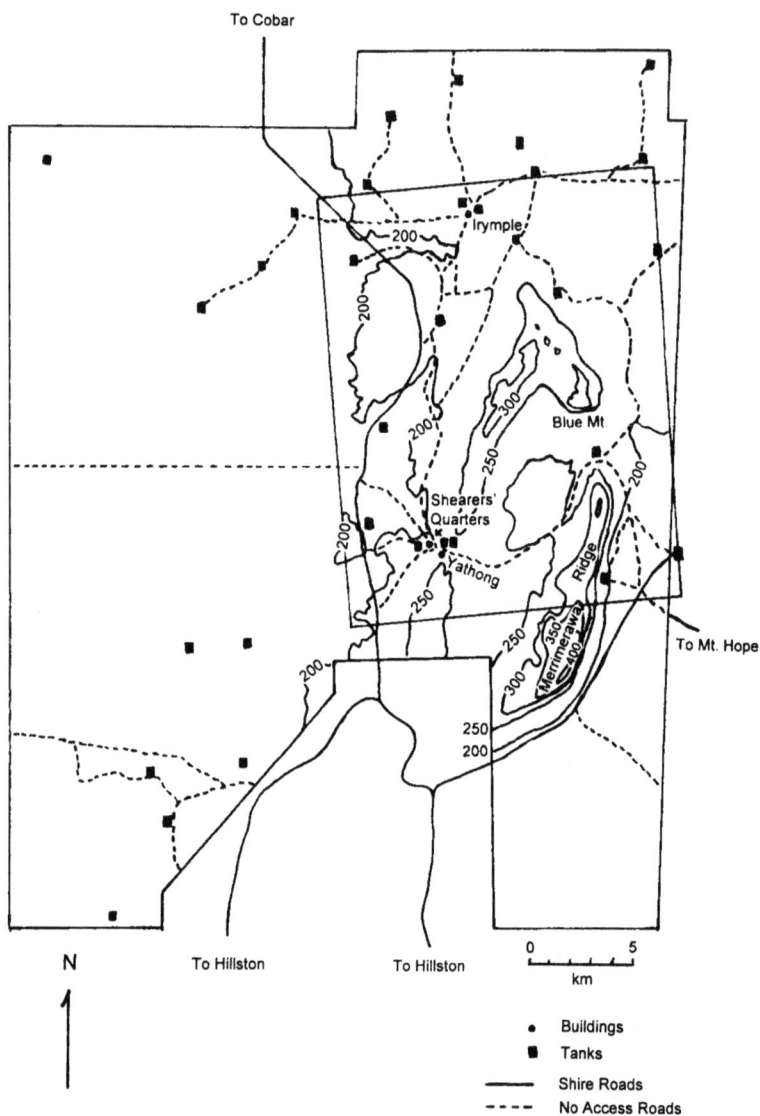

Figure 2.2. Map of Yathong Nature Reserve showing topography (50-m elevation contours above mean sea level), roads, earthen tanks, and buildings. A fire road (not shown) encircles the boundary. The rectangle on the eastern side shows the 288 km² intensive study area. The "no access" roads are dirt tracks closed to the public.

Road probably averaged less than a car per day. Meeting someone passing through was infrequent enough that it was common practice to stop and converse. In addition to shire roads, a system of dirt tracks, closed to the public, allowed reasonable access to most parts of the reserve (fig. 2.2).

A house and outbuildings at the Yathong Homestead were occupied by the reserve manager, Harry Warren, and his wife and two children. The arrival of our group, a family of four Americans and two Australian research assistants, constituted a human population increase of 150 percent. The population density of Yathong is typical of this region of the outback. The house at Irymple Homestead was still present and served as headquarters for research personnel from National Parks and Wildlife Scientific Services and CSIRO Wildlife and Rangelands Research Division. The Glenlea Homestead site in the southwestern part of the reserve was marked only by strewn wreckage.

The Shearers' Quarters for Yathong (fig. 2.3) are about 2 km west of Yathong Homestead, and the buildings include the shearers' quarters

Figure 2.3. Aerial view of Yathong Shearers' Quarters. The earthen tank is to the upper left. Water is pumped by a windmill into a raised metal tank near the earthen tank and conveyed by gravity to a second holding tank near the washhouse. The long building is the quarters building, the building to the right, the washhouse, and the building at the bottom, the cookhouse. Small outbuildings to the right are the dunny (outhouse, above) and fuel shelter (bottom).

themselves, a washhouse, and a cook shed. Shearers' quarters tradition-
ally housed crews of sheep shearers who traveled from station to station,
spending a few days at each, shearing the sheep and sacking the wool
each year. For this study, the Quarters played a similar role—as housing
for visiting researchers, volunteers, and work crews. The dilapidated
cookhouse was unused, and we renovated the building as headquarters
for this study.

Geology and Soils

The most prominent feature on Yathong is a northeast to southwest ridge
system lying along the eastern side (fig. 2.2). The major spine of the
system is Merrimerawa Ridge, a plate uptilted from the west, with a
precipitous eastern scarp that rises up abruptly from 190 m to 420 m
above sea level. These ridge systems are composed of reddish rock of
nonmarine sedimentary origin dating from the upper Devonian. On the
west side, the main ridge declines to around 260 m elevation over 2 km,
and then grades imperceptibly downward to the west. A lower parallel
ridge lies about 7 km west of the main ridge, reaching only about 320 m
along the ridge and 332 m at the outlying Blue Mountain (local hills not
to be confused with the Blue Mountains that are part of the Great Dividing Range) to the east. From the foot of this ridge, the landscape tilts
almost imperceptibly downward to around 180 m at the western boundary. The western half of the reserve is essentially flat.

The ridge system is drained to the west by Keginni Creek, an inter-
mittent stream that runs for several days following major rains, but is
dry most of the time. Keginni Creek drains into a sink where the water
disappears into the ground in an area of sand dunes on the western part
of the reserve. Away from the ridges, alluvial soils are of fine red silts
and sands with little development. Where stream courses are watercut
down to bedrock, soil profiles show almost no horizontal development.
These red soils constitute the whole of Yathong soils away from the
ridge tops. In the western part of the reserve, aeolian (windblown) dunes
of coarser red sands overlie the alluvial soils. These dunes are linear in
shape, about 4 to 5 m high, and lie parallel to one other in an east-west
direction, about 100 m apart (fig. 2.4). "Yathong" is an Aboriginal word
meaning sand hills.

Climate

Weather records on Yathong date only from 1983, when the first resident
manager, Harry Warren, began keeping records on rainfall, and then

Figure 2.4. The road to Cobar (Belford Road) showing windblown sand dunes characteristic of the western half of Yathong. The vehicle is the four-wheel drive vehicle loaned to the project by CSIRO. Note the vegetation, which is typical mallee habitat type.

added other weather parameters in 1985. We obtained long-term patterns from other nearby (by outback standards) weather stations, especially Cobar. Cobar, 120 km to the north and most representative of Yathong, provided a 104-year weather record (from 1882 to 1985). Yathong is best represented by records from Cobar. Nymagee, although closer, has a higher rainfall more typical of locations much farther east.

Climate in this region is characterized by moderately cold winters and extremely hot summers. Because Australia is in the southern hemisphere, the seasons are reversed, with January being the hottest month and July the coldest. It takes some time for a North American biologist to become accustomed to the sun being in the northern sky, and orientation is confused until one's internal compass gradually becomes reset.

Weather records from Cobar amply illustrate why outback Australia is a dry place. Mean annual rainfall is about 367 mm, whereas mean evaporation is 1956 mm, some 5.3 times as great (fig. 2.5). On average, there is a moisture deficit in all months, with the greatest deficit in summer months. The sun shines most of the time. Mean monthly number of hours of sunshine per day from 1963 to 1981 at Griffith, 220 km to the south, were 10 hours in January and 5.7 in July (Erskine and Smith 1983). The moisture deficit, in conjunction with high solar radiation and the lack of natural streams and lakes, makes Yathong a dry, parched

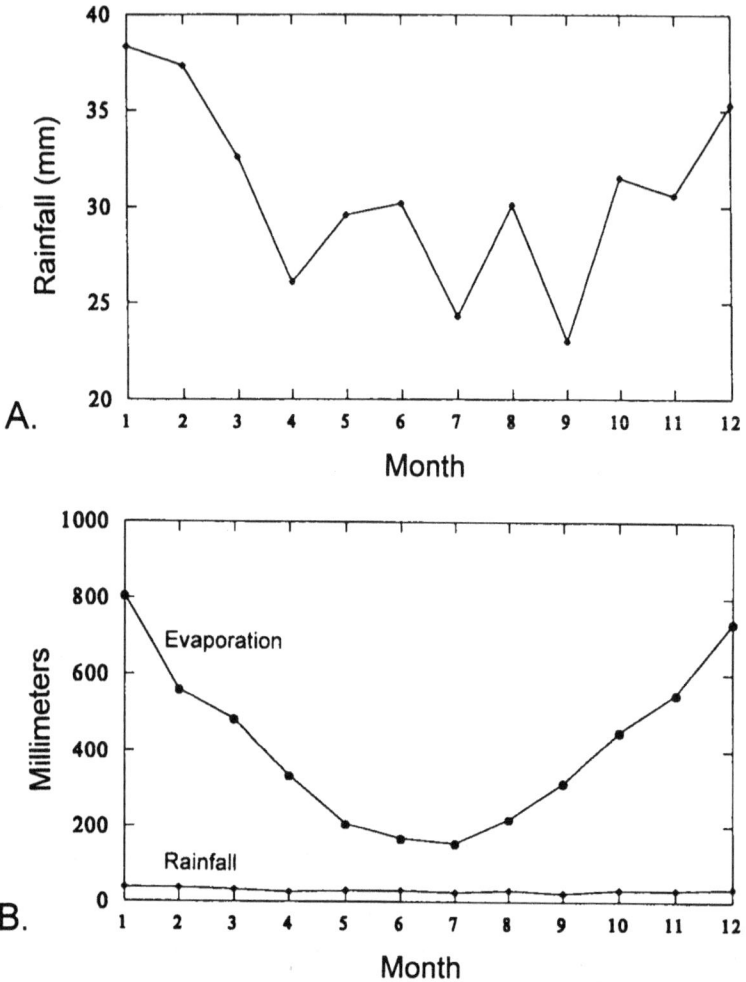

Figure 2.5. Mean monthly rainfall (A), evaporation with rainfall (for comparison) showing moisture deficit (B), solar radiation (C, next page), and temperature (D, next page). All but solar radiation are from Cobar and are based upon a 104-year record. Solar radiation was measured at Griffith and represents the mean for 13 years, 1968–1981.

environment most of the time. Significant precipitation comes in brief, sporadic rains and the water surplus produces flooding (fig. 2.6). The soft, unconsolidated soils turn into a quagmire and travel becomes impossible.

   Temperatures at Cobar (fig. 2.5) follow a typical seasonal pattern. When compared with Harry Warren's records, the mean monthly tem-

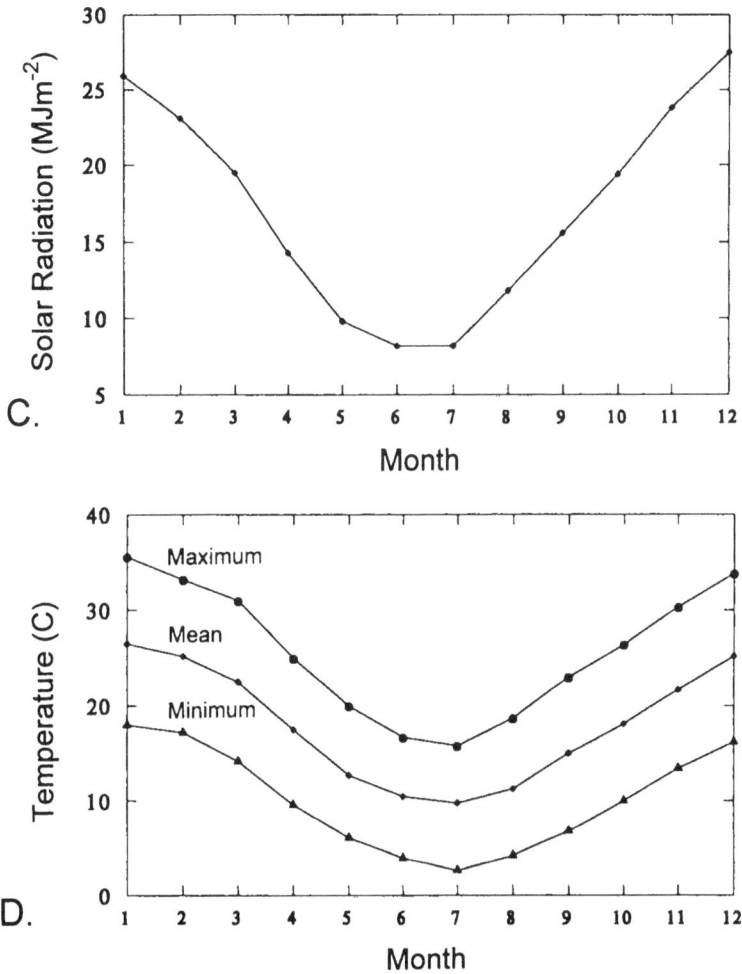

Figure 2.5. (continued)

peratures for overlapping years showed Yathong to be a degree or two warmer than Cobar, both for maximum and minimum. Although Cobar is closer to the equator, and thus is expected to be warmer, it is at a higher elevation (265 m) above sea level than Yathong Homestead (220 m).

Even within Yathong, the temperature varies greatly by microsite (see

Figure 2.6. Matched photos of the road between Yathong and Irymple Homesteads, about a half a kilometer from Keginni Creek, showing typical flooding conditions in the nearly level landscape.

chapter 3). Air temperatures commonly differed by 5 to 6 °C (even at night) between Shearers' Quarters and the Yathong Homestead, just 2 km away. Shearers' Quarters are in an open, low exposed site (200 m) while Yathong Homestead is in the trees and abuts a large rock outcrop. When freezing temperatures (0 °C) were recorded at Yathong Homestead, temperatures close to −10 °C were recorded by scientific thermometer at Shearers' Quarters, and our kitchen water pipes would freeze. The coldest temperature recorded at Shearers' Quarters was −10 °C, and the highest air temperature, in the shade, was 43 °C, again a degree or two higher than at Yathong Homestead.

Rainfall is the climatic variable of overriding importance in the Australian outback. It is rainfall that drives the remainder of the biological system (Caughley 1987a) and determines productivity for kangaroos, sheep, and virtually every other living thing in the ecosystem. Thus, an understanding of the variables contributing to the amount of rainfall is fundamental to understanding the fluctuating patterns of vegetation and kangaroo populations on Yathong.

The outback is notorious for its highly variable and unpredictable rainfall. Erskine and Smith (1983) illustrated the cover of their paper on weather at Griffith with a pair of dice in the air. Robertson et al. (1987) reported that there was no predictable pattern of rainfall amount at Menindee. Still, there are seasonal patterns useful to an understanding of rainfall in western New South Wales.

As noted previously, Yathong lies in a zone of overlap between summer rainfalls coming from the north and winter rainfalls from the south. Summer rains originate with the tropical monsoon storms that occur in the northern part of the continent. These cyclonic patterns produce thunderstorm activity that extends south across the continent as far as western New South Wales. At Yathong, this results in spectacular, usually localized thunderstorms, with much lightning, high winds, and occasional hail. These storms arise suddenly; in just a few hours, the day can change from intense heat to a storm with heavy dark thunder clouds and spectacular lightning. Torrential rain commonly occurs in the center of these storms, with several inches falling in 15 to 20 minutes. Rainfall is confined to a swath across the landscape, tracing the path of the storm. The edges are quite abrupt, and over a distance of 100 meters, one can go from completely dry to rainfall so heavy that one can hardly see. Such storms pass quickly, and in an hour, the sun may reappear in the company of enormous double rainbows.

For someone who has not experienced such a storm, it is hard to imagine its beauty and drama. The drab, bleached-out colors and unrelenting, oppressive heat of the sun through a seemingly endless succession of summer days are dramatically transformed. In a flash, torrents of rain wash away "dry and hot." The tremendous energy and spectacular lightning of such storms engage all the senses. The smell of rain on parched earth is unique and unforgettable. Passing, in the blink of an eye, from baked earth to water everywhere stretches one's credulity. People, too, are transformed. Minds focused for endurance are suddenly released to leap and bound. Even adults are compelled to go out and play in the mud, to capture the moment, for soon enough, the sun will reclaim the landscape.

In contrast, winter rains come from the south in broad, regional weather patterns. Rainfall is more prolonged, less intense, and extended over two or three days. Winter storms are associated with moderately strong winds and cold temperatures.

Long-term rainfall records from various locations in western New South Wales demonstrate a decline in mean rainfall from east to west (fig. 2.1). Further west, beyond the 300 mm isohyte, rainfall frequency distributions over the years follow a roughly normal distribution. In these areas, the amounts of summer and winter rainfall are approximately equal. East of the 300 mm isohyte, rainfall frequency distributions are more uniform, but jagged because of chance.

Yathong lies in this zone where, on average, summer rainfall exceeds winter rainfall. Rainfall by month at Cobar was greater in summer (Apr–Sep $\bar{x} = 204$ mm) than in winter (Oct–Mar $\bar{x} = 163$ mm). Also, standard deviations were greater in summer than winter. Summer rainfall by month was skewed toward lower values, whereas winter rainfall was nearly normally distributed about the mean. Thus, annual rainfall in the region around Yathong is determined primarily by summer rainfall, and this amount is more likely to be low than high. Total annual rainfall shows an irregular pattern because of the impact of variable summer rainfall. Whereas the average annual rainfall is higher than that west of the 300 mm isohyte, the variance between years is very great. West of the 300 mm isohyte, annual rainfall is lower on average, but winter rainfall equals or exceeds summer rainfall, so the frequency distribution of annual rainfall more closely approximates a normal distribution; that is, rainfall is less variable further west, which is the reverse of what one would expect.

Because weather at Yathong was not recorded until 1983, Cobar records are used to illustrate the rainfall history in the years preceding the study. A major drought occurred over much of southeastern Australia in 1982, when only 101 mm of rainfall fell at Cobar (fig. 2.7). An estimated

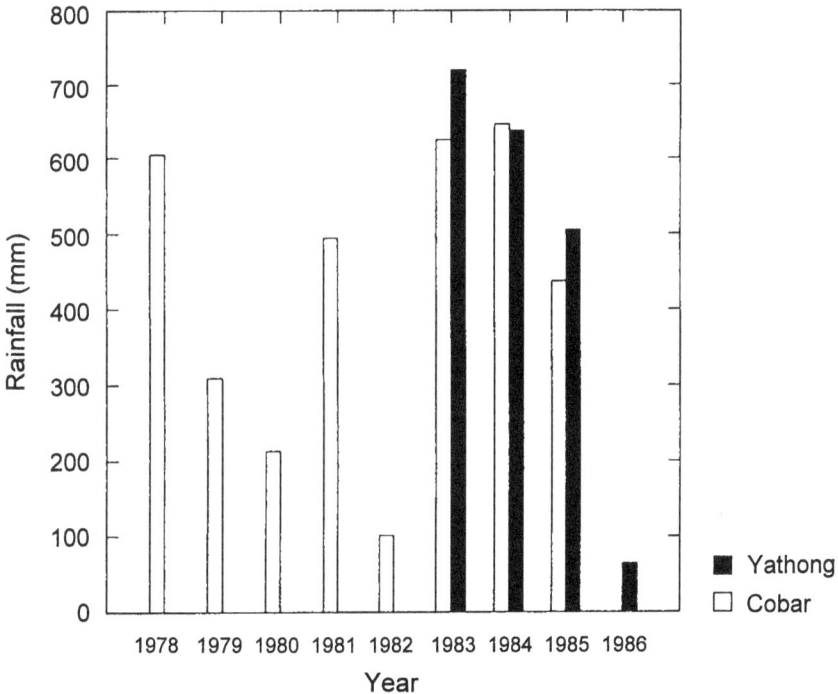

Figure 2.7. Rainfall (mm) by year at Cobar from 1978 to 1986 and at Yathong from 1983 to 1986.

43 percent of kangaroos died in western New South Wales (Caughley et al. 1984a), as did 50 percent of the kangaroos at Kinchega National Park near Menindee (fig. 2.1) (Caughley et al. 1987a). Kangaroo mortality was about 40 percent in South Australia and southern Queensland (Caughley et al. 1985b), illustrating the extent of this drought. Robertson (1986) reported that western grey kangaroos suffered higher mortalities than did red kangaroos.

This drought had an enormous impact on kangaroos at Yathong as well. Harry Warren reported that many kangaroos died during this period, and their skeletons were still abundant around water tanks when this study began. Pictures taken by Harry during the drought show herbaceous vegetation totally absent and shrubs greatly reduced (fig. 2.8). At Yathong, this drought was broken in 1983 by rainfalls of 35 mm in February, 145 mm in March, 225 mm in April, and 136 mm in May. These heavy rainfalls filled all of the earthen tanks with water, and fostered a rapid recovery of the vegetation. The 1982 drought was unusual

A.

B.

Figure 2.8. Matched photos of Shearers' Quarters tank in the 1982 drought after rainfall that flooded the tank (A) (March 1983, photo by Harry Warren) and in July 1985 (B). Note small round tree at left, windmill, and bush in left foreground.

regarding the great extent of eastern Australia that it affected. Most droughts are more variable over space, occurring in some places but not others. Time series analyses of the Cobar records suggest that a drought of this severity occurs locally about once in every 20 years.

Rainfall and Vegetation Growth During the Study

Records from both Cobar and Yathong show above-average rainfall in 1983, 1984, and 1985. Low rainfall occurred in 1986 (fig. 2.7). More detailed weather records and the availability of green feed for Yathong as measured during the period of continuous field work are given in figure 2.9. A long period of low rainfall occurred from December 16, 1985 to August 1, 1986. Total rainfall for 1986 was 64 mm. This dry period coincided with the heat of summer, and its influence on availability of green feed for kangaroos was also apparent (fig. 2.9).

Growth of green feed was significantly correlated with rainfall ($r = 0.56, P = 0.001$), and cross-correlation demonstrated a lag of one month ($r = 0.64, P < 0.001$) of green feed production following rainfall, as was previously reported by Wellard (1987) for Kinchega National Park. Moss and Croft (1999) reported somewhat longer time lags for Fowler's Gap (near the western boundary of New South Wales about 100 km north of Broken Hill) of two months for grasses and five months for forbs. We wonder if the difference might be related to temperature.

Temperatures (maximum, minimum) at Yathong were not correlated with green feed ($r = 0.12, P = 0.53, r = -0.02, P = 0.94$). However, this result masks some complex relationships between temperature and plant growth. Maximum temperatures in the summer and minimum temperatures in the winter are associated with high-pressure systems, which produce little or no rain. Thus, the most extreme values of temperature at either end of the scale tend to be associated with periods of little or no rain. Second, the lagged relationship of rainfall to green food production means that temperature has considerable time over which to exert influence. Direct observations of green growth showed that cold temperatures in winter delayed plant growth even when the soil was moist. However, cold temperatures extended the time over which soils retained moisture. Because winter temperatures are not so extreme as to completely prevent growth, green forage can be available at any time of year if soil moisture is available to support growth. This winter growth can be extremely important to the nutritional status of kangaroos. Thus, low temperature contributed to rainfall effects on plant growth by reducing

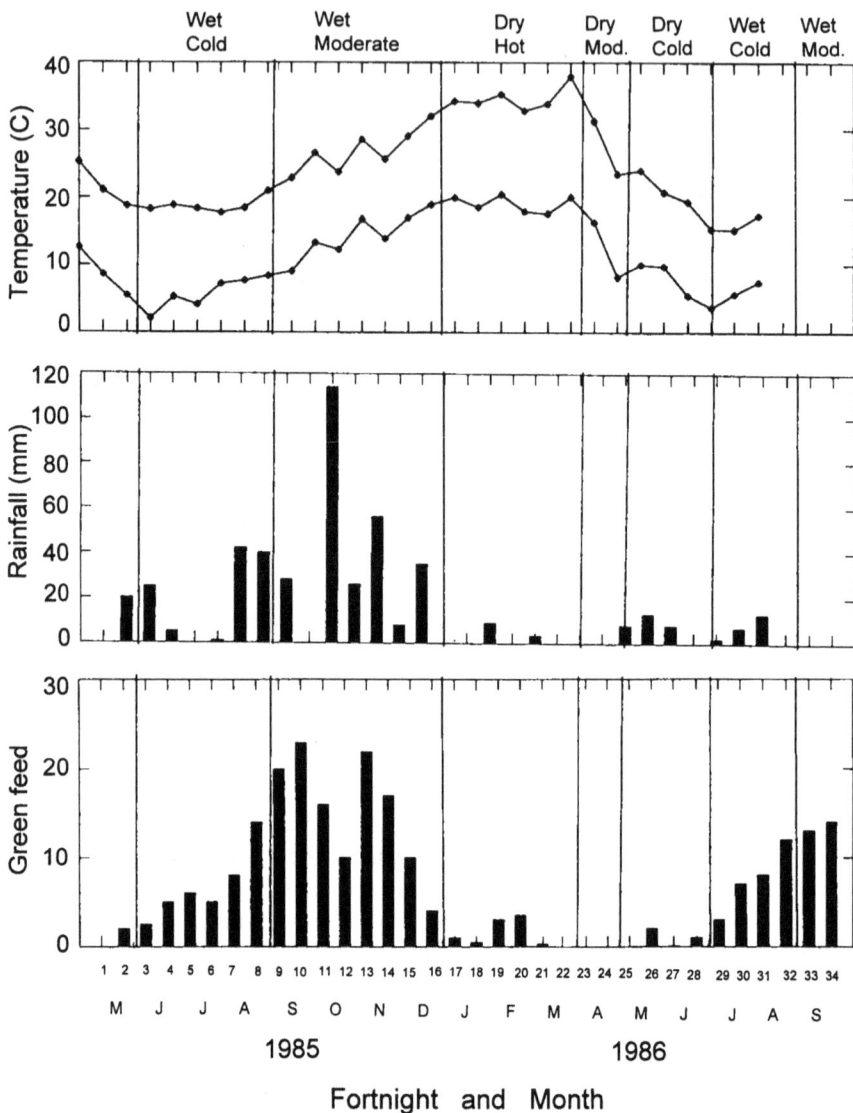

Figure 2.9. Temperature and rainfall recorded by this study at Yathong (Shearers' Quarters) and relative green feed availability as estimated visually and on photo point transects. Environmental periods are shown by vertical lines.

the rate, and hence, amount of growth, but also extended the period of growth and availability of green feed to kangaroos.

Conversely, the high temperatures of summer, although conducive to plant growth, resulted in rapid drying of soil moisture. This was particularly true if there was not a residue of mulch to insulate the soil surface from the intense solar radiation characteristic of this season. Mulch is sparse on Yathong because of fires, termites, and other factors. Consequently, relatively high rainfalls in November and December produced substantially less green plant growth than comparable rainfalls in August. Indeed, very high rainfall in early October 1985 (fig. 2.9) yielded no more plant growth than the August rains, partly because of soil saturation but also because the aseasonal high temperatures of this period resulted in less effective retention of moisture. Given the complexity of these relationships, it is not possible to examine the relationships statistically because of sample size limitations. Nevertheless, it is apparent from quantitative and qualitative evidence that rainfall is the key variable influencing green plant growth, but that seasonal temperatures modify that growth by reducing effective moisture in summer, and by physiologically slowing plant growth in winter.

Flora

If viewed on a continental scale, Yathong lies at the very western edge of the vegetation zone characterized by open woodlands. Just to the west, the shrublands, with only scattered small trees, predominate. According to the "Potential Natural Vegetation" map of Bridgewater (1987), Yathong lies in the "*Symphyomyrtus*" woodland (*Symphyomyrtus* is a *Eucalyptus* subgenus) with a shrub or grass understory, whereas a short distance west is considered *Acacia* (mulga) shrubland. On Yathong the woodland understory is predominantly grasses, with shrubs being uncommon.

In this study, vegetation of Yathong was categorized using an unpublished vegetation type map, produced by Ian Parker of CSIRO Wildlife and Rangelands Research Division in Gungahlin. He classified vegetation into seven major types, with subtypes within many of these. For purposes of this study, types and subtypes were combined into eight types found in an intensive study area in the eastern part of the reserve (fig. 2.10). Plant names and descriptions are taken from Harden (1990–1992) and Costermans (1983). Brief descriptions of the major features of each type from observations in this study follow.

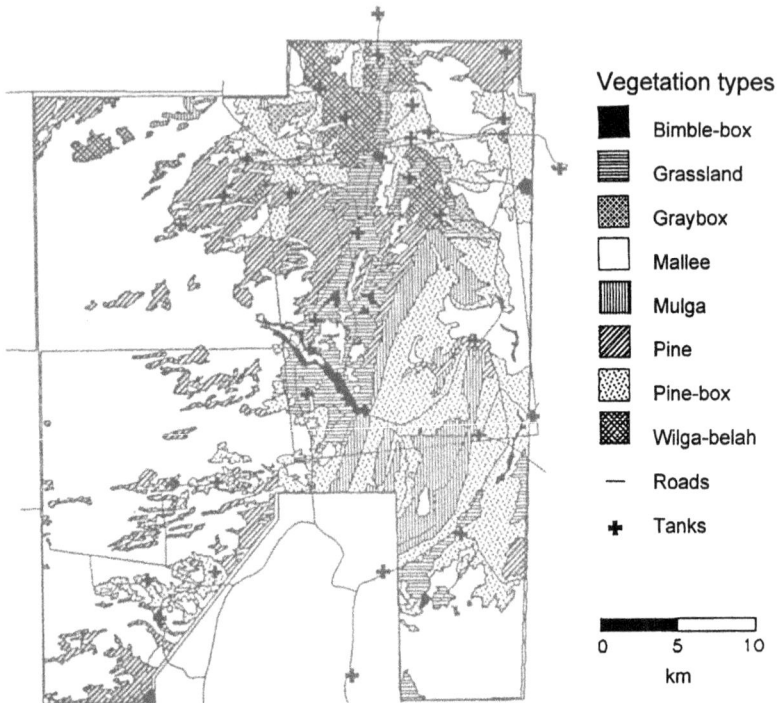

Figure 2.10. Vegetation type map of Yathong Nature Reserve after map produced by Ian Parker of CSIRO. Roads are shown as solid lines and earthen tanks by crosses. The rectangle on the eastern side of Yathong shows the 288 km² intensive study area.

Mallee

Mallee refers to an array of low-growing *Eucalyptus* tree species that share a common growth form. All have multiple stems arising from a common underground lignotuber crown on the root system. The stems angle outward like flowers in a vase, and the crown forms a parabola of foliage. In the mallee type (fig. 2.4), these trees occur in thick stands with nearest neighbor distances of about 6 m, which gives a density of about 100 individuals/ha (Cottam and Curtis 1956), and form a canopy cover of around 50 percent at a height of 3 to 4 m. Despite a typically sparse understory vegetation, the multiple stems and low canopy result in low visibility. Mallee occurs on relatively poor, usually slightly alkaline soils on Yathong, including level, low wet sites, sand dunes, and rocky sites. Mallee covers virtually all of the sand dune areas of the western flats of Yathong outside the intensive study area.

Predominant species of mallee on Yathong are red mallee (*Eucalyptus socialis*), dumosa mallee (*E. dumosa*), yorrell (*E. gracilis*), and grey mallee (*E. morrisii*). The sparse understory commonly includes Jersey cudweed (*Gnaphalium luteoalbum*).

Mulga

This vegetation type is characterized by the occurrence of open stands of mulga (*Acacia aneura*), a small tree (about 5 m in height) with a distinctive silver-grey appearance. Mulga is virtually the only tree occurring in this type, with a typical spacing of 10 to 15 m. On Yathong, this type is restricted to the rocky ridges with their coarse rubble and thin soils. There is sparse herbaceous understory with low productivity.

Bimble Box

Bimble box (*E. populnea*) (fig. 1.4, 1.5), a small to medium-sized tree (about 15 to 20 m in height) similar to quaking aspen (*Populus tremuloides*), is the dominant and nearly only tree in this type. Although bimble box is a common upland tree in higher rainfall areas to the east and south, on Yathong bimble box grows only in stands along temporary water courses, principally Keginni Creek. Step-point methods (see chapter 3) gave estimates of around 25 percent canopy cover, and nearest individual methods (Cottam and Curtis 1956) gave an estimated density of 20.1 trees/ha. The understory is composed of herbaceous species that grow well on floodplains.

Grey Box

This type is a relatively open woodland characterized by the presence of grey box (*E. microcarpa*), a medium to large tree of 25 m in height, that occurs on one small area of Yathong. It grades into the pine-box and wilga-belah types, and contains many of the same tree species occurring in those types. There is a fair herbaceous understory consisting of typical grassland species.

Pine-Box

This type is composed of scattered trees of white cypress-pine (*Callitris columellaris*) and gum-barked coolibah (*E. intertexta*). These open woodlands have a well-developed grassland understory. Gum-barked coolibah is always a widely spaced tree, whereas the density of cypress-pine is highly variable. White cypress-pine is a fire-adapted species that germinates following burning, apparently because fire releases the seeds. After burning, it initially forms dense, dog-hair thickets that are virtually impenetrable and produce a total visual screen. Step-point transects in one mid-successional-stage stand indicated 33 percent canopy cover, and nearest individual estimates gave a density of 200 trees/ha. Stand density is ultimately reduced through self-thinning and fire, to which younger, weaker saplings are particularly susceptible, so that larger, mature cypress-pines end up widely spaced.

In the 1930s, Chinese laborers were hired to ring-bark (girdle through the bark around the tree) mature cypress-pines to kill them in many areas on Yathong. The objective of cypress-pine reduction was to create more grasslands for sheep. Some sites where cypress-pines were removed from fertile soils now support productive grassland vegetation. Other sites characterized by poor soils, however, show low production of grasses despite the removal of cypress-pines.

Wilga-Belah

This vegetation type is a mixed woodland characterized by the prevalence of wilga (*Geijera parviflora*), a tree similar to weeping willow (*Salix babylonica*), and belah (*Casuarina cristata*). Wilga is a rounded, umbrella-shaped tree, about 6 to 8 m in height, with pendulous branches hanging to ground level (fig. 2.11). Kangaroos seek shelter during heavy

Figure 2.11. Wilga tree showing rounded crown and pendulous branches under which kangaroos often seek shelter from rain. Note carcass of dead radio-collared male red kangaroo that was dragged from under the canopy.

rains under the overarching canopy of these trees. Wilga was favored by graziers because, along with the much less common kurrajong (*Brachychiton populneus*) and rosewood (*Heterodendrum oleifolium*), it was among the few native tree species with foliage palatable to sheep. Branches could be cut from the canopy for emergency fodder during drought.

Belah, despite its long, pine needle–like foliage, is an angiosperm related to the oaks. It is often a large tree, and grows in many communities on poorer sites. Warrior bush (*Apophyllum anomalum*) is a common associate. Trees in this type tend to grow in ragged clumps, with open grasslands too small to map scattered in between.

Pine Associations

This type is a catchall category consisting of intergradations of the pine-box and wilga-belah types. It is the catholic mix, rather than any given species, that characterizes the type. As with the parent types, it features considerable open grassland between clumps of trees. Step-point transects gave a canopy coverage of 8.4 percent, and tree densities of 9.6 trees/ha.

Grassland

This type, as its name suggests, consists of open perennial grasslands, or grasslands with widely scattered trees (fig. 2.12). Although density of trees varied greatly by site, random step-point transects gave a tree canopy coverage of <0.1 percent, and an average density of 1.1 trees/ha. Some extensive areas cleared by the former station owners are still completely open. The most common grass that characterizes the type is spear grass (*Stipa variabilis*), a bunch grass that forms thick stands that grow to 25 to 50 cm in height, depending upon rainfall and soil quality. This species is often locally dominant in the winter rainfall zone (Beadle 1981). Scattered among the spear grass clumps at much lower densities are white-top (*Danthonia caespitosa*) and brush wiregrass (*Aristida behriana*).

Because of frequent fire, flooding, a high density of European rabbit warrens, and the long history of sheep grazing, there is considerable disturbance and microsite variation within the grassland type. Barley grass (*Hordeum leporinum*) grows in thick stands in former sheep holding pens (fig. 2.12), and in heavily grazed paddocks there are scattered individuals of wild turnip (*Brassica tournefortii*), blue crowfoot (*Erodium crinitum*), spring emex (or three-corner jack, *Emex australis*), burr medic (*Medicago polymorpha*), spear thistle (*Cirsium vulgare*), and clumps of star thistle (or Maltese cockspur, *Centaurea melitensis*) and saffron

A.

B.

Figure 2.12. Grassland vegetation type. Top photograph shows the general character of the landscape. At the right foreground is an extensive patch of mallee, with a small patch of wilga-belah at the extreme right center. Bottom photograph shows full earthen tank, with a pine association type in the foreground. The light-colored vegetation is the drying barley grass characteristic of old sheep paddocks.

thistle (*Carthamus lanatus*). Around old buildings, fences, and earthen tanks, the shrub black bluebush (*Maireana pyramidata*) and annuals such as blackberry nightshade (*Solanum nigrum*), camel melon (*Citrullus lanatus*), paddy melon (*Cucumis myriocarpus*), purple-flowered devil's claw (*Proboscidea louisiana*), Russian thistle (*Salsola kali*), slender dock (*Rumex brownii*), and horehound (*Marrubium vulgare*) are common. Also found around earthen tanks are tobacco bush (*Nicotiana glauca*) and wild tobacco (*N. suaveolens*).

Following burns, the whole countryside may be covered by common white sunray (or white paper daisy, *Rhodanthe corymbiflora*). Golden everlasting (or yellow paper daisy, *Helichrysum bracteatum*) is more sparsely scattered. In some low areas, Patterson's curse (*Echium plantagineum*) forms nearly pure stands punctuated by scattered individuals of great mullein (*Verbascum thapsus*). On the dry cracking mud of low areas, common heliotrope (*Heliotropium europaeum*) and Bathurst burr (*Xanthium spinosum*) are common. In other floodplains, an overstory of saffron thistle shelters species such as common sowthistle (*Sonchus oleraceus*), prickly lettuce (*Lactuca serriola*) and weeping lovegrass (*Eragrostis parviflora*).

Fire

The vegetation of Yathong (as that of much of Australia) is adapted to frequent wildfire (Gill et al. 1981; Ford 1985; Bridgewater 1987). The predominant grass at Yathong, spear grass, is the principal carrier of fire throughout the region. When this bunch grass dries, it becomes highly flammable and, due to its relatively high biomass, it carries fires extremely well. Presumably this is an adaptation that gives this species a competitive advantage over other less fire-adapted herbaceous species, and slows encroachment by woody species that are susceptible to fire in the seedling stage, as reported for *Triodia* grasslands in other parts of Australia (Suijdendorp 1981).

Fires in this region are primarily wildfires set by lightning strikes. In earlier times, aboriginal peoples routinely set fires to influence vegetation and animal life (Hallam 1985), but unacculturated aboriginal people no longer occur in the Yathong area. Graziers did not set fires and, in fact, feared them. Once started, wildfires were virtually impossible to suppress. The major means of fire control in this region was to set backfires, but ultimate suppression usually depended upon the coming of rainstorms. Fires consumed herbaceous biomass that, even when dry, was needed by graziers to support their sheep herds. Fires also threatened buildings, and in the time before steel fence posts, destroyed wooden-post fences, which burned to ground level or below. Posts had to be hand set, 15 feet apart, with four to five holes drilled in each to hold the smooth fence wires. Fence construction over the vast areas enclosed by a sheep paddock was a laborious and expensive task.

On Christmas day, 1984, a lightning storm started three separate fires near Cobar. These fires burned south some 260 km, and between January 1 and 8, 1985, burned about 80 percent of Yathong Nature Reserve. Thus, at the start of this study, burned areas consisted of bare ground with a thin, blackened residue of ash on burned bunches of *Stipa* grass. Many of the areas burned on Yathong originated as backfires, with roads forming fire breaks. The presence of many roads separating burned areas on one side from unburned areas on the other (supporting a tall stand of dry bunch grass), allowed us to compare kangaroo use of burned versus unburned grasslands.

Mammalian Fauna

Yathong Nature Reserve supports a varied herpetological and a spectacular ornithological fauna, but the native mammalian fauna is depauperate. The three species of kangaroos studied—red, eastern grey, and western grey—were numerous and widespread. The fourth species of kangaroo, the euro, was present but rare and confined to the rocky ridges. Four confirmed sightings were made over the 14 months of continuous field work: two adult males (one an old, battle-scarred veteran with shredded ears), and a female with a young-at-foot. Other than the kangaroos, native mammals were rare. Over the course of this study, we observed a total of one echidna (*Tachyglossus aculeatus*) and two species of bats (*Eptesicus vulturnus* and one unidentified species).

Introduced mammals, however, are extremely abundant on Yathong. European rabbits (*Oryctolagus cuniculus,* fig. 2.13) occur in pest proportions and their warrens (communal burrow systems) are ubiquitous in the open areas that constitute their major habitat. Rabbits seem to require good visibility and do particularly well in burned areas where cover is removed and new green growth is favored. In dense spear grass stands, warrens appeared to decrease. Absence of cover makes it difficult for predators to get close enough to catch rabbits before they escape down the burrows.

Feral goats (*Capra hircus,* fig. 2.13) are common on Yathong, despite periodic musters (roundups) by light plane and motor bike. So long as the boundary fences are not goat-proof (goats easily push through or

Figure 2.13. Introduced herbivorous competitors with kangaroos; European rabbits (A) in a warren. Note unpalatable thistles that mark warrens because rabbit feeding eliminates other vegetation. Feral goats (B) with wilga-belah vegetation type in the background.

under stranded, smooth-wire fences), there is little prospect of eliminating them from Yathong. Local graziers regard goats as "spare cash." Goats roam wild, occupy the poorer areas of the station, require no capital or labor investment, and their populations build up quickly. Thus, they are available for muster and sale as the need for cash arises. Mustering of goats on Yathong creates a sink that quickly attracts goats from surrounding stations. It is common to see hundreds of goats on Yathong only a few months following a muster.

Figure 2.14. Regeneration of kurrajong tree in decaying top of a dead gum-barked coolibah tree. Note pine-box vegetation type in the background.

Most of the impact of herbivory on woody plants observed on Yathong was the result of feeding by goats rather than kangaroos. Invariably, signs of removed twigs were associated with goat tracks or droppings. Goats have virtually eliminated reproduction by one palatable species of tree, kurrajong. We observed only two young specimens on all of Yathong, each growing in the dead, decaying centers of broken off gum-barked coolibah snags, about 3 m above ground level (fig. 2.14). During dry periods, goats even browse the toxic tobacco bushes that grow around earthen tanks.

Feral pigs (*Sus scrofa*) are common on Yathong, particularly in moister areas such as the bimble box stands along Keginni Creek. Fortunately, their numbers are low, and damage from their rooting activities is localized, partly due to control efforts by National Parks and Wildlife Service. As the vegetation dries out, however, pigs and goats visit earthen tanks regularly, where their wallowing in the mud and consumption of aquatic vegetation muddies the water with colloidal particles that do not settle out. Water quality declines precipitously. Although kangaroos and emus (*Dromaius novaehollandiae*) also visit the earthen tanks during dry periods, they drink from the edges without degrading water quality.

Major predators on Yathong are introduced foxes and domestic cats,

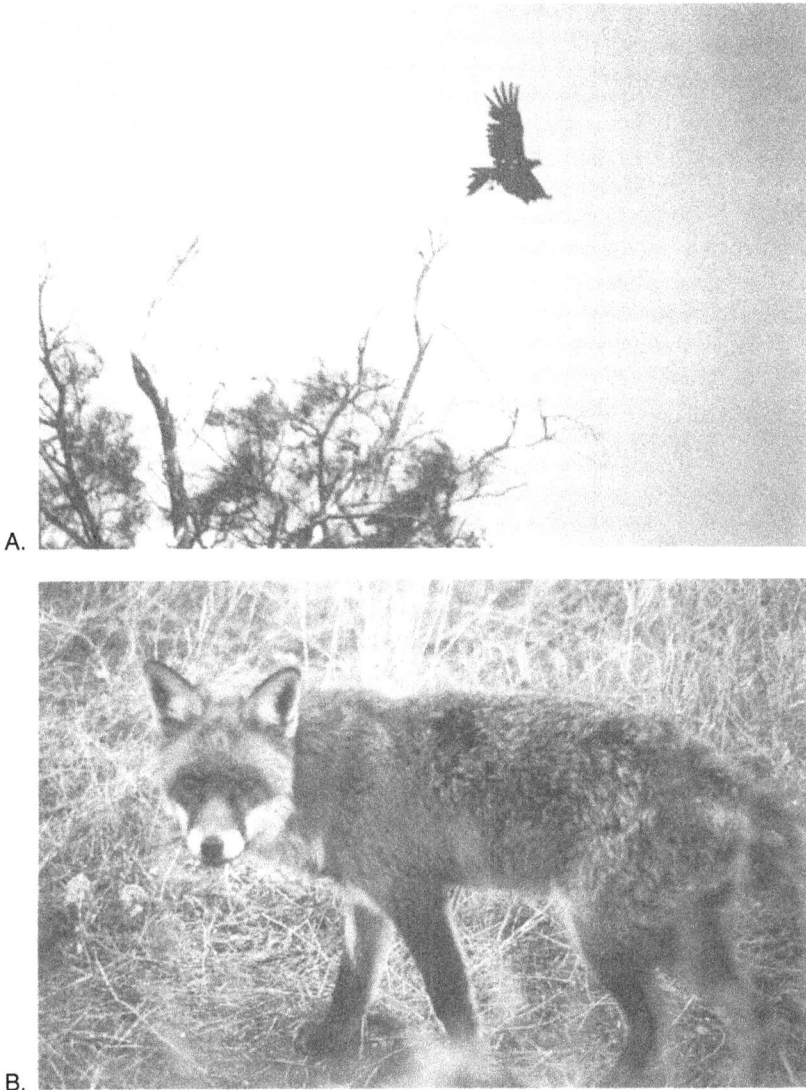

Figure 2.15. Major predators on Yathong; native wedge-tailed eagle (A) and introduced European red fox (B).

and native wedge-tailed eagles (*Aquila audax*) (fig. 2.15) and king brown snakes (*Pseudechis australis*). Attempts at control of foxes and cats by the National Parks and Wildlife Service have hardly dented their populations. European red foxes (*Vulpes vulpes*) were initially intro-duced to Australia in the 1860s for sport (Coman 1983). They are extremely abundant on Yathong. Two, three, or more were observed daily during field work.

Domestic cats (*Felis catus*) arrived with the earliest European settlers and quickly became feral. Aboriginals interviewed by Johnson et al. (1989) in the western deserts of Australia believed cats had always been there. Feral cats appear as abundant as foxes on Yathong, and these two predators no doubt have had an adverse impact on native small mammals, lizards, and ground-nesting birds.

Despite the abundance of rabbits, many foxes and cats are thin and scraggly during cool periods with green feed, which suggests they are not efficient at preying on healthy rabbits. During our stay, three foxes died by Shearers' Quarters. During late summer, rabbits severely infected with myxomatosis become dull and lethargic and are found along roads and under buildings away from warrens. These infected rabbits are easy prey, and furnish foxes and cats a great burst of food at this time. It seems likely that this abundance of vulnerable rabbits accounts for the high population densities of foxes and cats on Yathong.

Finally, the nonnative domestic house mouse (*Mus musculus*) infests buildings and other structures. These mice populations as well as myxomatosis-infected rabbits attracted cats and foxes to Shearers' Quarters. Attracted as well were numerous poisonous snakes, much to our concern for the safety of our small children.

3

METHODS

An ecological field study of this kind employs an array of methods designed to yield information on one or more specific topics. In this chapter, we present our methods in detail so that the reader may judge whether they are reasonable and sufficient to support the analyses and conclusions that follow in subsequent chapters. Our major methods were systematic in that they were conducted on a fixed schedule according to established procedures. Systematic methods included morning and night spotlight surveys, radio telemetry, photo points for vegetation change, and fecal collections for nitrogen determination. After initial trials to test the feasibility and logistics of our approach, and considering available resources, we set time schedules and collection procedures and developed standardized data collection forms. Initial trial data were not included in the analyses. Only data gathered after standardization are reported here.

Standardization was employed to eliminate variance due to method. Observers underwent intensive training at the start of the project, and training reinforcement was systematically incorporated to avoid drift among observers as time progressed. This way, any changes in results could be attributed to changes in kangaroo behavior and environmental variables.

A second set of methods was ad hoc, in that they pertained to unpredictable events and took advantage of opportunities that presented themselves. For example, concentrations of kangaroos on select feeding sites occurred at times. When this occurred, we sampled vegetative characteristics at these sites and compared them with randomly located sites in the same vegetation type. Direct observations of infrequent kangaroo behavior were also made in this manner. Any time sexual, agonistic, or mother-offspring behavior was observed, current project work was put aside and the behavioral bout recorded.

Every attempt was made to satisfy the assumptions of the statistical models by which the data were analyzed. Samples were chosen randomly (i.e., every observation had an equal chance of being selected) wherever possible, and where random selection was unworkable, the criteria for selection were clearly stated. Distributions and equality of variances were tested for normality, and transformed if necessary. When normality could not be achieved, nonparametric methods were applied.

Where relatively simple comparisons were tested, alpha was set at the conventional 0.05. In complex models with many variables, alpha of 0.01 or lower was used to reduce the likelihood of accepting relationships due to chance. We tried to be conservative, both statistically and biologically, by not making too much of small differences.

Finally, the cumulative results from all methods and available literature were examined for patterns. Did the results from independently collected data sets agree? Did the results from this study agree with other studies conducted in different areas by other researchers? Were predictions across variables consistent? For example, if a given species of kangaroo selected for vegetation types with high concealment cover, then it logically followed that the perpendicular distances at which this species was observed should have been less than for a species that selects for more open habitat. Examination for patterns is the best means available in a field observational study for establishing that variable and response represent cause and effect relationship rather than correlation by chance (where many correlations are attempted), or a spurious correlation where the related variable is not the causative variable.

The general statistical construct was to test the null hypothesis — that is, no difference (or change). Thus, rejection of the null hypothesis was grounds for accepting that there were changes over time of day, fortnight, month and season; or differences between species, sexes and age classes; or differences by habitat type, burn status, and so forth. Correlations between kangaroo behavior and time, season, and weather and forage conditions were then examined by multivariate statistical procedures. Statistical analysis was performed with the statistical package SYSTAT (Wilkinson et al. 1992) except where noted.

Environmental States

In temperate climates, the four calendar seasons—spring, summer, fall, and winter—serve well for data analysis. Calendar seasons have limitations for outback Australia. Although temperature during the study showed a typical seasonal progression, rainfall did not fit any observable pattern (fig. 2.9). Therefore, environmental states based upon combinations of temperature and availability of green versus dry forage were defined in lieu of calendar seasons. Temperature regimes were selected as "hot," representing high summer temperatures, "cold," representing winter temperatures, and "moderate," representing transitional temperatures between the hot and cold extremes. Forage was categorized as either green or dry. Because kangaroos search out green forage, the breakpoint between green and dry forage times was not placed at the midpoint of the transition, but instead at the disappearance point of green forage. The selection of these environmental states in relation to specific changes in temperatures and forage on Yathong is shown in fig. 2.9. The combination of "wet hot" did not occur during the study period.

Systematic Surveys

Four routes, covering a total of 88 km, were surveyed by vehicle each fortnight (two weeks), both in the early morning and by spotlight at night. Usually, one of the four routes was covered per day. A single observer, any one of three trained individuals, conducted each survey; at times a recorder accompanied the observer to record data only. The systematic survey routes allowed gathering of a wide array of information on kangaroo sociality, behavior, and activity, and on environmental variables such as weather conditions and seasonal effects on habitat quality and selection. In addition, we recorded statistics useful for line transect and mark-recapture estimates of population size and density.

At the start of the study, we randomly selected the route and day within a fortnight on which to conduct the surveys. However, this approach quickly proved unworkable because rainfall made certain (or sometimes all) routes impassable on (or even near) the selected day. Consequently, a fixed schedule was established such that from the beginning of each fortnight, the routes were surveyed in the reverse order of their likelihood of being impassable. Using this schedule, even with frequent rain delays, we were able to complete all routes within the fortnight.

Morning routes began at daybreak as soon as it was light enough to observe kangaroos. Night spotlight routes began at first full darkness. Spotlighting was done with a handheld, halogen bulb light that emitted a powerful beam of about 700,000 candlepower, using the general approach of McCullough (1982). Both sides of the road were searched, the side opposite the driver by holding the light over the roof of the vehicle. Kangaroos were readily visible at night due to their light pelage and moderately bright pink eyeshine.

At the beginning of each survey, the year, Julian date, and general weather conditions were recorded on standardized data forms. When kangaroos were sighted, the observer immediately noted the time from a dash-mounted digital clock and the vehicle speed from the speedometer. Vehicle speed varied from 5 to 40 km/hr, depending on the time needed to visually search both sides of the road, given the vegetation cover. The observer then estimated the perpendicular distance between the individual kangaroo (or the center of a group) where first sighted and the center line of the road. Observations with perpendicular distances exceeding 500 m (few in number) were arbitrarily excluded. Southwell and Weaver (1993) reported that comparable truncation resulted in no bias or loss of precision in their line transect estimates of a population of known size.

Observers were trained to estimate distances on markers (5-m intervals to 20 m, 10-m intervals from 20 to 100 m, and every 100 m thereafter) placed perpendicularly to the road next to the Shearers' Quarters. Observers refreshed their estimation skills frequently, both day and night, when driving to and from Quarters, and when visiting the dunny (outhouse).

Once the vehicle was stopped, the species, group size (social group was determined by close proximity and behavior), sex, and age composition, and presence and absence of eartags or radio collars were established with the aid of binoculars (fig. 3.1). All data were recorded on prepared data forms. Alarm behavior was recorded using four categories: (0) calm, continuing with normal activities; (1) alert, watching the vehicle; (2) hop, moving away from the vehicle at normal speed; and (3) bound, moving away at high speed. If a kangaroo remained calm or alert, the distance from the vehicle at which it did hop or bound (or if it never left the original site) was recorded as the vehicle moved past.

Location of the group was determined on x- and y-coordinates of a 1-km^2 gridded topographic map, and recorded to the nearest one-tenth km. Thanks to numerous roads, fences, vegetation type boundaries, and

Figure 3.1. Observing a kangaroo on a daylight systematic survey.

topographic cues, this level of accuracy was possible. Major habitat type was determined from the CSIRO vegetation type map of Yathong. The microhabitat of the specific site, a few meters surrounding where the kangaroo was first sighted, was estimated visually and recorded. For example, the kangaroo might have been observed in a patch of grassland in a major habitat type that was wooded.

Tree density surrounding the kangaroo was classified by these categories: (0) open grassland, trees greater than 100 m distance apart; (1) scattered trees, 25 to 100 m between trees; (2) woodland, 10 to 24 m between trees; (3) dense woodland, 9 or fewer m between trees; or (4) mallee. Shrub density was recorded as (0) no shrubs, (1) a few scattered shrubs, (2) moderate shrub density, and (3) thick shrubs obscuring vision. It should be noted that shrub stands are extremely uncommon on Yathong, and most observations were in sites with no shrubs.

Herbaceous cover was recorded as: (0) low green pick (new growth) less than 10 cm tall; (1) medium green, 11 to 25 cm tall; (2) tall green, greater than 25 cm tall; (3) tall drying, greater than 25 cm tall; (4) tall dry grass, greater than 25 cm tall; (5) tall dry grass with green pick underneath (criteria of 0 and 4 combined); (6) tall dry grass plus medium green growth (1 and 4 combined); (7) tall dry with tall green

(2 and 4 combined); and (8) tall green with tall drying (3 and 4 combined). Additionally, the site was classified as either burned (in the January 1985 burn) or unburned.

Light conditions were categorized as (0) full daylight after sunrise, (1) morning light, before the sunrise and/or light filtered through a heavy atmosphere (i.e., light that would cause color distortion on color film), and (2) night, which was coincident with spotlight surveys. In addition, the time of actual sunrise, when sunlight first fell on the landscape, was noted; this varied in the short term depending on route and topography, as well as seasonally.

Temperature was recorded in °C from a thermometer mounted on the radio antenna, but shielded from the wind of the moving vehicle. Wind speed was categorized as (0) calm, (1) light breeze, (2) moderate wind, and (3) strong wind, and wind direction as (0) north, (1) northeast, (2) east, and so on to (7) northwest, and (8) none. Cloud cover was categorized as (0) clear, (1) scattered clouds, (2) heavy clouds, and (3) completely overcast. Rainfall was categorized as (0) none, (1) drizzle or mist, (2) scattered showers, (3) light rain, and (4) heavy rain. Fog was categorized as (0) none, (1) light fog, and (2) heavy fog. Fog was very rare at Yathong.

Population Estimation

Mark-Recapture Population Estimate

A sample of 29 kangaroos was captured for radio telemetry studies (see Radio Telemetry below). Capture efforts were scattered broadly over Yathong, primarily along the same roads used for systematic surveys. This sample of known marked animals allowed the application of mark-recapture methods to derive an estimate of population size. We agree with Southwell (1989) that mark-recapture is too expensive for routine monitoring. However, where radio telemetry is being applied for other purposes, mark-recapture population estimations become possible, and accurate knowledge of the radio-collared sample allows several problematic assumptions of the method to be overcome.

In this case, resighting of marked animals, rather than recapture, was used to calculate population size. To maintain statistical independence, our research assistant, Stephen Thornton, who determined radioed kangaroo locations, never conducted systematic survey samples, whereas observers on the systematic surveys did not carry receiving equipment once the kangaroo sample was captured and the mark-recapture population estimation work begun.

Marked kangaroos were highly visible, both day and night. The radio collars were constructed of white, rubber-impregnated canvas belting, and this material was readily visible to the naked eye in both daylight and night spotlight observations. Each marked kangaroo was tagged in each ear with 32-mm-diameter aluminum eartags covered by reflective material color-coded by combinations in left and right ears that allowed individual identification. These eartags were easily observed in daylight, and reflected brightly at night by spotlight, far exceeding the intensity of kangaroo eyeshine. This raises the possibility that marked animals were differentially observed at night, but all were observed well within the distances at which most kangaroos, both marked and unmarked, were observed, so bias seemed unlikely. No eartags were lost over the period of the mark-recapture estimate observations. Because of these highly visible markers, it was extremely unlikely that tagged animals that were present during the systematic surveys were inadvertently overlooked.

The presence of radio transmitters allowed us to know exactly the number of living marked animals in the study area over time. Thus, the number of marked animals could be adjusted over time for mortality and movement out of the study area. For practical purposes, the kangaroo population could be treated as a closed population (see details in chapter 4), so Bailey's (1951) formula for a closed population was applied:

$$N = \frac{M(n + 1)}{(m + 1)},$$

where N is the population estimate, M is the number of animals marked, n is the sample of resighted animals, and m is the number of marked animals in the resighted sample. Standard error of this estimate is given by:

$$SE = \frac{M^2(n + 1)(n - m)}{(m + 1)^2(m + 2)}.$$

The results of McCullough and Hirth (1988) on white-tailed deer suggested that calculation of N should not be made on sample sizes of m less than 10, and inverse sampling with $m = 30$ was preferable. Initially, we intended to use inverse sampling with a preset $m = 30$, but mortality of marked animals and movement from the study occurred prior to reaching $m = 30$. Therefore, it seemed less problematic to use Bailey's model. In reality, the differences due to which model is applied are small compared to the problems with meeting the assumption of equal catchability and reobservability. Details of the analysis and evaluation of the assumptions are presented in chapter 4.

Line Transect Population Estimation

Line transects have previously been used to estimate kangaroo densities by Coulson and Raines (1985), Coulson (1993b), Coulson et al. (1990), Southwell (1989, 1994), Southwell and Weaver (1993), Southwell et al. (1995a, 1995b, 1997), and Clancy et al. (1997) with good results. However, these were mostly done by walking daylight transects, which was totally impractical in our large study area. Also, most of these studies were conducted in eastern Australia in more mesic areas with high kangaroo densities. Nevertheless, the good conformance of the known populations to estimated populations reported by Southwell (1994) supports the reliability of our methods. With tame kangaroos at very high density, he found some individuals were overlooked on the transect line, a remote likelihood in our low-density, relatively open habitat population. Southwell also reported that his wild populations displayed evasive behavior. By using a vehicle (as Southwell recommended) we had little problem with evasive behavior, and our recording of alarm codes allowed us to evaluate evasion as a variable. Similarly, Clancy et al. (1997) found that helicopter surveys resulted in less avoidance behavior than surveys on foot; the same is probably true for vehicles, because kangaroos are typically less alarmed by an observer in a vehicle than one on foot.

Perpendicular sighting distances from the systematic surveys were analyzed by the program DISTANCE (Laake et al. 1993) to derive population density. The center of the road was treated as the 0 point and kangaroo observations were treated by habitat type and season as a disappearing function of the distribution of perpendicular sighting distances.

The method assumes that all animals occurring on the line (0 perpendicular distance) are observed. This assumption was almost certainly met because of the high visibility on roads at Yathong. It is further assumed that the number of animals declines with distance away from the line, because some animals present are not observed.

Because kangaroos often occur in groups as well as singly (the latter were coded as "groups" of one), and given that statistically individual kangaroos in groups are not independent observations, the unit of observation for the line transect analysis was the group. Hence, DISTANCE was used to calculate density of groups. Southwell and Weaver (1993) found that using individuals instead of groups increased precision and did not bias the estimated population from the known population. Although instructive, we believe that the increase in precision may be an artifact of the larger sample size of individuals than groups, and that this approach is not advisable for the typical case where the true population is unknown.

DISTANCE fits various mathematical models to the disappearing function of groups, and gives statistics helpful in selecting the most appropriate fit. From the area under the best-fit equation and the length of the transect, the density of groups can be calculated. Density of individuals was calculated by multiplying density of groups by mean group size. The total population can be calculated from the individual density multiplied by the area studied. Anderson and Southwell (1995) reported that analysts using DISTANCE underestimated kangaroo populations by 10 to 12 percent. They could not determine the reasons for underestimation, so our estimates may similarly be underestimates.

Ideally, lines for the line transect method are selected at random. The scale of Yathong and the logistics of conducting the surveys made this impossible, so existing roads were used. Southwell and Fletcher (1990) used a similar approach to estimate abundance of whiptail wallabies (*Macropus parryi*). Whereas roads on Yathong are not strictly random in location, the generally gentle topography resulted in road placement being across most vegetation and topographic features, except for the main spines of the ridge system. These ridges were not included in systematic surveys because they were inaccessible to four-wheeled vehicles.

Therefore, we believe that the use of roads did not strongly bias the resultant estimates of kangaroos. Furthermore, the empirical test conducted on deer by Fowler (1985) suggests the use of roads per se did not greatly influence the outcome of density calculations, and additionally, that curvature of roads did not bias the result so long as it was not too extreme. Curvature of roads on Yathong (fig. 2.2) included in the systematic surveys was minimal because of the low relief.

Detailed analysis of model assumptions, mathematical fits, calculations of estimates, and other factors are presented with the results in chapter 4.

Species Composition

Numbers of three species of kangaroos observed in the systematic surveys were calculated as a percent of the total kangaroos on Yathong.

This percentage was compared to that derived from independent population estimates of the three species. Mixed species groups were analyzed for their composition and season of occurrence.

Sex and Age Composition

Kangaroos were classified as adult male, adult female, and young-at-foot. Adult males could easily be distinguished from adult females by size, differences in color patterns and body form, and differences in behavior. In most adult males, sexual dimorphism is well developed, as noted in chapter 1.

It is young males of the same size class as adult females that present problems in recognition. They represent a small proportion of the population; because their growth is relatively rapid, they quickly become larger than females. Furthermore, young males in the size class of adult females for all species have heavier, more muscular shoulders and forearms, and more robust skulls, particularly the rostrum. They frequently occurred alone, or if in a group, they tended to be separated from the main cluster, and were not followed by offspring. If the female-young group moved off, the young male usually moved separately, and often left the group entirely. All of these were identification cues for young males. If the kangaroo sat up facing the observer, the presence or absence of a pouch was usually apparent, and occasionally, one could see the scrotums of males, although these were difficult to detect in young males under observation conditions. In red kangaroos, young males are typically quite reddish in color, and are readily distinguished from the variegated gray, red, and white females. In fact, the red kangaroo is named for the color of the male, and "blue flyer," the local name for female red kangaroos, is more descriptive. However, there is considerable individual variation in color (Oliver 1986; Dawson 1995), with some males showing considerable gray and some females almost completely red. Nevertheless, color and the other characteristics noted above were sufficient to separate nearly all young male red kangaroos from adult females.

In eastern and western grey kangaroos, we could detect no color or pattern differences between young males and females, and consequently, morphological and behavioral cues were used. Because these young males were more difficult to distinguish from adult females than were red kangaroos, some errors of misclassification of young males as adult females probably occurred in these two species.

Young-at-foot are offspring that have left the mother's pouch, yet remain with the mother, nursing periodically from one nipple. Ordinarily, there is a pouch young attached to another nipple. Young-at-foot are almost always in close proximity (usually within 3 to 4 m) to their mothers and hop beside them when the group moves.

Juveniles (i.e., weaned offspring) were classified separately from young-at-foot. Juveniles are offspring that are weaned but remain in the company of the females. The transition from weaning to adulthood (sexual maturity) is rapid, and the identifiable juvenile stage is brief. Consequently, juveniles were uncommonly identified in the systematic surveys. They could be identified with certainty mainly in cases in which they associated with a female with young-at-foot, so the three could be compared simultaneously. Juvenile kangaroos were recorded so infrequently that, for data analysis, those with female groups were arbitrarily combined with young-at-foot, and juvenile males were combined with adult males.

Adult females with large pouch young, whose presence could be determined either by extruding head, feet, or tail, or by pouch distention, were identified and recorded. However, because most females have pouch young most of the time, except during severe drought, these data should be considered conservative. Only relatively large pouch young that could be readily determined at any distance were recorded in order to eliminate observational distance as a variable. Thus, pouch young that were recorded were probably in the last two months of pouch life before permanent emergence. The proportion of females with large pouch young indicated the relative stage of reproduction in females of the three species. Because visible young or distended pouches indicate that the young will soon leave the pouch to become young-at-foot, and because birth and copulation coincide with young leaving the pouch, the proportion of females with distended pouches is a useful gauge of reproductive activity and population increment (see Tyndale-Biscoe 1989 for a review). Although not equivalent biologically, for comparison of kangaroos with ungulates, it is useful to consider female kangaroos with large pouch young equivalent to placental mammals in late pregnancy, and young leaving the pouch as being approximately equivalent to birth (Millar 1977).

Sex ratios were calculated by species over time by dividing the number of males by the number of females observed in the systematic surveys to give the ratio of males:female. Data from morning and night spotlight surveys were compared, and other information on behavior was explored to attempt correction for differential observability between the sexes that biased the sex ratio.

Reproductive state of females was determined by dividing numbers of females with large pouch young and of young-at-foot observed in the systematic surveys by the total number of females observed to give the two ratios, large pouch young:female and young-at-foot:female.

Habitat Use from Systematic Surveys

Each group of kangaroos observed was assigned to a major vegetation type according to the vegetation type map. Therefore, the relative distribution of the three species by two sexes of kangaroos could be compared to vegetation types chosen. From these results, differential use of available vegetation types by species and sex could be derived. Preference for vegetation type could be derived by comparing the use observed in the systematic surveys against availability determined from map areas of each vegetation type. The area sampled by the systematic surveys was determined by two methods. The first entailed plotting the x- and y-axis grid coordinates of observed kangaroos, and using the outer boundary of points (away from the road on each side) of kangaroos observed. Because the distance at which kangaroos were observed varied with the concealment cover of different vegetation types, the outer boundaries were narrower in wooded types than in open grassland, and narrowest in the thick cover of the mallee type.

A second measure of area available by vegetation type was determined by the functional strip width as derived from the line transect analysis (the distance that included one-half of the area under the fitted curve times two to include both sides of the road) by vegetation type. These two measures of area sampled by the systematic surveys were then digitized to derive area by vegetation type. Then, the proportions of kangaroos by species and sex observed by vegetation type in the systematic surveys compared to the proportions of area of those vegetation types sampled could be tested by chi-square statistics on contingency tables. In the same manner, proportions selected by various species and sex categories could be tested against each other for differential use by sex and species. Finally, vegetation types sampled by systematic surveys were compared to vegetation types within the total study area to evaluate the representativeness of the sampled area for the study area as a whole.

Habitat characteristics within vegetation types selected by kangaroo species and sex categories could be examined by comparing microhabitat, tree density, shrub density, herbaceous stage, and burn status to elucidate the choices of these variables made by each category of kangaroo.

The same analysis was done on morning versus night spotlight systematic surveys to evaluate diel patterns of habitat use. Because it is known that kangaroo activities vary by time of day, it is useful to see if they carry out those different activities in different habitat types. For example, do kangaroos move from one type to feed to another type to rest? Is this related to exposure or heavier cover types, or do most types contain sufficient diversity to allow choice of suitable microsites for different activities? Does diel type selection vary with environmental state (e.g., hot, dry, or moderate temperature in combination with availability of green versus dry forage) or by burn status or current weather conditions? The number of variables that might influence activity and habitat use is extremely large.

Alarm Behavior

The probability of the different kangaroo species and sexes being alarmed by the vehicle used for the systematic surveys was of interest, because it is likely that behavior toward one kind of threat reflects behavior toward other kinds. Alarm is assumed to be an inherent trait of species and sex that may be modified by experience and local weather conditions. It is presumed to be an adaptation to threats over the evolutionary past, principally due to predation by natural predators and aboriginal humans (Jarman and Coulson 1989; Robertshaw and Harden 1989; Tunbridge 1988, 1991). Of course, since the arrival of Europeans, modern firearms have reinforced alarm behavior as graziers attempted to reduce grass and water consumption by kangaroos. Yathong's status as a Nature Reserve since 1978 has eliminated most current threats, although some poaching along public roads through the Reserve may continue. Nevertheless, the period of time that has elapsed since the onset of protection is too short for kangaroos to have lost a fundamental behavior. In any event, interspecies and sex comparisons should be valid; the relative excitability (or conversely, lack of excitability) to an identical disturbance (vehicle) can be compared across species and sexes.

Differences between morning (daylight) and night spotlight response of kangaroos to disturbance also are of interest, and relate to diel use of habitats. By the same token, species differences in activity in daylight versus darkness may well reflect inherent excitability, with the more excitable species being more active during cover of darkness. The degree of nocturnality of each species was indexed by dividing the number observed in the morning surveys by the number observed in the night spotlight surveys, with numbers being corrected for different sighting distances by the results of the line transect analysis. Other than the different sighting distances due to spotlight power limitations (i.e., in daylight, kangaroos can be seen at greater distances), the sampling effort in morning and night spotlight surveys was equal. The index of nocturnality, derived in this manner, would be greater than one if the species were observed more frequently in daytime, less than one if observed more frequently at night, and equal to one if there were no difference.

The tendency to flee also might influence the perpendicular distances at which kangaroos were observed on systematic surveys, much as Southwell (1994) reported for wild kangaroos on walked routes. Increased perpendicular distances would change the population estimates derived from line-transect analyses. By quantitatively assessing the degree of alarm behavior, the likelihood of bias can be evaluated for the various sex and age classes of kangaroos. Also, analysis of alarm behavior is helpful in evaluating the assumptions of the line-transect population estimation method being met, particularly as it relates to the problem of animal avoidance of the observer.

Influence of Weather on Kangaroo Behavior

Weather influences the behavior of most large mammals. Therefore, weather variables were recorded at the vehicle at the time of each kangaroo observation during the systematic surveys. Temperature was recorded as a continuous variable, while other weather variables were recorded by categories. Therefore, log-linear models were used to analyze these categorical data. Weather variables were examined for their influence on kangaroos of each species and sex for alarm behavior, habitat selection, nocturnality, sighting distances, use of burned versus unburned areas, and other factors.

As discussed in chapter 2, long-term weather records from Cobar and other locations in western New South Wales, and several years of records taken by Harry Warren at Yathong, were available. However, these records lacked sufficient detail for use with the radio telemetry activity data to examine for the immediate influence of weather variables on kangaroo behavior. To fill this need, a continuously recording weather station was set up at Shearers' Quarters, and collected data from December 20, 1985 through July 19, 1986. Instruments included a recording barometer (operated in the computer room, the most dust-free and temperature-stable room) in our living quarters, a recording hygrothermograph placed in a roofed (for shade) and screened lean-to on the side of the cook shed, and a recording wind speed and direction recorder mounted on a 2-m staff about 70 m from the buildings. Although a continuous-recording rain gauge was available, it was not operated because rainfall was so irregular in timing, amount, and location over Yathong. Whereas the other weather parameters taken at Shearers' Quarters could be assumed to be reasonably representative of those actually impinging on the scattered radio-collared kangaroos, rainfall could not. Therefore, when rainfall occurred, rainfall amounts at Shearers' Quarters were taken periodically (usually every 3 to 4 hours) from a nonrecording rain gauge, and written records were kept on general distribution of precipitation over Yathong.

Interspecific and Intersex Relationships

Spatial dispersion of the kangaroos by species and sex was expressed by the evenness with which they were distributed over the km^2 grids sampled on the systematic surveys. Niche breadth (*FT*) of Smith (1982) as recommended by Krebs (1989) was calculated as:

$$FT = \left[\sum_{j=1}^{n} \right] \sqrt{p_j a_j}$$

where *FT* is niche breadth, p_j = proportion of individuals found in or using resource state *j*, a_j = proportion that resource *j* is of the total resources, and *n* = total number of possible resource states.

Approximate 95% confidence interval of *FT* is given by:

$$\text{lower 95\% c.i.} = \sin\left(x - \frac{1.96}{2\sqrt{y}} \right),$$

and

$$\text{upper 95\% c.i.} = \sin\left(x + \frac{1.96}{2\sqrt{y}} \right),$$

where $x = \arcsin(FT)$ and y = total number of individuals studied = ΣN_j.

Values of *FT* can range from 0 to 1, with 0 being highly concentrated and 1 being uniformly distributed. Such concentrations may or may not be related to habitat selection, for concentrations can occur on the boundaries of two or more vegetation types, or may occupy only a small portion of a large vegetation type.

As recommended by Krebs (1989), niche overlap between the species and sexes within species was calculated by Morisita's (1959) measure C by the formula:

$$C = \frac{2\Sigma p_{ij} p_{ik}}{\Sigma^n p_{ij}[(n_{ij} - 1)/(N_j - 1)] + \Sigma^n p_{ik}[(n_{ik} - 1)/(N_k - 1)]}$$

where

p_{ij} = proportion grid i is of the total grids used by species (or sex) j,

p_{ik} = proportion grid i is of the total grids used by species (or sex) k,

n_{ij} = number of individuals of species (or sex) j that use grid i,

n_{ik} = number of individuals of species (or sex) k that use grid i,

N_j, N_k = total number of individuals of each species (or sex) in sample ($\Sigma n_{ij} = N_j$, $\Sigma n_{ik} = N_k$)

Differences by sex were examined to evaluate resource partitioning between the sexes. Given the great sexual dimorphism in kangaroos, constraints due to scale may well impact the sexes differently. For instance, larger body mass may allow males to have a wider temperature tolerance, and they may be able to subsist on lower-quality forages as reported for ungulates (Demment and Van Soest 1985) and kangaroos (Dawson 1989, 1995). Newsome (1980) reported different diets for the sexes in red kangaroos, and McCullough (1985) and Beier (1987) reported the same for white-tailed deer. Also, in many ungulate species, males and females are spatially separated outside of the breeding season, and may occupy different habitats as well. McCullough et al. (1989) showed that the sexes of white-tailed deer overlapped only about 56 percent, and when overlap increased, separation by habitat type was greatest. Similarly, a study of niche relationships among seven coexisting native North American ungulates by Y. McCullough (1980) showed that the differences in spatial use by the sexes in some species exceeded the difference in spatial use between species. Thus, when differences by sex were taken into account, this ungulate community of seven species occupied an even greater number of niches. Similar differences in habitat selection by sex were reported for eastern grey kangaroos (Kirkpatrick 1966; Grant 1973; Jarman and Southwell 1986), western grey kangaroos (Johnson 1983a), and red kangaroos (Johnson and Bayliss 1981; Johnson 1983a).

The analytical approach to niche relationships of kangaroos on Yathong was to explain as much of the total variance in kangaroo behavior as possible by species and sex, and to identify the variables influencing behavior and their relative importance. We wanted to know why kangaroos made the choices and behaved the way they did. On a conceptual level, this approach assumes that whereas some choices and some behavior may be random, an important portion is adaptive, the consequence of a long history of natural selection. In a harsh environment like the outback, it seems reasonable to assume that adaptation is more finely tuned, for the margin of error is narrower in severe environments where extremes are great.

Radio Telemetry

Radio telemetry served two purposes in this study. First, it was a means of obtaining information on home ranges, movements, and activity patterns of a sample of individual kangaroos. Radio telemetry was the most efficient and accurate method of obtaining these data. Second, radio telemetry produced unbiased estimates of kangaroo use of vegetation types, diurnal versus nocturnal activity, etc. to compare with parallel data obtained from the systematic surveys. These results were important to determine if limitations of direct observation and any observer influences on kangaroo behavior led to biases in the systematic survey data. This knowledge was fundamental to interpreting systematic survey results and determining whether species and sex differences from that method were valid only on a comparative basis (precise but not accurate), or were correct (both precise and accurate).

Radio collars were manufactured by Telonics, Inc. (Mesa, Arizona). We used the deer-sized package (MOD 500), with an expected battery life of about three years. Collars were made in the configuration developed by David Priddel (personal communication) for kangaroos at Kinchega National Park. A common problem with standard collars is that continual flexing of the cable antenna where it emerges from the collar causes the cable to break at that point; the signal strength subsequently drops to less than half that of the intact antenna. The Priddel pattern has an extra strap of collar material sewn perpendicularly to the collar encasing the antenna cable where it emerges. The enclosed cable flexes less, and this eliminated breakage of antennas.

Capture of Kangaroos

Kangaroos were captured for radio-collaring by "stunning" (Robertson and Gepp 1982). Stunning is the preferred (and most humane) method of capture for kangaroos because they do not respond well to drugs and are very prone to postcapture myopathy (Shepherd 1983). Consequently, our license from the New South Wales National Parks and Wildlife Service authorized capture only by stunning. This method proved efficient in another study (Priddel 1983, 1987).

Stunning is a capture technique that rightly prompts skepticism. Basically, the kangaroo's senses are disoriented, and then it is run down and tackled. A crew of seven is desirable, consisting of a vehicle driver, spotlighter, shooter, and four chasers. Ideally, two of the chasers are good sprinters, while two others are big and strong. The crew traverses the area in a large vehicle with the spotlighter seated on a rooftop rack and the shooter opposite the driver. The chasers are in the rear where they can get out of the vehicle rapidly. When a kangaroo of the appropriate species and sex is located by the spotlight, the vehicle approaches it as closely as possible without alarming it, preferably less than 50 m. (But we caught some kangaroos at 80 m.)

After the vehicle stops, the chasers dismount and position themselves two at each corner behind the vehicle, with a sprinter in the forward position. The shooter positions himself to shoot out of the front seat window or across the vehicle hood if the kangaroo is on the driver's side. The rifle is a .22 caliber rimfire fitted with a telescope sight, firing a high velocity (greater than the speed of sound) bullet. Strict safety procedures are followed to avoid accident.

The spotlighter continues to focus the spotlight directly on the animal to be captured. When the shooter and chasers are set, the shooter mimics the alarm call of the young-at-foot, causing the kangaroo to rise up to full alert position on its hind legs and tail. The shooter then fires a bullet between the ears of the kangaroo, an inch or two above the skull. By causing the kangaroo to rise to its full height, any unexpected movement will be downward away from the path of the bullet, for the kangaroo cannot hop without first gathering itself. Obviously, the job of shooter requires a skilled marksman. At the sound of the shot, the chasers sprint toward the kangaroo at full speed, just outside the beam of the spotlight (fig. 3.2). One of the sprinters makes the initial tackle, attempting to grab the kangaroo around the middle of the body from the back side. Once the kangaroo is grasped by the tackler, the other three chasers grab the tail, legs, and head to physically restrain the animal (fig. 3.2). The animal is then quickly blindfolded, which helps curtail its struggles.

Once the kangaroo is secured, the previously prepared (to minimize handling time) radio collar and eartags are affixed to the animal. Sex, and, if female, presence of pouch young, are determined. If a large pouch young is present, the pouch is secured with masking tape to prevent loss of the young after release.

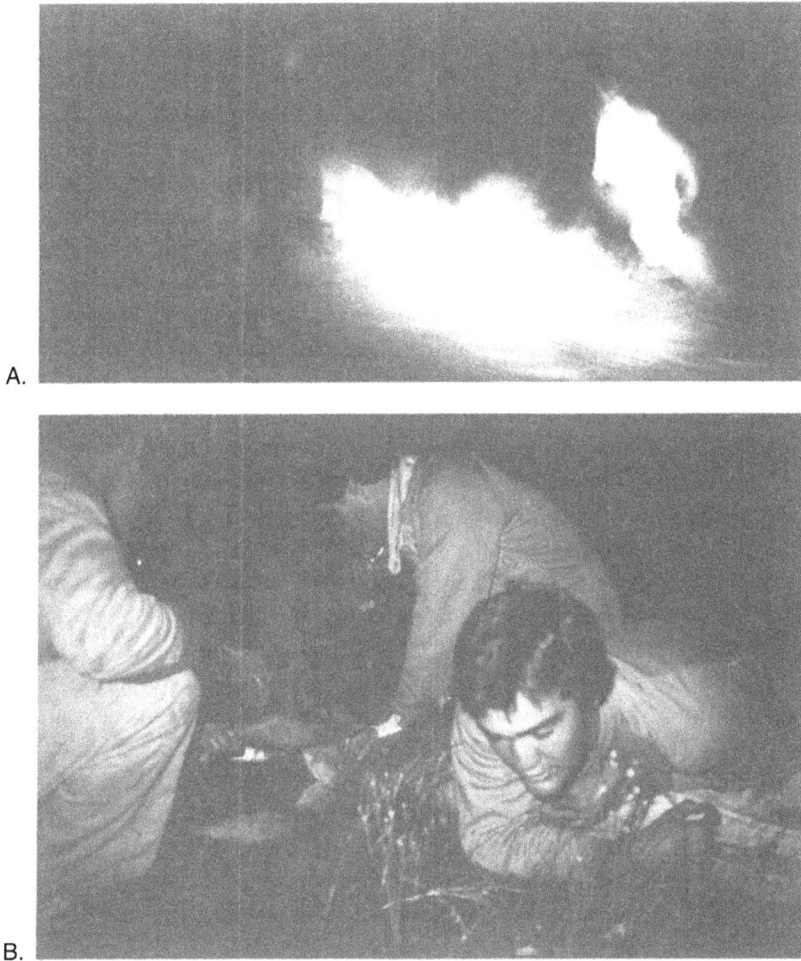

A.

B.

Figure 3.2.   Two catchers about to capture a kangaroo by stunning (above, photo by Marcia Teitgen). Below, restraining a large male red kangaroo for affixing a radio collar.

During release, one of the crew securely grasps the kangaroo by the tail while the others, on a prearranged signal, release the legs and head, remove the blindfold, and move back toward the vehicle. The crew member grasping the tail waits until the kangaroo regains its feet and recovers its balance. He continues to hold the tail (which necessitates hopping along behind large, strong males) until the kangaroo is oriented away from the vehicle and crew, and then releases the animal.

We made a total of 32 successful captures using this method. No kangaroo was struck by bullets or injured during handling. Average time from tackle to release was less than five minutes, with an extreme of about eight minutes. Injuries to the crew were mainly scrapes, bruises, and sprained ankles from sprinting in the dark amidst tree branches, downed limbs and rabbit warrens, and from wrestling strong animals on rough ground covered with an impressive array of plant stickers and spines. Several serious cuts came from western grey kangaroo claws, including one puncture completely through one crew member's finger that required medical treatment and stitches.

Chasing and tackling kangaroos is not for the faint of heart or slow of foot. A successful catcher must shut all else out of mind, focus on the starting gun, and at the sound of the shot, sprint recklessly toward the kangaroo. Tackling a kangaroo is exhilarating, partly because of the risks, partly because of the chase, but mainly because of the incongruity of it all. Capture efforts have the ambience of a contact sport.

It is not known with certainty why stunning works. It is presumed that the light blinds the animal, reducing effective vision, and the bullet between the ears perhaps produces a sonic boom that disorients hearing. If chasers cross the light beam or run noisily, the kangaroo simply runs away. Well-stunned animals slump over and remain motionless for three or four seconds. They then start to recover, and move away in disoriented fashion. It is important to be close to the kangaroo at the start, because to make a tackle, the chasers must reach the animal while it is still disoriented. Stunning works best on extremely dark nights but doesn't work at all on moonlit nights. A heavily clouded, overcast sky, with humid air and no wind is optimal. Under these conditions, kangaroos sometimes are tackled without their having moved at all. The fact that very dark nights are best suggests that vision is the sense most profoundly disoriented.

Two kangaroos were captured on July 24, 1985 by a crew of only two catchers. One was a red kangaroo female we named Scarlett O'Hara, who went on to furnish large amounts of information. The other, a young male red kangaroo, disappeared. We found his collar a few days later beside Belford Road and suspect that he was poached.

Thirty kangaroos were captured by a full crew between December 10 and 17, 1985. One of the advantages of stunning is that the desired species and sex can be singled out for capture. We wanted five radioed individuals of each sex of each species, and this was achieved for all species and sexes except eastern grey females, for which we managed to capture only four. We quickly filled the requirement for red kangaroos, and then western grey kangaroos. Most of the time was spent in trying to complete the sample of eastern grey kangaroos. Three kangaroos were captured but not radioed, two because they were too small and one because it was the wrong species. The latter was captured to settle a good-natured dispute about whether the individual was a western or eastern grey. Our call—western grey—proved correct.

The sample of 29 radioed animals was reduced by one in early January 1986 by the death of a young red kangaroo male about a month after capture. This individual was unusual in that upon release he did not depart, but remained in the vicinity and violently fought the collar. He repeatedly leaped into the air and dashed himself on the ground. We left the scene quickly in hopes that he would settle down. We observed him several hours later and he was still fighting the collar. The next day he was observed about 1 km away from the capture site, and while not violent, he was still working on the collar with his forepaws, trying to slip it off his head. He was found dead 31 days later, having displayed abnormal behavior until then. We did not use any data from this individual.

This young male red was the only kangaroo for which capture may have contributed directly to mortality. Although several other kangaroos momentarily fought the collar upon release, in a matter of seconds they ran away. After several days, they stopped trying to slip off the collar, and so far as could be determined from observations, behaved normally thereafter. They formed social bonds, bred, reproduced, and from all appearances became unaware of the collars. We had no indication that the radio-collared sample behaved differently from unradioed individuals, and we believe the results obtained on the radioed animals were an unbiased sample of the population. Data collection on the kangaroos captured in December 1985 began in January 1986, which allowed two weeks to a month for kangaroos to recover from any possible effects of capture and to become accustomed to the presence of the collars and eartags. Our assistant, Stephen Thornton, arrived on the project on January 5, 1986, and from that time on, he assumed full responsibility of field data collection on radio-collared kangaroos.

Home Range and Movements

Locations of radio-collared kangaroos were determined on a fixed schedule by triangulation from known points reached by car, motorbike, and on foot. After January 5, 1986, each individual was located once in the early morning, once in the evening, and once at night during each week. Occasionally, individuals could not be located, and were not sampled on this schedule. When an individual could not be found on the routine schedule, we made an effort to relocate it as quickly as possible. We found most individuals within several days by searching cross-country on a motorbike. Lost animals usually had moved more than 5 km, but signals could be detected over considerable distances because of the level terrain. The longest disappearance involved the female red kangaroo, Scarlett O'Hara, who was lost on September 23, 1985. We finally rented an airplane on October 21, and located Scarlett some 17 km from her previous haunts.

Locations were made by two or more triangulations from known points on a topographic map at as near to 90° angles to the site of the individual as possible. Azimuths were read with a precision, direct-sighting compass. Most azimuths were taken from within 1.5 km distance of the animal. Thanks to gentle topography and open vegetation, conditions for radio telemetry were nearly ideal. We had no problems with signal bounce.

Ordinarily, we avoided walking in on radio-collared animals to prevent abnormal movements due to disturbance by the observer. However, if a mortality was suspected, or particular life history information was needed, we approached radioed individuals but attempted to keep them unaware of our presence. In those cases, the expected locations based upon triangulation were very close to the observed location (usually within 50 m). Similarly, dead radioed animals and slipped collars were always found in close proximity to their triangulated position. Although we cannot quantitatively establish the margin of error in triangulated locations, we believe most would fall within a 50-m radius of the true location.

Locations were recorded on the x- and y-axis of a 1-km^2 gridded topographic map to the nearest 100-m unit. This scale would envelop most of the error in locations. Furthermore, any possible error would be trivial, given the scale of home range size and movements shown by the kangaroos. Some individuals moved widely, had to be triangulated at longer distances, and produced greater error in location. Nevertheless, their home ranges were correspondingly large, so the relative error would still be small. Conversely, some individuals were extremely sedentary. Triangulation distances in those cases were short, locations were precise, and absolute error was small. Direct observations of radio-collared individuals made incidental to other project work were also recorded. Locations of these individuals could be reliably placed due to the detail of the topographic map we used and the availability of recognizable features in the landscape.

In calculating home range sizes, movements to new locations that were within two previous home range diameters and involved return and continued use of the previous home range were considered home range extensions, and the new areas explored were included in the overall area of the home range. Movements of individual kangaroos that exceeded two previous home range diameters, and involved nonreturn to the former, were considered separate, serially occupied home ranges. Inclusion of the locations into a single home range would extremely distort the home range area, and inflate the actual use of space by the animal.

Home range locations were taken from January through September 1986, giving nine months of continuous data on all surviving kangaroos except Scarlett O'Hara, for whom data over 15 months were available. Following completion of continuous field work, until late June 1988, Stephen Thornton made trips to Yathong twice each year, in January and July, to relocate radio-collared animals. These locations were taken to determine whether or not home ranges established during the nine months of continuous radio telemetry persisted over the longer two and one-half year period.

Home range areas were calculated for the nine months of continuous field data with the programs McPAAL (Stüwe and Blohowiak 1985) and CALHOME (Kie et al. 1994). Five different calculation methods were used, and their outcomes compared and evaluated to see whether the assumptions were met. Some methods proved to be of little use, however, and no one method completely encompassed all of the interesting aspects of home range. Therefore, several methods were used to describe spatial use by individual kangaroos (see chapter 5).

Habitat Use by Radio-Collared Kangaroos

As described earlier in this chapter, distribution of kangaroos by species across vegetation types for the population as a whole was obtained from systematic surveys. However, these distributions were potentially biased because observability of kangaroos varied with vegetation type. The sample of radio-collared kangaroos could be used as a cross-check for such biases, because radioed animals could be located without being directly sighted and/or disturbed, and their grid locations assigned to vegetation types according to the vegetation type map. Radio telemetry data essentially lacked bias.

Data were analyzed by species and sex categories and by environmental state. The different species and sex categories and time periods in the radio-collared samples were tested (null hypothesis) for significance with x^2 (chi-square) tests of contingency tables. Vegetation types with less than five percent of the observations were deleted from the test.

Activity Patterns

Most visual methods of observing the activity patterns of wild animals are biased by difficulty of observation in different habitats and during hours of darkness, and by the lack of extended continuous records on given individuals. To overcome these problems, an automated radio telemetry system (Beier and McCullough 1988, 1990) was employed to continuously record activity patterns of kangaroos. Each radio collar had a mercury "tip-switch" built into the circuitry that opened and closed to change the signal pulse interval in response to the kangaroo's head position. The interval between pulses was 0.99 seconds with the head up and 0.66 with the head down.

Signals were received and processed by Telonics equipment, which consisted of a receiver, scanner, and digitizer. Signals were recorded on a dual-channel paper chart recorder (Rustrac, Galton Industries, Greenwich, RI). The switching frequency and timing of the entire system was controlled by a small controller–data logger computer (Polycorder, Omnidata International, Logan, UT) with a precise internal clock. The system was soon dubbed KROS, for "Kangaroo Radio Observational Station."

KROS worked as follows. The radio frequencies of one or more individuals to be monitored were entered into the scanner's memory. The scanner was controlled by the computer, which was programmed to turn itself on at a specified time, and to send a voltage signal to switch the scanner to the next frequency stored in memory. When more than one kangaroo was being monitored, the switch between individual frequencies was made at precise five-minute intervals. When a single kangaroo was monitored, a second, bogus frequency was stored in the scanner, and the computer was programmed to switch to the bogus frequency (no signal) for five seconds, and then switch back to the real frequency. This put precise time markers at five-minute intervals on the paper chart traces of a single kangaroo, which was necessary to overcome slightly variable rates the chart paper advanced due to battery power, ambient temperature, and so on. Although these advance errors were small on a five-minute time scale, they often became large when cumulative over the three days to a week the system was run without checking.

Signals were fed to the digitizer that converted the signals to numeri-

cal values of signal strength or amplitude (channel one) and pulse interval or period (channel two). Each channel was recorded on the paper
trace of the chart recorder, which was geared to a slow rate of travel,
giving about one week of continuous recording. This resulted in two
traces on the chart paper of about 12 mm length for each five-minute
sample of a given kangaroo. The switches on the digitizer were set such
that with signal loss, the pulse interval dropped to zero. Thus, traces due
to signal losses (that occurred for example, when kangaroos moved behind obstructions or beyond the range of the receiver) could be distinguished from real signals from kangaroos that were inactive.

Because the individual radio-collared kangaroos were spread over an
immense area, KROS had to be moved about to the locations of the
radioed kangaroos to be monitored. For portability, the receiving antenna was constructed of 1.5-m lightweight, interlocking television antenna sections, topped by a 2.5-m omnidirectional whip antenna. Either
four or five antenna sections were erected and supported by three guy-
lines of nylon rope, giving a tower height of 8.5 m with four sections or
10.0 m with five sections. The higher tower was used in situations where
greater topographic relief interfered with signals, or where individual
kangaroos had very large home ranges.

An antenna lead carried the signals from the omnidirectional antenna
to the receiver-recording system, which was housed in a weatherproof,
insulated picnic cooler box. The instruments were trickled-charged from
a heavy-duty 12-volt rechargeable automobile battery, housed in a second cooler box, with weatherproof leads to the receiver equipment in
the first box. Both cooler boxes were covered by a nylon tent rainfly to
protect the equipment from the extreme heat of direct sunlight, and from
rainfall (fig. 3.3). With experience, KROS could be set up in about
15 minutes, and taken down in 10 minutes.

Given the number of instruments, switches, wires, connections, battery packs, and other components, this system worked remarkably well,
producing interpretable activity traces for over 90 percent of the five-
minute traces. Backup components plus jury-rigging kept the system op-
erating. KROS could run, unattended, for a week (the length of the recording paper charts), but it was usually checked every three days.

KROS was moved to different sites around Yathong to equally sample
each individual radioed kangaroo. When possible, more than one individual (up to three total) was simultaneously monitored. A running log
and checklist of individuals monitored and yet to be monitored was kept
to be sure that each species, sex, and individual was equally sampled
each fortnight, to prevent biasing the results due to patterns of a given

Figure 3.3. Checking the movable activity recording system (KROS). Note the pine vegetation type in the background.

individual. The ideal, an equal sample of each individual, was not strictly met because of occasional movement of individuals out of range, and equipment failures. Nevertheless, equal sampling was closely approximated and the gaps in the record were essentially random. There is little doubt that the activity patterns obtained were representative of the radio-collared sample. Furthermore, the low variation among individuals of a given sex and species category indicates that the data were representative of the population at large. This assumption could be partially cross-checked by comparing the results from radio telemetry to observation patterns of kangaroo sightings in the systematic surveys.

Chart papers were hand-labeled at the start of each session with Julian date, time, radio frequencies and serial order of individuals being received. Time and individual thereafter could be determined by the number and order of five-minute traces in the record. When KROS was rechecked, the time and current frequency being monitored (read from the scanner) were hand-recorded on the current trace. This allowed backchecking to the previously recorded start or check time to verify that the number of five-minute periods was consistent with the starting and ending times, and the frequency being monitored was the correct one. On several occasions, electrical storms apparently produced sufficient voltage to scramble the frequencies in the memory of the scanner, and data were lost for the remainder of the chart.

Back at Shearers' Quarters, all paper charts were interpreted by the same research assistant, Judy Graham; thus, there was no observer variance in the interpretations. Each usable five-minute trace was scored as active or inactive based on the criteria developed by Beier and McCullough (1988). If both signal strength and pulse interval, in either head up or head down position, were virtually constant, the individual was scored as inactive for that five-minute period. Occasional "spikes" in a trace that was otherwise constant were ignored, because direct observations of bedded kangaroos showed that they occasionally moved around in their beds or rose up to scratch themselves.

All other trace patterns were recorded as active, characterized by continually shifting amplitude, and pulse intervals that varied between head up and head down. Active kangaroos change head position frequently, so many intermediate values were logged as the recorder's pointer moved up and down. If the kangaroo changed behavior from active to inactive within a five-minute trace, it was scored by the behavior that represented greater than half of the trace. Early in the study, direct observations on Scarlett O'Hara were used to develop criteria for interpretation of the traces. Backchecking the traces of radio-collared individuals observed during other project work showed that they had been correctly scored as active or inactive. Among active kangaroos, no attempt was made to discriminate specific kinds of activity, because direct observations of Scarlett O'Hara showed that several different activities produced similar traces.

Data on individual animal, Julian date, time, and active or inactive status were entered into a computer file, and activity patterns were examined by individual, sex by species, species, and all kangaroos combined. Percent activity was calculated by dividing the number of traces scored as active by the total number of traces in the sample time period multiplied by 100. Sample size (number of traces) depended upon the length of time over which the data were summed. For analysis of daily activity patterns, kangaroos were combined by species and sex categories to calculate percent activity by hour each month. Other analyses involved combining percent active by day and time over different environmental states, and correlations with weather variables recorded at Shearers' Quarters. Mean values by hour of weather variables other than rainfall were entered as continuous variables. Rainfall was entered as a categorical variable according to the following categories: (0) no rain, (1) light mist, (2) scattered showers, (3) widespread, heavy rainfall.

All differences between sexes, species, environmental states, and other variables were tested by analysis of variance (ANOVA), or where assumptions about distributions would not be met, the nonparametric equivalents. The hypothesis tested was the null case, i.e., no difference between the two or more categories being compared at alpha ≥ 0.05. The influence of weather variables on kangaroo activity was analyzed by multiple regression. Percent activity and weather variables were transformed where necessary to achieve equality of variances. The different species and sex categories were examined separately by environmental states.

The influence of weather was determined after removing the effects of Julian day and environmental state. The approach was to first run the independent variables in a stepwise regression with alpha to enter or delete set at 0.15 (Wilkinson et al. 1992). Important variables were then run sequentially in a multiple regression model with quadratic terms and interactions between variables to derive a model that maximized adjusted R^2 (multiple coefficient of determination) and minimized P (probability of significance). This model was then used to estimate the importance of the significant individual weather variables (or interactions) according to their individual F-ratios. Because of the large number of variables included and regressions run, alpha was set at 0.01.

Direct Observations of Kangaroo Behavior

We conducted direct behavioral observations of the three species as time permitted. We usually observed from the vehicle as a blind, because this alarmed kangaroos much less than a person on foot. None of the kangaroos was habituated, so the presence of the observer vehicle likely biased somewhat their normal behavior. Therefore, timing of observational bouts and calculating time spent in different activities were not done. To minimize observer interference, we conducted most daylight observations at 150 to 200 m distance with binoculars and a 20-power, window-mounted spotting scope.

Night observations were made by using a spotlight or night vision goggles. Aided by binoculars, observations by spotlight could be made up to 200 m. On moonlit nights, passive-light ("starlight") night vision goggles were effective up to 150 m in conjunction with binoculars. On darker nights, there was insufficient light to support use of binoculars, and this reduced the effective range of night vision goggles to a maximum of about 75 m. By spotlight, we could take notes in the diffusion of light emanating from the beam. With night vision goggles, we wrote using a supplementary infrared light source. A microcassette tape machine was used to record notes when the observer ran out of hands coordinating night vision goggles and binoculars.

Night spotlighting disrupted the normal behavior of the kangaroos, as they periodically turned to stare at the light and observer. But kangaroos soon ignored the vehicle and occupant when being observed in the dark with only the night vision goggles. The vehicle could be driven effectively without any lights, even on the darkest nights. Indeed, we quickly learned that when wearing night vision goggles, tail light illumination during braking was blinding, rather like a flashbulb going off in one's face.

Kangaroo behavior was mundane most of the time. Not including the time spent watching the observer, kangaroos spent 98 percent of their time eating or lying down; most of the other two percent was spent moving between where they fed and where they bedded. Social interactions (agonistic interactions, care-giving, and sexual behavior) represented a fraction of a percent of a kangaroo's time; they usually occurred quickly and then were over. Therefore, we elected to not schedule systematic behavioral observational time. Instead, other project activities were pursued, and when fortuitous bouts of interesting behavior were observed, other work was momentarily interrupted and the bout was recorded. We timed such behavior with a digital wristwatch.

Vegetation Photo Points

We did not have time or labor to do detailed monitoring of vegetation growth and biomass production. Nevertheless, it was necessary to generally monitor the growth and drying of vegetation because of its great importance to kangaroos as food (Moss and Croft 1999). To do this, we established marked photo points in major vegetation types and photographed them with color film in the second fortnight of each month. Five photo points were established in grassland, three in upland areas, and two in floodplain areas. One photo point was established in the mallee, bimble box, and pine association vegetation types. At each photo point, a marker stake was placed on each side of the road, and orientation stakes were placed 10 paces from the road on either side. The photographer stood at the roadside stake, and placed the central, focus spot of the camera on the top of the second stake. Thus, the views were virtually identical between months. A comparable photo taken on the opposite side of the road resulted in two photos in opposite directions at each photo point each month.

In addition, copious field notes were kept on herbaceous vegetation phenology, i.e., timing, whether green or dry, flowering, seeding, and whether or not grazed by kangaroos. Photo point reference, in conjunction with these notes and herbaceous scoring of kangaroo feeding sites in the systematic surveys, gave a reasonable qualitative record of herbaceous vegetation dynamics, and these form the bases of green feed abundance plotted in figure 2.9.

Step-Point Vegetation Surveys

At times it was desirable to have more detailed comparisons of two contrasting circumstances for herbaceous vegetation. These involved comparison of (1) herbaceous plant cover between vegetation types at the same time, (2) plant cover between burned and unburned sites (on opposite sides of the road) within vegetation types, and (3) concentrated kangaroo feeding sites with randomly selected sites in the same vegetation type at the same time.

The step-point method (Evans and Love 1957) was used because it is easily adapted to relatively sparse herbaceous vegetation, and rapidly learned by unskilled observers to produce repeatable results within five percent variability for percent cover. University Research Expedition Program volunteers, working with project personnel, performed much of this work.

Areas to be sampled were mapped on topographic and vegetation maps. A systematic sampling scheme with a random starting line was used to locate transect lines. Once the first line was located, the others were systematically assigned, usually parallel to the first at 100-m intervals (occasionally, the angle was altered to avoid intersecting a different vegetation type).

Each observer was assigned a transect line by compass azimuth toward a distant landmark. Sample points were determined by pacing the transect line. At an appropriate number of paces (which varied with the size of area to be sampled, but usually was every fifth pace), the point directly under a pen mark on the toe of the right shoe was scored as bare ground, litter, or vegetated. Vegetated hits were classified as herbaceous forb or grass, and as new green shoots, green clumps, or dry clumps. The points on herbaceous vegetation were counted only if the mark fell within the basal area of the plant. Thus, using this method, tops of stems pressed down by the foot were discounted. While pacing, observers kept their eyes on the point of sight on the horizon to prevent influencing pace length or foot placement.

Each transect consisted of 100 points for the herbaceous layer, which included points that fell on vegetation, litter, or bare ground. Once each point was scored and recorded for the herbaceous layer, the shrub and tree layers were determined. If, when vertically projected, the point fell within the crown area of a shrub or tree, in addition to a herbaceous hit, the point was recorded on shrub or tree canopy layers. When the transect was completed, on the return trip to the point of origin, the observer determined and recorded the nearest tree to points at one-fifth the transect length (five observations/transect) in vegetation types with low tree density, and one-tenth the transect length (10 observations/transect) in types with high tree density. Observers lacking reliable estimation skills determined the distances to the nearest tree by pacing.

Percent cover for bare ground, litter, and herbaceous vegetation was calculated from the 100 hits at ground level. Percent canopy cover for shrubs and trees was calculated without reference to bare ground or litter.

Stem density of trees was calculated from distance estimates as

$$\frac{10,000}{(\bar{x}2)^2} = \frac{tree}{ha}$$

where $\bar{x}$ = the mean distance in m (Cottam and Curtis 1956).

Each step-point transect was considered an observation, and the mean and variance for the site were calculated from the sample of observations (i.e., number of transects). Differences in herbaceous cover between sites being compared were tested by analysis of variance (ANOVA) with values transformed if necessary to meet the requirements for equal variances. The null hypothesis was tested and alpha was set at 0.05. Differences between sites in shrub canopy cover, tree canopy cover, and tree stem density were tested by paired t-tests using the same criteria for acceptance.

Fecal Nitrogen

The value and limitations of fecal nitrogen as an indicator of diet quality have been debated in the literature (Robbins 1983; Hobbs 1987; Leslie and Starkey 1987). It is apparent that, for a variety of reasons, fecal nitrogen is not an unvarying proportion of dietary nitrogen, and that there are limitations in using it as an indicator of diet quality. However, if the comparisons are carefully chosen, we believe it is a useful indicator of relative diet quality within species across environmental states, and across species in the same environmental state. The justification for this view is given in chapter 9. It is also worth noting that nitrogen is the fecal component that currently is most useful for giving low-cost information about quality, as opposed to composition, of diet. As such, it warrants exploration in a nonintrusive study such as this.

Fecal collections were made each month in the second fortnight. For each of the three kangaroo species, 10 samples were collected from different individuals. All droppings were collected fresh from sites where kangaroos were observed immediately prior to collection. There was virtually no chance of misidentifying the species of kangaroo producing the dropping collected.

We attempted to make the fecal collections broadly across the intensive study area (fig. 2.2) to avoid local influences on the results. This proved to be difficult for eastern grey kangaroos, which at times could be found only in certain areas. In these cases, samples were collected over at least several separate sites. We soon learned that although kangaroos defecated under a variety of circumstances, a majority of defecations occurred during feeding. Searching for droppings around bedding sites, travel paths, or by tracking fleeing kangaroos was abandoned, and efforts were directed solely to individuals observed actively feeding.

All of the droppings of a single defecation were collected and placed in a paper bag that was labeled with species, date, and area of collection. Samples were air-dried (easily done at Yathong) and stored for analysis. Fecal nitrogen determination was performed by Ruth Windridge under the direction of David Tongway of CSIRO Rangelands Research Center in Deniliquin according to the Kjeldahl method of Nelson and Sommers (1972, 1973, 1980). Initial statistical tests showed that variation between pellets within a group was usually within the variation in the analytical technique. Duplicate subsamples ($N = 17$) had a mean difference for combined species of 0.06% (range 0.14–0.00) compared to five tests of six pellets within samples for eastern grey kangaroos that had standard deviations ranging from 0.06 to 0.03; two tests of six pellets for red kangaroos ranging from 0.16 to 0.09; and three tests of western grey kangaroos ranging from 0.10 to 0.06. Therefore, for most dropping samples, only a single pellet was used to determine nitrogen, and this was treated as the sample; sample size was the number of droppings, which equaled 10 for each species for each month.

Finally, a series of pellet groups was measured for mean dimensions, so pellet size, which is broadly correlated with individual kangaroo size, could be correlated with nitrogen content. Where sex or age of the kangaroo producing the dropping was known unambiguously, this information was recorded to verify the correlation between size of dropping to size and sex of kangaroo.

From an understanding of intraspecies variation, the interspecies comparisons could be made more parsimoniously, and with greater assurance that the species was the source of differences, rather than animal size, sex, or analytical technique. Differences between species were tested by ANOVA with a factorial model with month and species as "treatments" and individual fecal samples by species as replications.

4

POPULATION SIZE AND COMPOSITION

Kangaroo populations in outback Australia fluctuate markedly due to periodic drought (Caughley 1987b, Cairns and Grigg 1993). Populations increase during periods of rainfall and then crash with the next drought. Mortalities exceeding 50 percent of the population are common (Caughley et al. 1985b). Populations at Yathong suffered this magnitude of mortality during the 1982 drought.

Given this long-term fluctuation in numbers, estimates of population size at any point in time represent a fleeting, ever-changing snapshot in the continuing saga of rain, grass, and kangaroos. Rainfall is linked in a complex way to El Niño–Southern Oscillation and La Niña events (Nicholls 1991). With the onset of such events, the short-term rainfall and drought pattern is somewhat predictable. Drought accompanies El Niño–Southern Oscillation events, and high rainfall coincides with La Niña events. However, the interval between subsequent El Niño–Southern Oscillation events is not predictable, nor are the intervening weather patterns, themselves highly variable. Thus longer term patterns of meaning for kangaroo populations are, for practical purposes, unpredictable. Vegetative growth and kangaroo numbers, with appropriate lags, are determined by rainfall, and are whiplashed in time and out-of-sync, usually being out of phase with the current state of rainfall, which ordinarily comes in pulses.

Fully understanding such a dynamic system would require study over a very long period of time, for one need not only measure the three variables in a time series (even years may be too long a unit of time, given the aseasonal character of rainfall), but one must also have a sequence of booms and busts in order to examine for nonlinearities within functional relationships (i.e., lags are not constant). Obviously, such a project is not feasible for an American on sabbatical every seventh year. Even Australians have found this task daunting, and have fallen back on using a commonly measured variable—rainfall—and modeling the system with short-term data on vegetation and kangaroos (Caughley 1987b).

Fortunately, our task at Yathong was somewhat simplified. We had no expectations beyond obtaining more than one small piece of the larger puzzle. But as we discovered, the position of that piece in the larger puzzle was determined by the 1982 drought. That drought greatly reduced kangaroo numbers broadly over Australia (Caughley et al. 1987; Cairns and Grigg, 1993) including at Yathong, and the rainfall records thereafter indicated the timing of rainfall pulses that drive the system. Given the short time between this major drought and our study, the maximum rate of kangaroo population growth, the occurrence of rainfall pulses, and the descriptions of local residents, we deduced that Yathong populations were in the recovery phase. The major question was how far into recovery were the populations, and were they approaching their maximum limits during good rainfall periods?

Population analysis was based primarily on systematic survey data. The survey routes broadly sampled the intensive study area (fig. 2.2), and should be representative of that part of Yathong. They are not, however, representative of the extensive mallee areas on the western side of the reserve. Known biases of the systematic surveys were corrected, where possible, by unbiased data from the radio-telemetry locations.

Sex Ratio

The gross (uncorrected) sex ratio for all systematic survey data over 13 months was 0.31 (males:1 female) for red kangaroos ($N = 4627$), 0.45 for eastern greys ($N = 538$), and 0.51 for western greys ($N = 3457$). Over a period of months, the observed sex ratio of all species changed (fig. 4.1). High variation in the sex ratio of eastern grey kangaroos was no doubt largely due to small sample sizes, and only the gross downward trend is likely to be real. Although sex ratio trends of red and western grey kangaroos were more consistent, small variations are probably due to sampling error, and only the major trends are likely to be reliable.

Given that no significant adult mortality was detected, we do not believe there was a real change in true sex ratio over the study period. The seasonal change in observed ratios was due to behavior, particularly the association of males with females during periods of higher breeding activity (see chapter 8) and differential use of habitats by the sexes (see chapter 10), both of which influenced relative observability.

One approach to correcting the obvious sex ratio bias is to use the ratios obtained during breeding periods when association between the sexes was highest, and concealment cover of different habitats was least likely to bias observability. Mixed-sex groups were highest (see figs. 7.2 to 7.4) for red and western grey kangaroos in July and August, and these periods coincided with observed breeding behavior. Although the sample size for eastern grey kangaroos was less, their breeding season seemed to occur at about the same time as for the other two species.

Sex ratios of all three species were higher early in the study (fig. 4.1), indicating a clear overrepresentation of males in the systematic surveys as compared to the rest of the year. Ratios for August and September (when adequate samples for all three species were available) were 0.50 for red, 0.63 for eastern grey, and 0.76 for western grey kangaroos. These ratios set the upper limit that the sex ratios could take. However, we believe they are biased to the high side, because males not only associate with females at this time, but generally move about in open areas where they are more likely to be observed. Thus, it is likely that this approach overcorrects, and errs in the opposite direction.

Habitat use is the major variable influencing the sex ratios observed from the systematic surveys. For both red and western grey kangaroos, there was a highly significant difference in the proportion of each sex observed between grassland and other habitats with greater concealment cover (red chi-square = 39.22, df 1, $P < 0.001$, $N = 4451$; western grey chi-square = 125.46, df 1, $P < 0.001$, $N = 3457$). Finer separations of habitats with concealment cover were not made because such partitioning resulted in small sample sizes for many habitat types.

Eastern grey kangaroos showed no significant difference in the proportions by sex between grassland and other habitat types (chi-square = 1.01, df 1, $P = 0.316$, $N = 538$). This result is due partly to the relatively small sample sizes. Nevertheless, eastern grey kangaroos of both sexes were highly selective of wooded habitats, which made observation by direct methods difficult, but reduced bias in numbers of each sex observed. For this species, no correction factor for habitat use was applied. The overall ratio of 0.45 from systematic surveys is likely a bit too low, and the ratio of 0.63 from the breeding season is too high. A ratio in the neighborhood of 0.50 male:1 female was probably reasonable for this species.

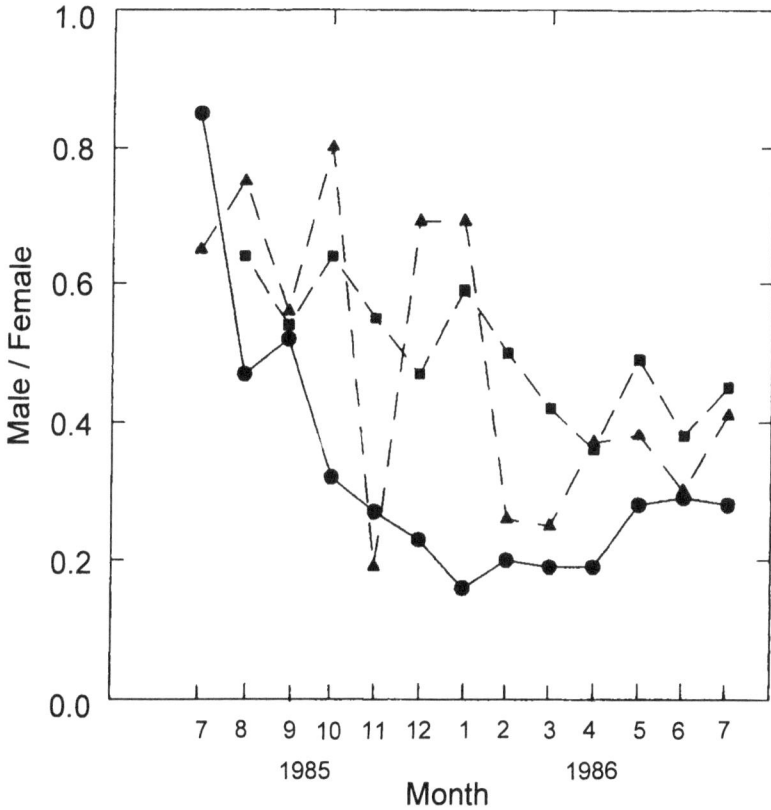

Figure 4.1. Sex ratios (males:1 female, uncorrected) for eastern grey (triangles), red (circles), and western grey (squares) kangaroos as derived from the systematic surveys at Yathong Nature Reserve.

Correction factors were applied to the observed sex ratios of red and western grey kangaroos as follows. The ratios from the systematic surveys were derived separately for grassland and for other habitats with higher concealment cover. These ratios were weighted by the relative use of grasslands from the unbiased radio-telemetry data to derive a corrected sex ratio. For example, for red kangaroos the ratio of male:1 female was 0.26 for grasslands and 0.41 for other habitats. Radio locations showed 28.6% of red kangaroo locations (both sexes combined) in grasslands and 71.4% in other habitats. Weighting the sex ratios from grassland and other habitats by these percentages of habitat use gave a corrected sex ratio of 0.36 for red kangaroos. Similar correction methods were applied to western grey kangaroos; the gross ratio of 0.51 was corrected to 0.55 by these methods. This is somewhat higher than the 0.46 ratio reported for western grey kangaroos in southwest Australia by Arnold et al. (1991). Thus, the sex ratio was the most unbalanced towards males in reds, with the two species of greys also unbalanced, but similar.

The juvenile sex ratio was not determined in this study, but the sex ratio of pouch young previously reported showed ratios near equality (Arnold et al. 1991), or unbalanced towards males (Johnson and Jarman 1983; Norbury et al. 1988). This suggests significant male-biased mortality of kangaroos during young-at-foot and juvenile stages, as reported by Norbury et al. (1988). Quin (1989) reported that male-biased mortality occurred at most life stages of eastern grey kangaroos. This results in a sex ratio of adults skewed towards females, as is common in North American deer (McCullough 1979; Connolly 1981). Adult mortality at Yathong was too low by itself to account for the imbalance in the sex ratio. Higher rates of mortality for juvenile males probably continue among young adults (Norbury et al. 1988; Dawson 1995).

Age Ratios

Age ratios were based upon the number of young-at-foot/number of adult females observed in the systematic surveys. Age ratios varied over the months of the study (fig. 4.2), but unlike variation in sex ratios, these variations surely reflect the appearance of successive cohorts of offspring. Given the seasonal nature of reproduction and presence of large pouch young (see fig. 8.2), young-at-foot are the oldest cohort of dependent offspring, displaced from the pouch by the mother to make way for the younger cohort of pouch young. The pouch young, in turn, will eventually be replaced by suspended blastulas of the third and youngest generation in red and some eastern grey kangaroos, and by conception and birth in those eastern greys without suspended development, and in western grey kangaroos.

Gross year-long age ratios from the systematic surveys were 0.36 young-at-foot:1 female for red, 0.39 for eastern grey, and 0.44 for western grey kangaroos. There was no significant difference in the age ratio of eastern grey (chi-square = 1.57, df 1, $P = 0.21$, $N = 517$) or western grey (chi-square = 0.19, df 1, $P = 0.67$, $N = 2698$) kangaroos observed in grassland versus other habitat types with greater concealment cover. Therefore, no correction for habitat was indicated. There was a significant difference in the age ratio for red kangaroos (chi-square = 8.08, df 1, $P = 0.004$, $N = 4627$) with the ratio being 0.38 in grassland and

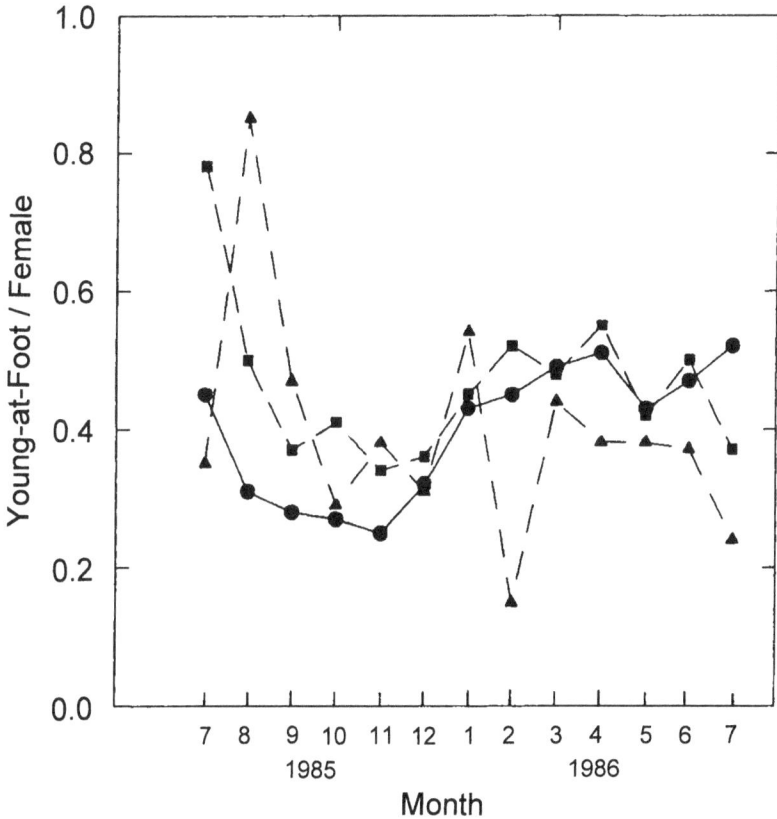

Figure 4.2. Age ratio (young-at-foot:1 female, uncorrected) for eastern grey (triangles), red (circles), and western grey (squares) kangaroos as derived from systematic surveys at Yathong Nature Reserve.

0.31 in more closed habitats. Because the other two species did not show this difference, it is unlikely that a bias against observing young-at-foot in closed habitats exists for red kangaroos alone; thus, the age ratio of red kangaroos was not corrected either.

The similarity of age ratios among the three species at Yathong suggests that reproductive rates were roughly comparable, although the year-long rates are somewhat misleading in that they include two cohorts of offspring. Figure 4.2 shows that the first cohort tended to be higher in the two species of greys than in red kangaroos, whereas the second cohort was similar in red and western grey kangaroos and somewhat lower in eastern grey kangaroos.

That about 60 percent of adult females of kangaroos at Yathong were without young-at-foot suggests a relatively low rate of reproduction. Part of the absence of young-at-foot is attributable to relatively weak synchronization of breeding in kangaroos. Still, the ratio bespeaks a population that is not expanding rapidly. The two generations of red kangaroos (fig. 4.2) were similar in magnitude, whereas the first generation of both species of greys was higher than the second. Given that the first was produced during relatively favorable rainfall conditions, and the second during low rainfall, the response of the two species of grey kangaroos was as expected. Why the first generation of red kangaroos was not greater is an open question. However, as noted in chapter 6, the red kangaroo, when wet, is subject to cold stress at Yathong, near the southern edge of their geographic distribution, and small young, presumably, would be most susceptible. Thus, the cold stress associated with high rainfall may offset the favorable food conditions produced by rainfall in the coldest (i.e., southernmost) parts of the red kangaroo's range. It would seem that warm season rainfall would promote greater success for this aseasonal breeding species than would cold season rainfall. This conjecture is discussed further in chapter 10.

Mark-Recapture Population Estimate

All marked kangaroos observed in the systematic surveys were identified and recorded. The ratio of marked to unmarked kangaroos was used to calculate an estimate of population size by the Petersen-Lincoln method (Caughley 1977). Marking occurred mainly in December 1985, and marked animals of the completed sample occurred in systematic surveys from January 1986 through the end of these counts in July 1986. Overall, 43 observations of marked kangaroos were made in the systematic counts. Over time, however, death of marked kangaroos and movements of some marked individuals out of the area under observation (see chapter 5) changed the effective size of the marked population.

In general, it is desirable to allow a relatively short time between marking and reobservation to minimize these problems. Even though the radio collars provided us the size of the marked population (because movements and death were known), accumulation of observations over time would require calculation of a weighted average of the marked population as sample size of marked animals declines. Therefore, it seemed preferable to confine the population estimate to data collected shortly after marking, while the full sample of marked kangaroos was still alive and within the study area.

Table 4.1. Species composition of kangaroos in the Yathong intensive study area obtained from combined systematic surveys and population size and density estimates based upon the total number of kangaroos estimated by the mark-recapture method.

	Systematic Surveys		Mark-Recapture	
			Population	Density
Species	N	Percent	Estimate	Estimate (/km²)
Eastern grey	778	6.9	263	0.9
Red	6358	56.4	2154	7.5
Western grey	4137	36.7	1402	4.9
Total	11,273	100.0	3819 ± 777 [1]	13.3

[1] Standard error.

Between January 4 and April 4, 1986, all kangaroos marked by radio collars and eartags (and not suffering immediate capture-induced mortality) were in the population. During this period, 3136 kangaroos of all species were observed in the daylight and night spotlight systematic surveys, which sampled the intensive study area of approximately 288 km². Because the number of marked kangaroos was too small to make estimates separated by species, all species were combined for the estimate. A total of 22 marked kangaroos was observed. Using the Bailey formula (chapter 3), the estimated number of kangaroos was 3819, with a standard error of ± 777. For 288 km² this gives an estimated density of 13.3 kangaroos/km². This total estimate of kangaroos can be broken down by kangaroo species using the species percentages in the combined systematic surveys (table 4.1). This resulted in an estimate of 263 eastern grey, 2154 red, and 1402 western grey kangaroos in the intensive study area.

The accuracy of a mark-recapture estimate based on resighting is strongly influenced by biases in observability of marked and unmarked animals (McCullough and Hirth 1988). Essentially, the method requires complete mixing of the marked sample with the unmarked population. Because animals do not move at random, but instead occupy home ranges or other nonrandom subsets of the environment (see chapter 5), it is important that the marked sample be a representative cross-section of the population to be estimated. Trapping methods are prone to capture unrepresentative samples because traps are attractive to some individuals that seek out the bait, and repellent to others that avoid strange objects in the environment. Also, the placement of traps often leads to a nonrandom and nonsystematic placement pattern. Biases in capture can severely distort estimates of population size, and, unfortunately, these biases cannot be corrected by increasing the sample of marked animals or of recaptured (resighted) animals. For example, McCullough and Hirth (1988) found that estimates of white-tailed deer, where 80 percent of the population was marked, were incorrect because of bias in capture and reobservation.

We could not test the capture bias in this study because the true population was unknown. However, intuitively, using an active method, such as stunning, that seeks out animals and takes them as encountered, seems unlikely to show the biases in capture associated with trapping methods. Furthermore, the capture occurred along the same roads as the resighting, and the road system was spread broadly over the intensive study area. Consequently, these spatial attributes favored representativeness in both capture and resighting.

In contrast to capture, it is possible to evaluate the resighting sample by comparing it to the relative abundance of the species and sexes in the systematic surveys (table 4.2). In fact, both the sex and species distribution of the resighted marked animals and the systematic surveys were reasonably similar, suggesting general representativeness of the marked sample (table 4.2). The biggest discrepancy was in species comparisons of western grey kangaroos, which were overrepresented in the resighted sample, and red kangaroos, which were underrepresented. We

Table 4.2. Number of observed versus expected marked animals by sex and species. Expected rates are based upon percentage of the species in systematic surveys and weighted sex ratio for the three species combined.

	Eastern Grey	Red	Western Grey	Total	Percent of Observed	Weighted Sex Percentage[1]
Male	2	4	6	12	27.9	29.9
Female	2	14	15	31	72.1	70.1
Total	4	18	21	43		
Percent of marked animals observed	9.3	41.9	48.8			
Percent of animals in systematic surveys	6.9	56.4	36.7			

[1] Percent of the species in the systematic surveys (last row) by estimated percentage by sex for each species (eastern grey = 33%, red = 26%, and western grey = 35% males).

believe this difference was due to the small and consistent home ranges of western greys (see chapter 5), which increased the probability of re-sighting, whereas the large home range and inconsistent occupation pattern shown by red kangaroos reduced probability of resighting. Consequently, the numbers of western grey kangaroos may have been somewhat overestimated, whereas red kangaroos may have been somewhat underestimated.

Line Transect Population Estimate

Using the program DISTANCE (Laake et al. 1993), line transect analysis was performed on perpendicular distances of kangaroos obtained from systematic surveys. (See Burnham et al. 1980 for a full description of this method.) Only night spotlight samples were used in this analysis, as these were less biased by poor observability of kangaroos in heavier vegetation cover, and they better fit the sighting distance distribution functions. The night spotlight and daylight samples could not be combined because the distributions were different due to the limited observability distance of the spotlight counts. For red and western grey kangaroos, sample sizes were sufficient to derive estimates separately for the sexes (males versus females plus young-at-foot); however, eastern grey kangaroo sexes were combined in the analysis because the small sample sizes resulted in erratic results for each sex separately.

In all cases, the best-fitting model for distribution of perpendicular distance selected by DISTANCE was used. For combined eastern grey sexes and red males, this was hazard rate, for red and western grey females, uniform, and for western grey males, half-normal. No pooling of observations was required to obtain a reasonable fit as judged by chi-square statistics and by visual examination of function lines superimposed on histograms of perpendicular distances.

Population estimates derived by DISTANCE were similar to those obtained by mark-recapture (table 4.3). The sum of the separate line transect estimates for all kangaroos was a density of 12.83/km^2 as compared to 13.30 for mark-recapture. If we underestimated our populations by 10 to 12 percent, as reported by other users of DISTANCE (Anderson and Southwell 1995), the difference would be in the opposite direction, but still the results would be similar. All of the separate line transect estimates for sex by species were similar to those calculated for mark-recapture (table 4.3). Once again, eastern greys were least numerous, followed by western greys, with red kangaroos being most abundant. The relative magnitude of the estimates was comparable between the

Table 4.3. Population density estimates (kangaroos/km^2) from line transect analysis of night spotlight systematic surveys of kangaroos using the program DISTANCE, compared with mark-recapture estimates of density in the intensive study area.

	N^1	Line Transect $(\bar{x} + SE)$	Mark-Recapture
Eastern grey males	75	0.43[2]	0.30
Eastern grey females	143	0.88[2]	0.60
Red males	276	2.07 ($\pm$ 0.35)	1.95
Red females	955	4.94 ($\pm$ 0.94)	5.55
Western grey males	240	1.19 ($\pm$ 0.08)	1.72
Western grey females	531	3.32 ($\pm$ 0.32)	3.18
Total		12.83	13.30

[1] Sample size.
[2] The total estimate for eastern grey kangaroos (1.31 $\pm$ 0.53, N = 188) was broken down to sex by the estimated sex ratio of 0.5.

two methods of estimation. Density of each sex and species was multiplied by the area of intensive study (288 km^2) to get estimates of 124 males and 253 females and young-at-foot eastern grey (377 total), 596 males and 1423 females and young-at-foot red (2019 total), and 343 males and 956 females and young-at-foot western grey (1299 total) kangaroos.

Considering red and western grey kangaroos only, western greys constituted 39% of the total population by line transect, and 40% by mark-recapture. Thus, the possible overestimate of western grey kangaroos by the mark-recapture method suggested in table 4.2 does not seem to be borne out by the results from the line transect method.

One other independent cross-check of accuracy of red and western grey kangaroo estimates can be made using the sex ratio derived earlier. Recall that sex ratio was derived independently, as were the estimates of numbers of males and females by the line transect method. (Comparable estimates by sex were not done for eastern grey kangaroos because of small sample sizes.) The sex ratio of estimated numbers of red and western grey kangaroos did not match the ratio derived earlier (see Sex Ratio above). For red kangaroos, the line transect estimate ratio was 0.42, whereas the corrected ratio from systematic counts was 0.36; for western grey kangaroos, the comparable ratios were 0.36 and 0.55. If it is assumed that the corrected sex ratio from systematic counts was accurate, and if the line transect population estimates were accurate for red females, then the adjusted estimate would be 1935 red kangaroos, or 84 fewer than the unadjusted line transect estimates. Conversely, if one assumed the line transect estimate for red males was accurate, the adjusted line transect estimate would be 2252, or 233 greater than the unadjusted estimate. Comparable values for western grey females would be 1482 (183 greater) and for western grey males would be 967 (332 fewer).

The bounds of adjusted line transect estimates (1935 to 2252 for red kangaroos, 967 to 1482 for western grey kangaroos) are well within the standard errors of the unadjusted line transect estimates. This means that the discrepancy of sex ratios between line transect estimates and systematic counts are relatively minor, and if taken into account, the estimates do not change greatly. Because it is not possible to determine which of the three estimates (i.e., sex ratio from systematic counts, line transect estimates of females, or line transect estimates of males) includes error (or combination of errors), the unadjusted ratios were retained as reasonable.

Discussion of Population Results

Sex ratios of Yathong kangaroo populations were biased in favor of females. Eastern grey kangaroos had a sex ratio of about 0.50 and western greys, 0.55. The sex ratio was most unbalanced in red kangaroos, in which there were only 0.36 males for every female. Ratios unbalanced toward females are typical of large, polygynous mammals, including kangaroos (Dawson 1995). Also, ratios tend to be most unbalanced in species with the greatest sexual dimorphism, which was true in this study—the most dimorphic species, the red kangaroo, had the most unbalanced ratio.

Estimated age ratios were 0.36 for red, 0.39 for western grey, and 0.44 for eastern grey kangaroos. These ratios, in conjunction with the small number of kangaroo mortalities discovered in our study, suggest a growing population. Nearly all carcasses found were very old individuals that had survived the 1982 drought and died of causes related to old age, including those mortalities of radio-collared animals not due to capture effects (one red male and one eastern grey male). It seems likely that much of the population was born since 1982, and thus was in prime age classes. Maximum longevity reported for kangaroos is at least 27 years (Bailey 1992), but most populations are younger, with few individuals exceeding 10 to 15 years (Wilson 1975; Norbury et al. 1988; Quin 1989; Stuart-Dick and Higginbottom 1989).

From all appearances, the population at Yathong has recovered to a moderate density from the severe reduction during the 1982 drought.

The population estimates of this study suggested a density of around 13 kangaroos (all species combined)/km². For comparison, the density maps of Caughley (1987a) give densities in the Yathong area in 1980–82 (predrought) of 1 to 5 eastern grey, 5 to 10 red, and 5 to 10 western grey kangaroos, for 11 to 25 total kangaroos/km², or roughly comparable.

The estimates of kangaroos by the two different methods used in this study gave gratifyingly similar results. Although strictly speaking they cannot be considered as independent, because the resighting for mark-recapture and the distance estimates for line transect were taken on the same surveys, there was no reason to suspect that the probability of observing a marked animal had any relationship to the probability of animals occurring at various perpendicular distances from the transect. We believe that, for practical purposes, the two methods were independent, and that the use of the same surveys for both resighting and obtaining perpendicular distances was simply a matter of efficient use of effort.

It is further reassuring that the two methods gave comparable differences between the sexes and species. None of the results was contradictory with reference to relative abundance of the species and sexes. Any concern we may have about incorrect estimates is with eastern grey kangaroos. Because of their strong association with wooded habitat and their skittishness, we believe that they were probably underestimated. Still, there is little doubt that there are relatively few eastern greys compared to the other two species, and any conceivable underestimate would not change the total number of kangaroos on Yathong very much.

The area studied represents some of the best kangaroo habitat on Yathong. The densities found there are not representative of the entire reserve. Most of the western half of Yathong is extensive mallee habitat (fig. 2.10), and the only species that used that habitat commonly was the western grey.

If one takes the use of mallee by kangaroos (see table 10.1) and applies it to the density estimates by the line transect method (table 4.3), one obtains an estimate of density in mallee of 1.01 for western grey, 0.45 for red, and 0.05 for eastern grey kangaroos. Total density in mallee would be 1.51 kangaroos/km² or about 12% of the density in the intensive study area. We believe this estimate of kangaroos for mallee across Yathong is too high, as much of the mallee use by the radioed kangaroos in the intensive study area (fig. 2.10) occurred in small patches of mallee that were interspersed with superior habitat. We believe it is unlikely that any eastern greys, and extremely few reds, used the extensive stands of mallee typical of the western half of Yathong.

Mulga is another vegetation type nearly unused by kangaroos. It occurs on rocky ridge tops such as Merrimerawa Ridge (figs. 2.2, 2.10), and one rarely finds either tracks or droppings in those areas. Eastern grey females made the only recorded use of mulga by radioed animals, and then only 1.7% (see table 10.1). This would suggest a density of 0.01 kangaroos/km^2. Of course, the euro occurs there in sparse numbers that would not increase the total kangaroo density by much. Mulga constitutes only 5.6% of Yathong, and the density there is insignificant in the estimation of kangaroos on Yathong.

Although the overall trend in kangaroo numbers at Yathong is upward, it should not be assumed that all cohorts showed the same response. The success of any given cohort is overwhelmingly influenced by the rainfall pattern at the time it is produced. Thus, while most cohorts are successful during an increasing trend, some will not be, due to periodic poor conditions from lack of rain. Thus, the second cohort observed in this study was smaller for both species of grey kangaroos (fig. 4.2), whereas for red kangaroos it held steady.

When kangaroo populations at Yathong do reach food resource limits, stable numbers are not to be expected (Caughley 1987b, McLeod 1997). A more realistic expectation is that amount of rainfall will impact differently on successive cohorts to give dynamic fluctuations in population size that track, but seldom match, carrying capacity. This population behavior was termed "centripetality" by Caughley (1987b).

It is notable that Bayliss (1987) reported combined densities of red and western grey kangaroos at Kinchega National Park of about 15/km^2 during drought and 55/km^2 during good rainfall periods. These estimates seem extraordinarily high given the nondrought estimate of around 13/km^2 at Yathong in this study. Ordinarily, one would expect that a more productive area like Yathong, with around 317 mm of average yearly rainfall, would support more kangaroos than Kinchega, which has an average of 236 mm per year. But note, also, that the estimates of kangaroo density by Caughley (1987a) in the Yathong area in 1980–82 (prior to the 1982 drought) ranged between 11 and 25, also less than those reported by Bayliss (1987) for Kinchega.

Coulson (1993b) estimated densities of western grey kangaroos by strip counts in a study area within Hattah-Kulkyne National Park on the Murray River in northwestern Victoria, and reported estimates of 18 to 58/km^2. He also cited the 1983 unpublished work of D. G. Morgan, who reported densities of 24.8 western grey and 1.8 red kangaroos per km^2 over the whole park. This park has rainfall similar to Yathong (320 mm per year) and has grassland, mallee, woodland, and lakebed vegetation. However, its lake and stream systems suggest a more benign environment than the mean rainfall would suggest. It also has a patchy and interdigitated distribution of vegetation types, which increases carrying capacity for kangaroos (Short et al. 1983; Coulson 1993a). In accord with a more productive environment, the home ranges of western grey kangaroos at Hattah-Kulkyne (Coulson 1993a) were approximately one-half the size of those at Yathong (see chapter 5). These differences from Yathong may well account for the higher density of kangaroos at Hattah-Kulkyne National Park.

Our estimates of a density of $1.3/km^2$ for eastern grey kangaroos would be about as expected from the estimate of $5.3/km^2$ reported by Southwell et al. (1997) for the heartland of their range in eastern Australia. Yathong lies near the western end of the continuous distribution of eastern greys (i.e., not including western extensions along river courses), and low density would be expected on the margins of their habitat.

The difference in kangaroo density between Yathong and Kinchega National Park is harder to explain. Estimating kangaroos is an inexact science, and part of the difference can be attributed to error in methods. Yet estimates at both Yathong and Kinchega have reasonable error terms, and the discrepancy seems too great to be accounted for by method error. One could postulate that kangaroo numbers at Yathong are in the process of building up, and eventually will exceed those at Kinchega; it is likely there will be some further growth at Yathong. Still, there are problems with this postulate. Note that kangaroo density at Kinchega during the 1982 drought (around $15/km^2$) was similar to or greater than the density of around $13/km^2$ in this study at Yathong after three to four years of recovery. Our impression from work at Yathong is that the high density reported at Kinchega is not likely to be reached at Yathong; four times the observed density does not seem attainable unless an extraordinary string of good rainfall years occurs in a row.

We suggest that there may be fundamental differences in the way the ecosystems of Yathong and Kinchega function that would account for a lower density of kangaroos in a richer environment. First, rainfall at Kinchega is quite uniform throughout the seasons (Robertson et al. 1987), whereas it varies greatly at Yathong. Thus, effective precipitation and total precipitation are likely to be more equal at Kinchega than at Yathong, where effective precipitation has a greater impact on kangaroo density. In other words, lows within years can have as great an effect on kangaroo density as lows between years. Growth of kangaroo numbers at Yathong is interrupted by both seasonal and annual short-falls. Other things being equal, this will dampen population growth more at Yathong than at Kinchega.

Second, the shrub layer is almost absent at Yathong, whereas it is well developed at Kinchega. During dry periods, the shrub layer furnishes alternate food of moderate to low quality for kangaroos at Kinchega (Barker 1987), which likely moderates the difference between the good and bad periods. At Yathong, dried grasses of very low quality are the best foods available during dry periods. As with uniform seasonal rainfall, less variable food quality may dampen out short-term impacts at Kinchega and allows kangaroo densities to track more closely the long-term rainfall trends.

Finally, and most important for overall carrying capacity of kangaroos, is the well-developed woodland component of the vegetation of Yathong (fig. 2.10). A substantial proportion of annual productivity is sequestered by trees. The woody portion of that productivity is beyond the reach of kangaroos, and is indigestible as well. A small amount of tree production—leaves at the bottom of the canopy—is within the reach of kangaroos, but much of this foliage has concentrations of secondary compounds that discourage consumption. Most of the leaf canopy falls at senescence, when it is dry, and contains little nutrient content for kangaroos. Eventually, it is decomposed mainly by termites or fire. Bacterial decomposition of litter is minor at Yathong because of the dry climate and frequent fire regime. By and large, plant productivity going to woody species on Yathong is lost to kangaroos. They are almost totally dependent upon the herbaceous layer.

We believe the system at Yathong behaves substantially differently from that reported for Kinchega National Park by Caughley et al. (1987a). Although kangaroo numbers fluctuate considerably at Yathong, we do not believe they show the long, random-walk trends of measured (Bayliss 1987) and modeled (Caughley 1987b) behavior of kangaroo populations at Kinchega. Still, McLeod's (1997) conclusion that the concept of a stable equilibrium carrying capacity does not apply to kangaroos in the outback would be true of Yathong. The population consequences of severe, periodic droughts, such as that in 1982, are common across the two systems. However, we suggest that kangaroo numbers at Yathong show less consistent long-term trends in numbers and lower amplitude in fluctuations due to the greater impact of short-term effects (e.g., variable seasonal rainfall) and lower overall magnitude due to the lower production of the herbaceous layer alone of Yathong versus the combined herbaceous and shrub layers of Kinchega. Furthermore, it is notable that fire—an important factor reducing plant biomass at Yathong—is not mentioned at all as an environmental variable at Kinchega by Caughley et al. (1987a).

5

HOME RANGE AND MOVEMENTS

Background

Place for animals is as important as it is for humans. Consider that it is simply inefficient to locate food, water, and other life requisites by searching randomly for them. Even in an environment as unpredictable as outback Australia, there is constancy of vegetation types, soils, and topography; additionally, forage conditions, once established, have a certain time expectancy. Thus, even food is likely to be found in locations where it was previously found.

By having a place of its own, as well as familiarity with the landscape, the individual kangaroo can take advantage of past experience to minimize the time and energy required to maintain life, grow, and reproduce. It allows retreat from predators along known routes that favor escape. Also, it is an advantage to live in a community of known social relationships. This reduces the time required to establish social hierarchies, find mates, and strive against competitors.

Most mammals occupy a known area of consistent size and shape that changes little once the individual reaches adulthood and becomes integrated into the social community. Burt (1943) first coined the term "home range" to categorize this behavior, previously only loosely understood from natural history studies. The home range concept has continued to be useful to describe space use, despite a number of cases that deviate from the classical idea in a number of respects. It is common to refer to core areas, where individuals spend a majority of time, and these are connected by travel paths between core areas. This space use can be thought of as a "web" of travel routes, with nodes at core areas where routes intersect. The perimeter of the web's outermost locations encompasses the home range, despite the fact that some part of the enclosed area is used little or not at all. Still, a key attribute is the predictability by which the animal uses space over time. Continuing interest in permutations in home range use has been fostered by radio telemetry studies, which provide a level of detail about movements over time that could not be acquired by observational approaches for most species, and by the development of computer programs to reduce spot locations into a meaningful statistical description of the home range.

Today, many case studies show that most mammalian species demonstrate a use of space that can be reasonably described as some permutation of a home range. Even so, biologists continue to be intrigued to discover whether there are species (or regional permutations) that display an unpredictable use of space. Such use would not be truly random in the statistical sense, because present location will influence the likelihood of future location, at least over the short term. Yet, there does exist the case of nomadism, in which movements are sufficiently erratic as to be undefinable in the home range context.

Kangaroos are of particular interest in this regard because early studies described them as nomadic, and some recent studies of radiocollared individuals showed movement patterns that were unpredictable, and apparently not tied to a specific locality. In outback folklore, kangaroos were essentially nomadic, showing only temporary attachment to any given site. Because drought conditions were frequent and unpredictable, kangaroos were believed to follow the rains; for them, storms seen on the distant horizon, or the smell of rain carried on the wind were cues to the lush green pick that was sure to follow. Kangaroos were thought to manage themselves much as graziers managed sheep flocks, by shifting themselves about according to forage availability.

This view was reinforced by the common observation that local areas that previously had few kangaroos would, after rains, have large aggregations of kangaroos feeding on the new green forage. These "mobs," as they were called, were presumed to be the gathering of many nomadic individuals, each independently determining its own course. Kangaroos were mostly seen either alone or in the company of few other individuals. The absence of sightings of large mobs moving en masse from one area to another suggested independent, nomadic movement.

Early observational studies failed to resolve the question. Frith (1964), over nearly a year, repeatedly sighted a recognizable group of kangaroos that remained within a 5-km circle. This would suggest that kangaroos were not as nomadic as generally believed; but of course, any given year might be different, and that year, drought may not have been severe enough to cause them to move. Bailey (1971) trapped and marked 143 kangaroos at water holes in northwestern New South Wales. A majority of resightings were made within 8 km of the trapping site, but during a drought, one animal was sighted 216 km away. Denny (1982) caught and tagged kangaroos at water holes in the same region. Over 50 percent of the resightings were made at the point of capture. However, one individual was recorded to move over 300 km to South Australia. Furthermore, Denny (1982) reported that young males showed the greatest movements, raising the possibility that long-range movements were not due to nomadism, but rather, to dispersal by a given sex and age class—a common phenomenon in mammals (Greenwood 1980), including kangaroos (Johnson 1989). Thus, large-scale kangaroo movements, rather than indicating nomadism, may, in fact, have been the ordinary movements of a large animal in a low-quality, highly patchy habitat.

Although these studies showed that there was variation in movement behavior within the population, the proportion of individuals that fell into the sedentary versus nomadic categories could not be determined. The ambiguity of these early results traces partly to the inherent limitations of using visual observation of marked animals. For example, many of the marked individuals may never be seen again. Resightings are biased toward the area of marking for two reasons. First, using common sense, one would look first (if not most frequently) where the individuals were known to have been. And second, concentric circles, if incremented by equally increased radii, encompass an enormously expanded area. Marked kangaroo density is higher in the first concentric circle than in subsequent ones, even if absolute numbers are equal across several circles; more marked individuals will be sighted in the areas of higher marked animal density, thereby biasing those observations made closest to the point of capture. In addition, observation or recovery of carcasses of marked individuals at long distances are sporadic events and causality of long-range movements cannot be inferred from the few data points obtained.

The observational study best controlled for biases was conducted by Priddel (1983, 1987) at Kinchega National Park and the adjacent Tandou Sheep Station near Menindee in western New South Wales. A sample of 261 red kangaroos and 170 western grey kangaroos was captured by stunning (chapter 3), a technique that results in less concentration of capture sites than trapping at water sources. Although resighting depended mainly on chance observations, systematic effort was made by walking predetermined compass lines. The longest movement recorded was 285 km by a red kangaroo female. The longest movement of a western grey kangaroo was 85 km by a male. However, most individuals of both species were sedentary. For red kangaroos, 17 of 22 individuals sighted within 1000 days following capture were within 7 km of their place of capture. Western grey kangaroos were similarly sedentary. However, some individuals were highly mobile, if not nomadic. Priddel (1987) estimated that less than seven percent of his marked individuals fell in that category. Croft (1991) reported some long-distance movements by red kangaroos in his study at Fowler's Gap. In comparison, Denny (1982) estimated around 20 percent mobile animals at Tibooburra in northwestern New South Wales.

Similar observational studies on the eastern grey kangaroo, in the mountains of the eastern part of its range, showed a mainly sedentary pattern. Kirkpatrick (1966) and Jarman and Taylor (1983) found eastern greys to be sedentary with small, overlapping home ranges, although in the latter study, one adult, reproducing female moved to a new area 17 km away.

The advent of radio telemetry has overcome the inherent limitations of visual observation of marked individuals. Now, one can not only determine the movements of all radioed individuals, but also get enough successive observations to piece together the movement patterns of each individual. Given a reasonable sample of radioed kangaroos, therefore, one can determine the proportion of the population showing various movement patterns. Furthermore, one can refine the spatial characteristics of the home range for sedentary individuals.

Priddel (1983, 1987) used radio telemetry to study use of space by red and western grey kangaroos in and around Kinchega National Park. Both species showed relatively small home ranges (mean = 7.7 km^2 for red and 6.9 km^2 for western grey kangaroos). Priddel concluded that red kangaroos had larger average home ranges than grey kangaroos, and males had larger home ranges than females. Oliver (1986) also reported small home ranges for red kangaroos, and that females were more sedentary than males. Using telemetry, Jaremovic and Croft (1987) confirmed the small size of eastern grey kangaroo home ranges in good habitat, which were usually less than 0.5 km^2. Dawson (1995) commented on the small home range of a female eastern grey kangaroo at Fowler's Gap at the western end of their range (7.7 ha). But in the western part of their range, eastern grey kangaroos occupy forested stringers along stream courses, a productive habitat that also, presumably, would impose narrow linear home ranges. Coulson (1993a) reported home range sizes of 1.98 and 2.53 km^2 for female and male western grey kangaroos, respectively, at Hattah-Kulkyne National Park in Victoria. Coulson et al. (1999) reported female eastern grey home ranges of 0.27 to 1.60 km^2, and a young male that moved 15 km away.

Our study of kangaroo home range and movement at Yathong Nature Reserve may be viewed in context with these results elsewhere. Yathong is not nearly as benign and stable an environment as eastern New South Wales, where most studies of eastern grey kangaroos have been conducted. Rainfall at Yathong averages around 350 mm/year as compared to around 800 mm/year in eastern New South Wales, an area that has much cooler summer temperatures as well.

Conditions become more harsh as one moves from east to west in New South Wales. At Tibooburra, where Denny (1982) worked, average rainfall is 150 mm (chapter 2)—on average, the most severe part of New South Wales. Menindee, the site of Priddel's (1987) study, has an average rainfall of 236 mm (Robertson et al. 1987; chapter 2). Yathong has higher rainfall and is in a zone where drought is less frequent and less extreme. Thus, eastern New South Wales, Yathong, Menindee, and Tibooburra represent a decline in precipitation and increase in drought severity.

It is generally believed that mammalian home range size bears some relationship to availability of resources, as conveniently indexed by metabolic needs (McNab 1963). It follows, therefore, that home range size should be some function of animal body size and quality of the habitat. If body size is held constant, then home range size should be greater in poor environments than in richer environments. This prediction may be examined by comparing species and sexes across the environmental gradients discussed above. Thus, one would expect that eastern grey kangaroos at Yathong would have larger home ranges than those in eastern New South Wales. And red kangaroos would be expected to have progressively larger home ranges as one moved from Yathong to Menindee to Tibooburra. Similarly, western grey kangaroos would be expected to have larger home ranges at Menindee than at Yathong. There is some suggestion that this relationship holds. Priddel (1987) found that red and western grey kangaroo home ranges were larger on Tandou Sheep Station, where grazing by sheep was heavy, than on adjacent ungrazed Kinchega National Park.

The probability of nomadic behavior should be a corollary of home range size. Because nomadism is a response to fluctuations in resource availability, one would predict that nomadism would be most prevalent at Tibooburra, less so at Menindee, and least at Yathong. Again, suggestive evidence is available—Denny (1982) estimated that about 20% of the kangaroos at Tibooburra were nomadic, against Priddel's (1987) estimate of about seven percent nomadic at Menindee.

Kangaroo movements and space use at Yathong were determined to evaluate these expectations. We successfully radio-collared a total of 28 kangaroos captured by stunning (see chapter 3); logistically, this was the maximum number of radioed animals we could monitor. Furthermore, the presence of three species by two sexes on Yathong presents the opportunity to compare body size and home range size to see what proportion of the variation in home range size can be accounted for by body size, and what is attributable to species (sexes combined), and sex within species. The sample included five individuals of each sex of each species, except for male red kangaroos and female eastern grey kangaroos, for which sample size was four each. Locations were made by triangulation without disturbing the animals (chapter 3).

Home Range Size

The various species and sexes of kangaroos showed different home range sizes (fig. 5.1, table 5.1). Home ranges of red males were largest (fig. 5.1D) and those of western grey females, smallest (fig. 5.1E). Red kangaroos had larger home ranges than the other two species (fig. 5.1, table 5.1). ANOVA tests of minimum convex polygon home ranges showed that red kangaroo home ranges were significantly larger than eastern grey ($F = 5.09$, $P < 0.025$) or western grey ($F = 13.19$, $P = 0.001$) home ranges. There was no significant difference between eastern and western greys ($F = 1.39$, $P = 0.25$). Overall, males of all species had significantly larger home ranges than females ($F = 6.41$, $P = 0.018$).

In a comparison of sex and species, male red kangaroos had significantly larger home ranges than females and males ($P < 0.05$) for each species. The differences in home range area of red males versus all other kangaroos combined was highly significant ($F = 13.695$, $P = 0.001$). Using the Fourier 95 and minimum concave polygon home range methods, eastern and western grey males were found to be significantly different from eastern grey females ($P < 0.05$). For the other home range methods (table 5.1), the differences between eastern and western grey

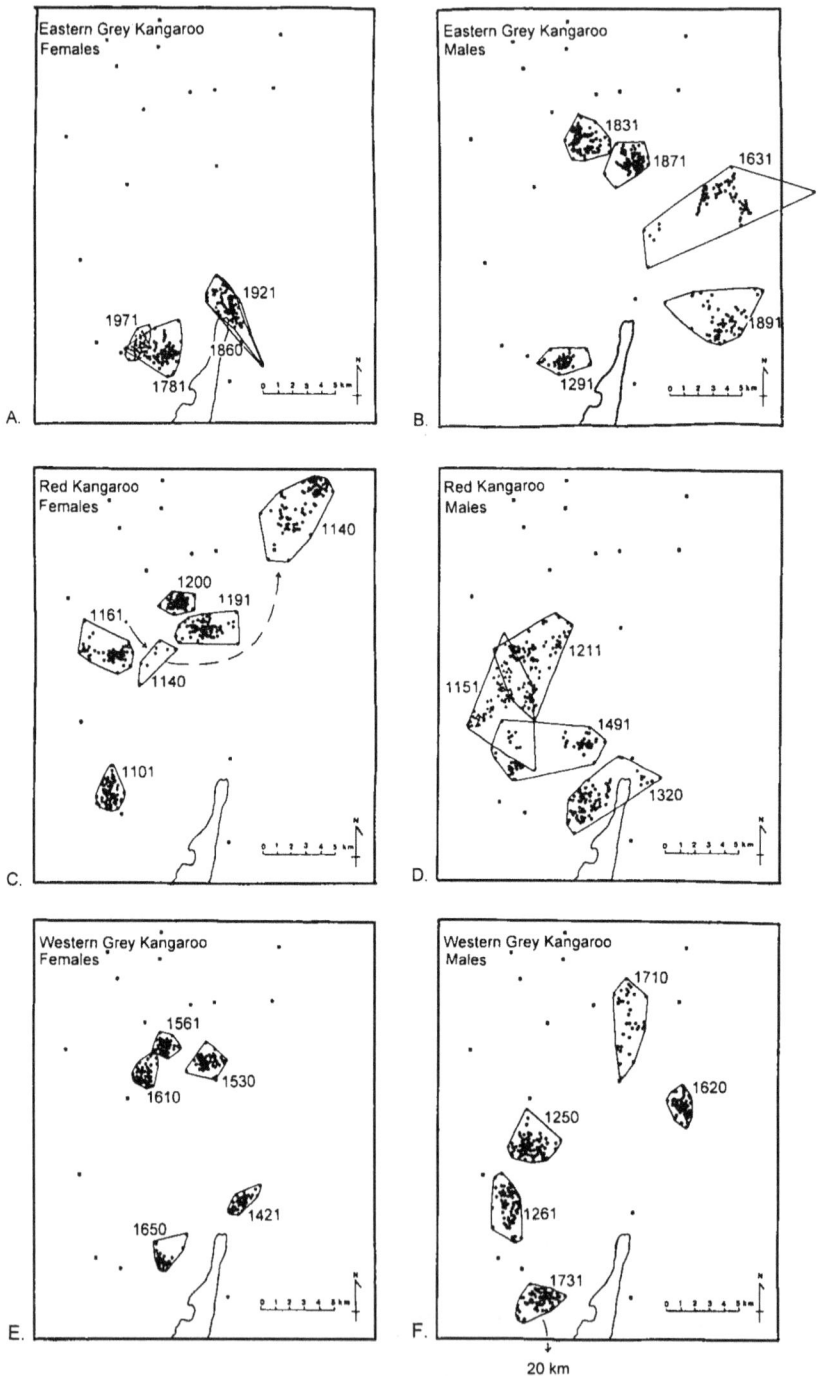

Figure 5.1. Home range areas of radio-collared kangaroos by the minimum convex polygon method. Squares indicate locations of earthen water tanks, and the outlined projection at the bottom of the panels is the northern end of Merrimerawa Ridge.

Table 5.1. Mean and standard deviations (in parentheses) of home range areas (km^2) of species and sex categories of kangaroos on Yathong Nature Reserve as calculated by McPAAL (minimum convex and concave polygons, Fourier series, and harmonic mean) and CALHOME (adaptive kernel).

Species	Sex	N	Minimum Convex Polygon	Minimum Concave Polygon	Fourier Series 95%	Harmonic Mean 95%	Adaptive Kernel 95%
Eastern grey	F	4	7.77 (3.38)	4.16 (1.44)	3.78 (1.67)	5.28 (1.66)	5.28 (2.64)
Eastern grey	M	5	13.56 (11.03)	7.63 (3.76)	9.45 (6.44)	12.61 (10.77)	11.77 (7.61)
Red	F	5	13.85 (12.31)	7.93 (5.73)	4.72 (3.22)	10.92 (9.14)	10.30 (8.33)
Red	M	4	26.14 (11.31)	12.70 (4.21)	18.79 (5.25)	19.97 (3.87)	20.82 (4.28)
Western grey	F	5	3.32 (1.23)	2.24 (0.85)	2.11 (1.22)	3.17 (0.95)	3.51 (1.58)
Western grey	M	5	8.97 (1.35)	6.93 (0.89)	9.89 (9.31)	7.62 (2.31)	8.61 (3.53)

males and western grey females were close to significant, and we believe the differences were real, despite the fact that they did not reach statistical significance at 0.05.

In addition to small sample size, variation within species and sex categories contributed to lack of significance of some comparisons. However, male red kangaroos—the category with significantly larger home ranges than all others—also exhibited the greatest variation in home range size. Oliver (1986) reported that male red kangaroos had larger home ranges than females, but Croft (1991) found similar home range sizes for both sexes of red kangaroos in western New South Wales. A major contributor to variance was age-related body size of the individual. There was a great range of body sizes in our radio-collared male red kangaroos, from a medium-sized individual to one that was very large (in local parlance, an "old man roo"). Home range area of the red males increased directly with their rank order in size. Similarly, for male eastern grey kangaroos, the smallest male, a young animal, had the smallest home range by far, and another young male was next smallest. These results agree with Jarman and Southwell (1986), who also reported that home range size was related to male size in eastern grey kangaroos in eastern Australia. However, our largest male eastern grey did not have the largest home range, although the size difference was small between him and two other large males.

Among male western grey kangaroos, one was a medium-sized individual, and all the others were large and not distinguishable in mass by appearance. Although this medium-sized male did not have the smallest home range, the size variation in home range among western grey males was notably small (8.71 to 10.78 km^2). Variation in female western grey kangaroos was similarly small (range 2.41 to 5.55 km^2). It may well be that this is characteristic of the species, at least in this region. The great similarity of home range characteristics for western grey kangaroos (fig. 5.1E and 5.1F) and the high predictability with which they could be located within an expected area was impressive.

Contrarily, the great variation in home range sizes of female eastern grey and female red kangaroos could not be explained by body size. All but Scarlett O'Hara (who was young and small when captured and collared) were adult females that varied little in size, and within six months, Scarlett was indistinguishable by size from the others. Among red females, one had a very large home range (33.89 km^2), and Scarlett, like her namesake, moved around a good deal (17.12 km^2 home range). Two others that had similar small-sized pouch young had extremely small home ranges (4.00 and 5.34 km^2). The first occupied primarily bimble box and open grassland types near Keginni Creek whereas the second occupied mainly wilga-belah and open grassland. Thus, habitat occupied, by itself, did not account for why such a small area was utilized by these two individuals.

Similarly, body size did not account for the variation in female eastern grey home ranges. The smallest female in the sample (who, nevertheless, had a pouch young) had the largest home range (10.45 km^2) whereas the female with the smallest home range (2.87 km^2) was large in size. In keeping with the species ordering of red kangaroos having largest and most variable home ranges, eastern grey kangaroos intermediate, and western grey kangaroos, the smallest and least variable, the range in home range sizes among eastern grey females was much less than that among red females. This comparison was complicated by the fact that two eastern grey females had virtually identical home ranges, and frequently were visually observed together as a social pair (Nos. 1860 and 1921, fig. 5.1A). However, at times they were observed or located by radio in separate areas. Jaremovic (1984) reported similar association between individuals in eastern Australia, where these associations would be more expected due to their more integrated social groups (Kaufmann 1975; Jaremovic 1984; Jarman and Southwell 1986).

Total expanse of area used by kangaroos was well represented by the minimum convex polygon method. However, this method did not account for the fact that much of the area included may have been used little or not at all while certain core areas (sometimes called "activity centers") were used heavily. Thus, the difference in home range size between species or sexes may not represent a larger amount of total core area, but rather a greater dispersion of separate core areas.

A simple index of dispersion of core areas is to divide minimum convex polygon area by minimum concave polygon area. Three species/sex categories had similar values (eastern grey male = 1.78, female = 1.87, and red female = 1.75), whereas one (red male = 2.06) had a substantially higher value and the other two were less (western grey male = 1.29, and female = 1.48). These figures suggest that red males have widely dispersed core areas whereas western grey kangaroos of both sexes tend to have home ranges with a central core area.

Most of the variance in the ratio of minimum convex polygon to minimum concave polygon home range size was due to sex ($F = 8.29$, $P = 0.008$) but not to species ($F = 0.97$, $P = 0.393$). Most of the effect of sex was attributed to males ($F = 23.67$, $P < 0.001$), although females came close to statistical significance ($F = 4.55$, $P = 0.054$).

The harmonic mean and adaptive kernel home range methods are particularly useful in analyzing the use of space by kangaroos because these methods can calculate area to include various percentages of the observations and enclose areas topographic in nature (Spencer and Barrett 1984). Thus, enclosure of a given percentage of observations may include several unconnected core areas, better reflecting actual use of space than a method requiring a contiguous distribution.

All individual radio-collared kangaroo home ranges were calculated by both the harmonic mean and adaptive kernel methods enclosing 10%, 25%, 50%, 75%, 95%, and 100% of the locations. The nonparametric adaptive kernel calculation relaxes some of the parametric assumptions of the harmonic mean method (Kie et al. 1996). Plots of area on percent of locations for each individual were essentially linear (fitting quadratic models increased R^2 only a few percent) up to about 75% inclusion; above 75% inclusion, the slope of lines increased rapidly and thus became strongly curvilinear (fig. 5.2). Although the linear segment of the "curves" was not forced through zero, all regression lines passed near or through the origin.

These results suggest that home range area expanded at a nearly constant rate up to about 75% inclusion of locations; in other words, usage and area were closely linked, suggesting a biological basis as, for example, food availability. Beyond 75% inclusion, area increased progressively greater than expected, suggesting gaps in usage of the increased area. Thus, the periphery of the home range seems to consist of isolated intrusions into the surrounding area, and intervening areas are not consolidated into consistent use (fig. 5.3). This characteristic of the fringe of the home range held (to various degrees) for all species and both sexes of kangaroos, and may well be typical of most large, herbivorous mammals.

The fact that, independent of sex and species, home range area expanded at a constant rate up to about 75% inclusion of locations, suggests that the slope constant (*b*) may be a useful statistic for comparing the "requirement" for area between sexes and species. This "area requirement constant" (ARC) for home range obviously encompasses minimum requirements (as the animals were surviving and reproducing), but may encompass excess requirements as well.

The ARCs show relationships by sex and species that parallel those of home range size. Constants were 0.015 for western grey females, 0.036 for eastern grey females, 0.041 for red kangaroo females, 0.043 for western grey males, 0.051 for eastern grey males, and 0.096 for red kangaroo males. Consequently, females had lower constants than males, and constants by species were lowest for western grey, intermediate for

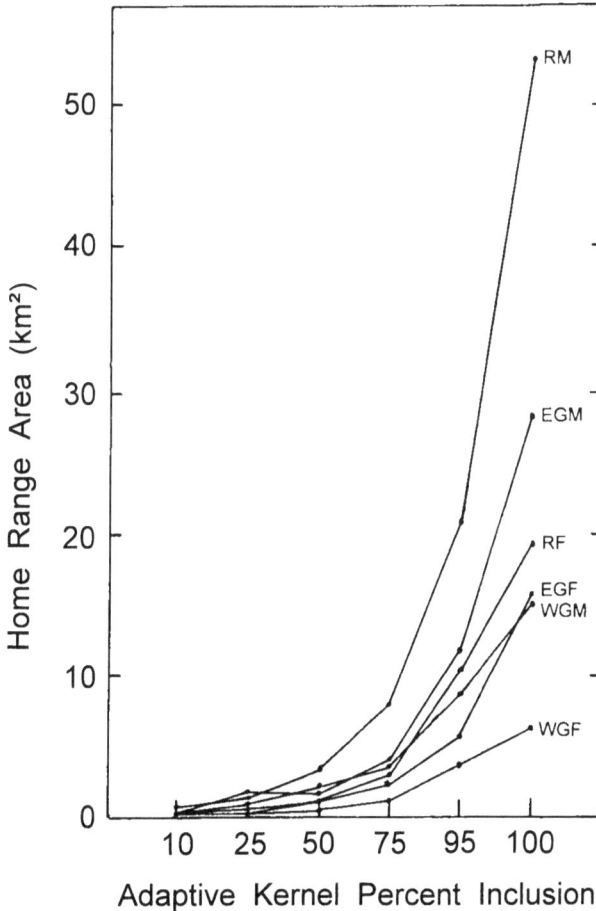

Figure 5.2.   Plot of home range area on percent inclusion as calculated by the adaptive kernel method with the program CALHOME. Mean values for species and sex categories are plotted with the individual values not shown because of the confusion of overlapping points.

eastern grey, and highest for red kangaroos. The 100% used area can be projected by multiplying the requirement constant by 100 (the constant a can be ignored because the regression line passed through the origin) to obtain the predicted total home range. Western grey females would have an ARC home range of 1.5 km^2, eastern grey females 3.6 km^2, red females 4.1 km^2, western grey males 4.3 km^2, eastern males 5.1 km^2, and red males 9.6 km^2. Comparison of these predicted home range areas

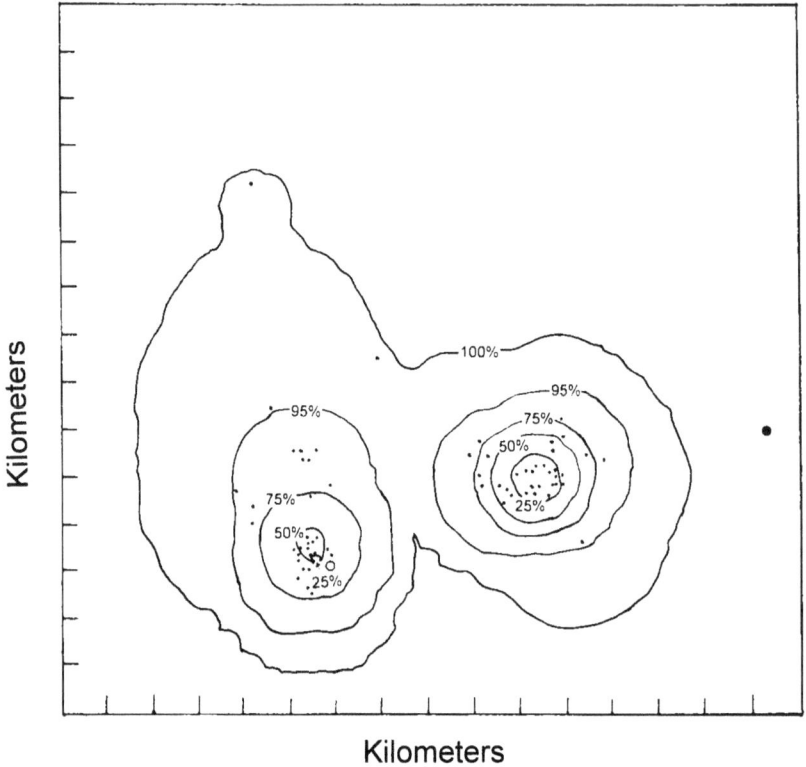

Figure 5.3. Home range of red male 1491 based on 74 radio telemetry locations. He was a large male who was captured (● on map) on December 11, 1985, 3.8 km east of the easternmost telemetry location, and died (○ on map) at an old age on July 3, 1986. The five contour lines indicate the percent area inclusions. This home range illustrates the pattern of several core areas, with scattered locations on the periphery.

with actual measured home ranges (table 5.1) shows that the predictions were substantially smaller, illustrating unused areas when the peripheral 25% of locations are included.

The minimum convex polygon areas for red kangaroo females were 1.78 times larger than those for eastern grey females, whereas ARC home ranges were only 1.14 times larger. Thus, when the effect of the unused area is removed, red females require only slightly larger areas than do eastern grey females, whereas observed locations for red females are scattered over almost twice the area.

The ARC home ranges resulted in even less distortion of the comparison between female red and western grey kangaroos. This comparison represents a study extreme because western grey females had small, compact home ranges whereas some red females had dispersed core areas and scattered perimeter locations. Red females had ARC home ranges 2.73 times greater than western grey females whereas minimum convex polygon areas were 4.17 times greater for red females.

Similar discrepancies held for comparison of males across species. The ARC home range for eastern grey males was 1.19 times larger than that for western grey males whereas the minimum convex polygon was 1.51 times larger. We attributed this difference to the more scattered use of home range perimeters by eastern grey males. Comparable values for eastern grey males in relation to red males were 1.88 and 1.93, respectively. This similarity in ratios indicates that males of both species have relatively unused perimeters, even though red males have home range areas nearly twice as large as eastern grey males. Red males had an ARC home range 2.23 times that of western grey males, compared to 2.91 times greater for the minimum convex polygon, again illustrating the relatively compact home range area of western grey males and the dispersed home range area of red kangaroo males.

McNab (1963) reported a positive relationship between body size and home range area, presumably due to the higher energetic costs of a larger body size. He suggested that home range size should be related to metabolic body size, so that mass should be raised to the 0.75 power. Therefore, the larger the body size, the greater the home range area required to support the species. However, subsequent studies have resulted in higher scaling exponents than 0.75. Harestad and Bunnell (1979) using data from 55 species of mammals, reported a scaling exponent of 1.08, and Swihart et al. (1988), using data from 23 mammals, reported a scaling exponent of 1.42, both significantly different from 0.75. Kangaroos in this study greatly exceeded the 0.75 exponent. For 100% ARC home ranges, the exponents were 1.49 to 1.74 for the species-sex combinations, and for home ranges measured by minimum convex polygon, they varied from 1.70 to 2.09.

Distances Moved Within the Home Range

Movements of kangaroos over the study period can be categorized as short-distance movements within the home range, occasional movements that expand the home range, and rare long-distance movements that result in establishment of a new, distinct home range. We first present results on movements within the previously delineated home range.

Movements were analyzed quantitatively by determining the linear

Table 5.2.   Minimum distances moved (km) between successive locations of radio-collared kangaroos between January 10 and September 25, 1986, when individuals were located every third day on average.

Species	Sex	N	Mean Distance	(SD)	Modal Distance [1]	Maximum Distance
Eastern grey	F	394	0.89	(0.85)	0.2–0.4	6.9
Eastern grey	M	410	0.87	(0.71)	0.6–0.8	5.6
Red	F	495	0.91	(1.71)	0.2–0.4	21.1
Red	M	371	1.34	(1.78)	0.2–0.4	18.8
Western grey	F	446	0.60	(0.47)	0.2–0.4	4.4
Western grey	M	428	0.93	(1.27)	0.4–0.6	16.4

[1] Modal distance in 0.2 km intervals.

(i.e., straight line) distances moved between subsequent location fixes of radio-collared individuals. The data analyzed included the period January 10 through September 25, 1986, during which each individual was located every third day on average. Actual distances moved could not be determined because kangaroos seldom move in a straight line, and may retrace their own paths on a daily basis. Thus, the data are indices to compare relative distances moved between sexes, species, and time periods (table 5.2).

Frequency distributions of movements by 0.2-km intervals showed that movement for all kangaroos was highly skewed towards short distances, with only a few percent of the movements being longer distances. Therefore, the mode is presented along with the mean. For western grey females, 93 percent of the movements were one km or less, and for eastern grey females, 76 percent were one km or less (table 5.2).

Other combinations of species and sexes showed similar frequency distributions, although different categories moved over greater or lesser distances. Distances moved by species and sex paralleled the results of home range area (tables 5.1 and 5.2), which was expected. Nevertheless, the two may not be correlated, because movement within the home range could be independent of the home range area. Thus, for example, male red kangaroos showed the highest mean distance moved in keeping with their largest home range area. However, they also had the highest standard deviation. Furthermore, they had a small modal distance moved, comparable to females of all species (0.2 to 0.4 km). This points to the fact that although male red kangaroos periodically move long distances from place to place, once in a place (a core area), they stay there and are relocated repeatedly in this core area. They then remain sedentary until the next movement to a different core area. Croft (1982) reported that red males shifted activity centers more frequently than females, and our results are consistent with his observations. This behavior of males probably also explains why Coulson (1997a) found that female red kangaroos were as likely to be killed on roads as males, whereas for the two species of grey kangaroos, males predominated in mortalities.

Eastern and western grey male kangaroos, by contrast, have small mean distances moved, but higher modes (table 5.2). Even though they have much smaller home range areas than red kangaroo males, they move about within their home range on a more regular basis. Thus, the home range area of male grey kangaroos is more fully occupied over time than that of red kangaroo males, which have large areas of the home range unused for long intervals. Also, as apparent from the previous section, large areas in the home range of red males are not used at all.

Females of all species, even red kangaroos (with their relatively large home range), showed small mean distances moved (table 5.2). Mean distances paralleled home range sizes, with female red kangaroos having longest distances moved and greatest standard deviation, and female western greys having least distances moved.

There was no consistent relationship of distances moved over the seasons, or by environmental states ($P > 0.05$, data log transformed to satisfy normality). Although there was no difference in distances moved between subsequent locations for males with time of day, for females of all three species, locations taken during early afternoon (1200–1500) showed substantially shorter distances moved than at other times of day. Early afternoon is the period of the day when kangaroos are relatively inactive (see chapter 6). This finding implies that males rest in no particular relationship to their areas of activity, whereas females are likely to rest nearer to their previous resting sites.

Direct tests of this early afternoon behavior were complicated by low sample size and the time elapsed between consecutive early afternoon locations. (Recall that locations were taken on average every three days, and every third location was a daytime location, most of which were taken in the morning. Thus, the average interval between early afternoon locations was 25 days.) Despite the long time between subsequent locations taken in early afternoon, the mean distance between subsequent locations for eastern grey females was 0.83 km, red females 1.09 km, and western grey females 0.45 km. Only red females, which had a higher likelihood of shifting core areas than the other two species (similar to males), had higher values than the mean values given in table 5.2. The "stay at home" western grey females had mean early afternoon distances of only 0.45 km, compared to 0.60 km average distances; this finding strongly points towards use of consistent resting sites, from which activities are pursued in various directions. Early in the study, we suspected that western grey females might be dead because of the close proximity of subsequent radio locations (usually in the mallee vegetation type), and that variation in locations was due to triangulation error. Indeed, on several occasions, we walked in on radio-collared western grey females only to confirm that they were alive. We found that western grey females often returned to the same exact resting site.

The short distances between early afternoon locations for female eastern grey and red kangaroos is probably also due to a central resting pattern. Males, due to their scattered core areas (red kangaroos) and active movement within home ranges (eastern and western grey), did not show centrally located resting areas. Neither eastern nor western grey male kangaroos showed any pattern in resting sites. The scale and variance for distances moved for early afternoon locations were indistinguishable from other times of day. Curiously, red kangaroo males showed some tendency towards bimodality in early afternoon locations with one set of values clustered at lower distances than at other times of day, and another set of values scattered at high distances. These results might indicate that for scattered distances, male red kangaroos shifted between core areas, whereas for clustered distances, there is a central resting area within given core areas. However, the data supporting this interpretation of male red kangaroo behavior are a bit tenuous. A larger sample size would be required to clearly establish that the frequency distribution of distances moved showed bimodality rather than an unusual scatter due to chance.

A final consideration about short-term movements is the individual variation among species and sex categories. Western grey females all seemed to be stamped from the same mold and, to a lesser degree, western grey males also tended to behave similarly. As previously noted, variation in movement behavior in eastern grey and red kangaroo males was largely attributed to size and age, with small, young males behaving more like females, and large old males showing much higher mobility. However, eastern grey and, particularly, red kangaroo females showed variation in movements that was not readily attributable to either size or age. Scarlett O'Hara, the youngest and smallest radio-collared red kangaroo female, was a notorious gadabout, as was red female 1191. Conversely, red females 1101 and 1200, both large and apparently at least medium-aged, had minimum convex polygon home ranges of only 4.00 and 5.34 km^2. Eastern grey kangaroo females had minimum convex polygon home range areas that varied from 2.87 to 10.45 km^2, again, in no apparent relationship to size, age, or habitat types occupied. These variations seem to be truly individual in origin, in that endogenous variables do not seem to adequately explain the differences observed.

Home Range Extensions

As noted in the methods section (chapter 3), long-distance movements consisted of range extensions and shifts to new ranges. Movements to new areas within two diameters of the previous home range area by the minimum convex polygon method were considered home range extensions. In most cases, our radio-collared individuals continued to use the previous home range area.

Although some range extensions in the study probably were true extensions, many of the apparent extensions probably were simply a reflection of the period of study. Most of the radio-collared individuals were fully adult animals when captured, with previously well established home ranges. As we accumulated data on the locations of these individuals, we developed a growing sense of the area occupied. However, because perimeter areas are less frequently used, a longer period of time was required to adequately observe use of the entire established area. We believe most of the apparent extensions we observed were of this nature. The full extent of the area utilized may be difficult to determine, however. For example, the two eastern grey females that occurred together and shared a common range were found most often near an earthen tank early in the study. When the tank dried up, their locations shifted to the southeast, with one location suspiciously in the direction of another tank. If they made brief excursions to that tank for water, it's possible that we may not have located them while at the tank. We suspect this was the case as eastern greys are the most water-dependent of the three species studied and no other water source existed within their home range. However, other individuals showed similar linear home range extensions into areas where no water occurred so we cannot assume that these two were visiting the tank.

Some of the increase in home range area may have been due to age. For example, Clancy and Croft (1989) reported that wallaroos expanded their home ranges over three years, with a declining rate of expansion over time. Some of our animals were relatively young, and it's possible they may have been exploring new areas as they grew older. Nevertheless, most locations—obtained each January and June in the two years following the period of continuous field work—fell within the previously established home ranges. Projecting this pattern backwards, we can assume that most of an animal's home range area is established fairly early in the independent life of the individual, as described below for Scarlett O'Hara. However, a probable case of true range extension was shown by western grey male 1710 (fig. 5.1F). Between December 1985 and February 1986, this large, old individual occupied a home range with an earthen water tank at its center. By March 1986, he had extended his range about 3 km to the south, and the March locations barely overlapped the earlier ones. On April 17, he was found dead by Mallee Tank, about 8 km south of the center of his range prior to March. This movement coincided with the absence of green feed and the drying up of the earthen tank in his original home range. This older male apparently could no longer survive such harsh conditions.

Establishment of New Home Ranges

Sudden, permanent shifts of individuals to new areas far removed from the previously occupied home range are of great interest, because they contradict our usual view of home range behavior. There were two notable cases of this kind observed in this study. The first, and best documented case, involved the redoubtable Scarlett O'Hara (aka 1140) (fig. 5.1C). She was a small red kangaroo female when captured on the night of July 24, 1985. She did not have a pouch young, and we believe she was a juvenile only recently separated from her mother. She was not in the company of any other kangaroos. Although we attempted to release her in a northwesterly direction (away from the vehicle), she persisted in facing southeast, and eventually, was released in that direction.

Six days later, on July 30, we located her in an adjacent grassy valley, 4.2 km from the capture site. We did not locate her radio signal in the capture area on the intervening days, and presumed she moved to a new site soon after release. On the next day, she was observed 2.8 km to the north, in the company of a large red kangaroo male. Between then and September 11, she was found in the same valley, occupying a minimum convex polygon home range area of 2.83 km². Between August 1 and August 10, she was alone. On August 24, she was in the company of another red female and on August 27, these two were joined by a large red male that was maintaining a tending bond with Scarlett (we named him "Rhett Butler," of course). We believe Scarlett mated for the first time during this consort period. On September 11, she was located alone in the same area.

Between September 12 and September 22, Scarlett could not be found in the area previously occupied, and subsequent ground searches for her radio signal proved fruitless. On October 21, Scarlett was finally located by airplane in the very northeastern corner of Yathong, some 17 km northeast of her previous home range. During the remainder of October, she remained there within an area of about three km^2 (fig. 5.1C).

In November and December, she extended her home range to the northeast onto the adjacent Stanniford Sheep Station. By the end of December, she had occupied 16.5 km^2 of the total 17.12 km^2 of this new minimum convex polygon home range (as determined at the end of the field study period, September 25, 1986). All subsequent January and June locations through 1988 were within this home range area. In the short term, she moved broadly over the home area. Because of her unusual history of movements, we walked in to observe Scarlett more than we did other radioed kangaroos. Sometimes she was found alone, and other times, in company with other red kangaroos of various ages and sexes.

A general idea of home range establishment can be derived from these observations of Scarlett. She apparently was captured as a lone juvenile not long after severance of the young-at-foot dependence with her mother. Her social bonds were fluid, for she was as likely to be found alone as with other kangaroos. The number and sex of individuals with which she was found in association changed repeatedly. She was never again found in the area of capture, and her first home range was scattered over a narrow, linear area. It was here that she apparently bred for the first time.

After about two months in this initial home range, she left it rather abruptly for a new home range, well removed from the first. This would seem to be a case of dispersal, but because her area of birth was not known, the interpretation is unclear. Nevertheless, the exploration involved in establishing a permanent home range is apparent.

These long-range movements came during a period when green feed and water were abundant and widely distributed, so a search for better resources did not seem to be the impetus. Similarly, social forces did not seem to be involved. No antagonistic behavior was observed, Scarlett showed fluid social contacts, and her moves were apparently made alone. Her behavior gave us the distinct impression of individual choice, rather than the force of constraints.

For about the first month and a half in the new home range, most of her locations were in a relatively small area. Over the following two months, her home range gradually expanded as she explored new areas on the perimeter. Thereafter, relatively little home range extension occurred, as most movements carried Scarlett from place to place within the previously explored area. Two years later, her last recorded location was still in the same area. Oliver (1986) reported similar changes in location by young red kangaroos as they established themselves. These results are consistent with the Edwards et al. (1994) report of dispersal by young females, and more sedentary behavior by adult animals.

A second case of long-range shift in home range involved western grey male 1731 (fig. 5.1F). He was a large, fully adult male when captured on December 15, 1985. Between then and September 7, 1986, he occupied a predictable home range area of 10.8 km^2 between two rocky ridges southeast of Yathong Homestead. For no apparent reason, he abandoned this home range on September 8. From September 8 through 10, we searched intensively for him by motorbike. His distant weak signal was picked up on September 10, and the following day he was observed alone and apparently normal. He had moved a straight-line distance of 20 km south of his previous home range. All subsequent locations, including those in January and July through 1988, were in this new area. He was occupying a particularly large home range in the new area; the few locations we had time to obtain being scattered widely over about 20 km^2. This new area was in a very large mallee stand where it was not possible to approach closely by vehicle, or even by motorbike, and so locations were taken at greater distances than usual, with consequent large error. Still, no conceivable error in locations could account for such a large area of occupation. It is tempting to speculate that such a large area was required in the new home range because of the low quality of mallee as kangaroo habitat, compared to the mainly pine-box habitat of his earlier range.

The causes prompting this shift in home range were unknown. The mallee where he took up occupancy was burned, and recent rains had resulted in a flush of green vegetation. However, green feed was widely available, including abundant feed on his former home range. Also, Caughley et al. (1985a) reported that kangaroos moved away from burned areas in mallee. Because this western grey was a large, fully adult animal that had successfully occupied an established home range, social factors do not seem to be a likely cause of the move.

Discussion of Home Range and Movements

The picture of use of space by kangaroos on Yathong is fairly clear. All of the radio-collared kangaroos in this study showed fidelity to a place in the landscape that can reasonably be described as a home range in the sense of Burt (1943). Even the two individual kangaroos that showed permanent shifts in home range areas showed typical home range behavior in both previous and new areas. Notable further evidence for red kangaroos—the species with greatest variance in home range use—was given by Priddel et al. (1988), who reported that all but one female of six adults (two males, four females) that were translocated 13 to 20 km away, returned to their previous home ranges.

Red kangaroos had the largest home ranges, eastern grey kangaroos intermediate home ranges, and western grey kangaroos the smallest home ranges. Females of all species had home ranges that were substantially smaller than those of males. Variance in home range size among individuals increased with mean home range size. Red kangaroos, with the largest home ranges, also had the greatest individual variation in home range sizes. In red males, variation was related to size and age, whereas for females, it seemed to be individual choice, independent of size and age. In general, the results of Priddel (1987) and this study are in agreement. Red kangaroos have larger home range areas than western grey kangaroos, females of both species have smaller home ranges than males, and a small subset of the population shows substantially greater mobility than the majority.

Eastern grey kangaroos had intermediate size home ranges, and intermediate variation between individual home range areas. Although the smallest eastern grey male had the smallest home range, the largest male had a relatively small home range. Two eastern grey males had notably larger home ranges than the other two. Finally, western grey kangaroos had the smallest home ranges and individual variation for both sexes was small. Western grey females were remarkably predictable in their use of space. Still, the home ranges (minimum convex polygon) of western grey kangaroos at Yathong were approximately twice as large as those reported by Coulson (1993a) for Hattah-Kulkyne National Park in northwestern Victoria, and Arnold et al. (1994) for Durokoppin Nature Reserve in Western Australia.

Mean short-term movements were more or less correlated with home range size. Species and sexes showed mean movements in proportion to overall home range areas. However, male eastern and western grey kangaroos had especially high modes of short-term movements, suggesting that over time they move about their smaller home range areas more regularly than other species and sex categories. Female kangaroos of all species apparently had central resting locations, whereas males did not. Priddel (1987) reported that western grey kangaroos had established, consistent resting sites whereas red kangaroos did not, but he did not report separately for the sexes. Differences in methods between the two studies obscure the comparison of kangaroo behavior.

Long-term movements at Yathong were not consistent across species or sexes, or by environmental states. Causal factors were obscure. Two permanent shifts in home range areas occurred during periods of high resource availability. Oliver (1986) reported red kangaroo dispersal due to environmental decline, and that dispersal distances increased with drought. Only the old western grey male that died might have followed that pattern in this study. Johnson (1989) reported that dispersal as adults was uncommon in macropodids. Our sample size was insufficient to determine dispersal patterns. Greenwood (1980) reported that, in general, mammals show male-biased dispersal, and Johnson (1989) and Stuart-Dick and Higginbottom (1989) reported the same for kangaroos.

Our results show that energetic requirements, alone, are not sufficient to account for home range size. First, such comparisons assume equal habitat productivity (Harestad and Bunnell 1979) and similar efficiency of digestion and assimilation. These assumptions may not hold. For instance, the Australian outback is likely to be lower in productivity than North American habitats studied by the above authors. Also, relative to metabolic weight, a species such as the western grey kangaroo, which specializes in digesting dry, coarse forage, can occupy smaller, more compact home ranges than can a green feed specialist like the red kangaroo, which can only find its patchily distributed food by moving over a dispersed home range containing a lot of infrequently used interstitial space.

Secondly, energetically based home ranges do not take social behavior into account (Damuth 1981). All three species in this study showed serial polygynous breeding systems, in which males wandered about in search of estrous females (see chapter 8). On the basis of size alone, male home ranges overlap 2.87 female home ranges for western grey, 1.42 for eastern grey, and 2.34 for red kangaroos, respectively. However, this comparison overlooks the fact that male and female home ranges are not superimposed exactly, and that the home ranges of both sexes overlap extensively. Therefore, sex ratio and minimum overlap of home ranges are minimum estimates of the number of females with which the average male comes into contact. Periodic aggregation in mobs brings an even greater contact between females and males. Probably the average male kangaroo comes into contact with 15 or more females without leaving his home range. Because most breeding is accomplished by the biggest males, they probably have even greater contact than the average male, particularly in red kangaroos.

Although minimum physiological requirements may be useful to set minimum home range size requirement, minimum requirements are not what influence actual home range size. Fitness is the logical measure of optimal home range size. Selection should favor an animal expanding its home range until the cost of movement, risk of predation, etc., balances the additional gain of resource quantity and quality, and opportunities to mate and reproduce. Because females emphasize production of offspring, whereas males emphasize competition for mates, the sexes also use space for different needs.

No kangaroo in this study could be classified as nomadic. Thus, Yathong Nature Reserve is apparently a sufficiently benign environment to favor permanent attachment to place. Although no protracted drought occurred during the study, there was a harsh dry period during which most of the earthen water tanks dried up and no herbaceous green feed was available. Moreover, kangaroos remained within their home ranges even as free water became unavailable. They did not move to areas where thunderstorms resulted in heavy rainfall, even when these storms were clearly visible on the horizon and the lightning and thunder were obvious to a human observer. It is inconceivable that the kangaroos on Yathong could not detect these rainstorms. Yet, none of the radio-collared kangaroos left their home areas for these rain areas, where green feed was sure to appear within a week. Finally, the two kangaroos that shifted home ranges did so not during a dry period, but during times when green feed was abundant and widely available.

Denny's (1982) finding that 20 percent of the kangaroos at Tibooburra, and Priddel's (1987) report that seven percent of the kangaroos at Kinchega National Park in western New South Wales were nomadic suggests that nomadic behavior may be related to harshness of the environment. Average rainfall at Tibooburra is about 150 mm per year, at Kinchega 236 mm, and at Yathong, 350 mm. This suggests that nomadism is a characteristic of kangaroos only in the most extreme conditions, and that such behavior disappears as one moves across isohytes of increasing rainfall from west to east in New South Wales. In Hattah-Kulkyne National Park, an area apparently supporting greater density of kangaroos than Yathong, no long-distance movements were recorded by Coulson (1993a).

However, an alternate explanation of the results must be considered: categorization of individual kangaroos as nomadic may be an artifact of short-term records on individuals. For example, if location records of kangaroos on Yathong were few in number or made over a short period of time, the two permanent shifts in home ranges could have been thought of as nomadic, as could the widespread movements of some individuals with large home ranges (these include three males and one female red kangaroo and one male eastern grey). Thus, as many as 7 of the 29 radio-collared kangaroos (24 percent) might have been categorized as nomadic in a less intensive study. Only a relatively long-term record with large numbers of locations allowed a better understanding of the use of space by kangaroos at Yathong. Nevertheless, mitochondrial DNA studies of the genetic structuring of red kangaroos reported by Clegg et al. (1998) showed that eastern populations living in higher rainfall and richer habitats showed more structuring that those in drier, more homogeneous environments. Thus, the more stable home ranges at Yathong than further west would be consistent with genetic results, and suggest that nomadic movement of the western population was real, and not an artifact of the observation method.

Note that minimum convex polygon home range over a three-month period calculated by Priddel (1987, table 7.1) and Priddel et al. (1988) at Kinchega was 7.74 km² for red kangaroos. These would be less than the comparable mean home areas by species (i.e., sexes combined) for Yathong of 19.31 km² for red kangaroos, despite the fact that Yathong is a richer environment. The home range sizes of red kangaroos at Yathong were similar to those reported for an arid area in Western Australia by Norbury et al. (1994). It is likely that part of the difference between Kinchega and Yathong was due to length of time over which locations were taken (three months at Kinchega, 10 at Yathong) rather than different behavior of kangaroos between the two areas. This is a particular problem with red kangaroos that tend to remain in given core areas for a period of time, then shift to a new core area some distance away. The more sedentary western grey kangaroos showed the expected relationship, with home ranges at Kinchega of 6.92 km² and 6.15 km² at Yathong.

This is not to imply that the animals reported by Denny and Priddel were not nomadic, but only to indicate that further work is required before a completely unambiguous understanding of kangaroo use of space in relation to long-term rainfall patterns can be derived. In this regard, Norbury et al. (1994) proposed that home range size and movements of red kangaroos are facultative in response to drought cycles in the unpredictable environments of outback Australia. As previously reported by Croft (1991), they found that adult males extended their movements when vegetation biomass was low. Females, by contrast, remained sedentary. The questions about kangaroo mobility in relation to environmental variation are of great interest, and clearly warrant further work.

6

ACTIVITY PATTERNS

Kangaroos, like most animals, follow schedules. These help keep life orderly and efficient by sequencing tasks and structuring time to take advantage of favorable conditions and minimize adverse environmental effects. The basic schedule is the daily one. It sets the routine over the 24 hours of alternating day and night. However, atypical weather conditions may influence kangaroos to depart from their usual daily routine. The daily cycle also changes with the seasons as the days grow shorter or longer, and temperatures rise and fall. Finally, activities change with phenological stage of the vegetation, which determines the length of time it takes to gather a meal. At Yathong, green or dry feed can occur at any season.

Activity data were gathered by radio telemetry, using the combined signal amplitude and period (pulse interval) to determine if animals were actively moving about, or resting (chapter 3). Because these data were collected remotely and automatically, there was no observer influence or visibility bias; and, because the electronic system recorded continuously, all times of day and night were sampled equally.

Seasonal Patterns

Seasonal rainfall patterns at Yathong are divided into summer (northern) patterns and winter (southern) patterns. These were described in detail in chapter 2. The percent of time spent active for all species and sexes of kangaroos changed greatly over the nine months of monitoring (table 6.1). Percent of time active ranged from around 40 percent in

Table 6.1. Mean (standard error) percent activity of kangaroos by species and sex, by month, and overall percent active and hours per day active for the study.

Month	Eastern Grey Female	Eastern Grey Male	Red Female	Red Male	Western Grey Female	Western Grey Male	All Kangaroos[1]
Jan	41.7 (5.6)	37.0 (5.5)	48.0 (5.9)	29.6 (5.1)	38.6 (5.6)	55.4 (6.0)	42.0 (1.4)
Feb	39.4 (5.6)	43.9 (5.3)	35.6 (3.9)	51.0 (8.4)	55.6 (6.7)	47.7 (5.9)	48.4 (1.6)
Mar	61.3 (5.9)	43.0 (5.5)	36.8 (5.4)	43.1 (6.3)	50.6 (6.4)	41.6 (5.8)	45.3 (1.2)
Apr	50.8 (6.3)	47.9 (4.8)	45.0 (5.1)	47.5 (5.0)	49.8 (5.8)	53.2 (5.8)	51.1 (1.1)
May	64.0 (5.4)	64.0 (5.4)	55.3 (5.2)	47.0 (5.0)	59.5 (5.6)	64.5 (5.9)	59.5 (1.1)
Jun	70.2 (3.9)	76.5 (5.8)	63.6 (4.7)	55.2 (4.2)	71.0 (3.9)	63.0 (5.5)	64.9 (1.0)
Jul	73.5 (4.8)	75.2 (4.1)	52.0 (5.8)	52.4 (3.8)	61.5 (3.4)	58.4 (4.9)	62.1 (1.1)
Aug	80.8 (4.6)	84.3 (4.9)	68.8 (6.5)	73.3 (5.5)	67.0 (5.8)	75.5 (6.1)	71.9 (2.0)
Sep	64.4 (5.8)	64.2 (5.1)	58.0 (4.6)	60.7 (5.7)	67.5 (5.1)	63.9 (4.4)	65.0 (1.1)
Overall % active[1]	60.5 (3.7)	51.7 (4.0)	52.6 (3.8)	52.4 (36.3)	54.5 (3.7)	58.4 (3.6)	55.5
Hours/day active	14.52	12.41	12.62	12.58	13.08	14.02	13.32

[1] All data combined.

the hot summer months (January, February, March) to a high of about 70 percent in August, the coldest month. Eastern grey kangaroos showed the greatest activity of all (near 80 percent) in the coldest month, August (table 6.1). The change of percent activity over months for all species and sexes combined was highly significant ($F = 59.54$, $P < 0.001$), as it was for all species and sexes individually ($P < 0.001$). These data indicate that the change in seasons grossly influenced all kangaroos in similar ways, independent of both species and sex.

Although seasonal changes in the environment influenced all sex and species categories in the same general directions, among species and sexes, different sizes, physiological requirements, and use of habitat affected the degree of change. Of 15 sex by species tests, six were statistically different ($P < 0.05$). These tests point out that each sex by species has activity patterns that may differ from other species and the opposite sex. Only red male kangaroos were not significantly different ($P > 0.05$) from any other sex or species category.

Comparing percent activity by month partially masks the associations of activity with environmental states, however, because months, although well correlated with temperature, do not indicate the occurrence of rainfall, and subsequent availability of green forage. This variation can be better represented by examining activity during environmental states based on a combination of temperature and green feed availability (fig. 2.9, chapter 3). These results (fig. 6.1) again show that kangaroos are similar in overall pattern, with percent of time active being greater during cold periods than hot or moderate periods. The higher activity during cold periods was independent of whether it was wet or dry (fig. 6.1, $P = 0.698$), suggesting that temperature per se was the most important variable. Presumably, the greater energy demand to maintain body temperature (Dawson 1995) was the reason for extending the time spent active, most of which involved feeding (specific activities were verified by direct observations; see chapter 8).

Percent of time active was most similar between sexes and species during the dry-hot and dry-moderate environmental states (fig. 6.1). The extremely high daytime temperatures and high solar radiation that limited activities at midday, were probably the main variables forcing similar activity schedules, as further discussed below under daily patterns. For other environmental states, activity between species and sexes was more variable. These states were less constraining on behavior and allowed greater latitude than did the extremes of the dry-hot period.

Despite their overall similarity, again, there were significant differences between the species and sex combinations. For all environmental states combined, every comparison between sexes and across species

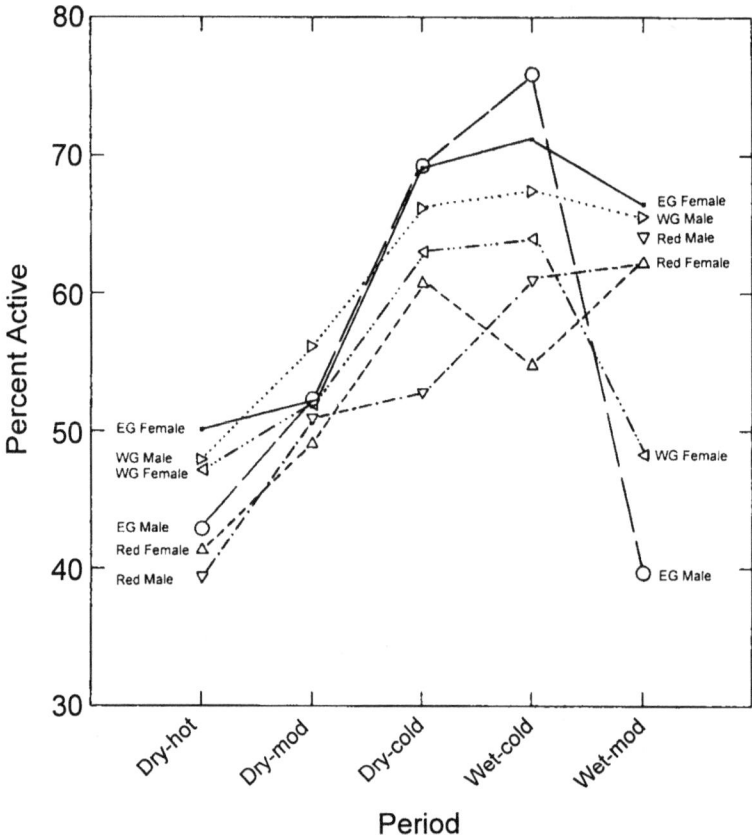

Figure 6.1.   Mean percent active by kangaroo species and sex by environmental period. Environmental periods are ordered in the time sequence they occurred during the study. Chi-square tests between species and sex combinations are given in table 6.2.

was significantly different ($P < 0.001$ in all cases). Very large sample sizes were available, and this contributed to the level of significance observed. When species and sex combinations were broken down by environmental state, with consequent reduction in sample size, the results of tests for differences were more variable, although most were significantly different (table 6.2). The dry-moderate temperature state showed the least differences between combinations of species and sex, with only 2 of the 15 combinations (both sexes of red versus western grey males) yielding significant differences at $P < 0.05$. Other patterns were weak. For dry-cold and wet-cold states, only red kangaroos showed a difference between the sexes (dry-cold); conversely, for eastern greys, the sexes were nearly identical for dry-cold (table 6.2).

Table 6.2. Pearson chi-square probability values (1 df) for tests of activity between sexes and species of kangaroos by environmental state.

	Eastern Grey Female	Eastern Grey Male	Red Female	Red Male	Western Grey Female
Dry-Hot					
Eastern grey male	<0.001				
Red female	<0.001	0.312			
Red male	<0.001	0.064	0.325		
Western grey female	0.055	0.005	<0.001	<0.001	
Western grey male	0.156	0.001	<0.001	<0.001	0.616
Dry-Moderate					
Eastern grey male	0.819				
Red female	0.283	0.220			
Red male	0.724	0.571	0.230		
Western grey female	0.953	0.873	0.286	0.697	
Western grey male	0.054	0.117	0.004	0.018	0.082
Dry-Cold					
Eastern grey male	0.908				
Red female	<0.001	<0.001			
Red male	<0.001	<0.001	0.008		
Western grey female	<0.001	<0.001	0.117	<0.001	
Western grey male	0.065	0.075	<0.001	<0.001	0.061
Wet-Cold					
Eastern grey male	0.068				
Red female	<0.001	<0.001			
Red male	<0.001	<0.001	0.413		
Western grey female	0.001	<0.001	0.001	0.116	
Western grey male	0.104	0.001	<0.001	0.002	0.103
Wet-Moderate					
Eastern grey male	<0.001				
Red female	0.029	<0.001			
Red male	0.332	<0.001	0.431		
Western grey female	<0.001	<0.001	<0.001	<0.001	
Western grey male	0.654	<0.001	0.098	0.568	<0.001

Daily Activity Patterns

Upon plotting activity patterns by hour for all three species, it can be seen that eastern and western greys were similar overall across the diel cycle (fig. 6.2), whereas red kangaroos showed less activity at night. All three species were similar during midday hours. Correlation between eastern grey and western grey kangaroo mean activities by hour was

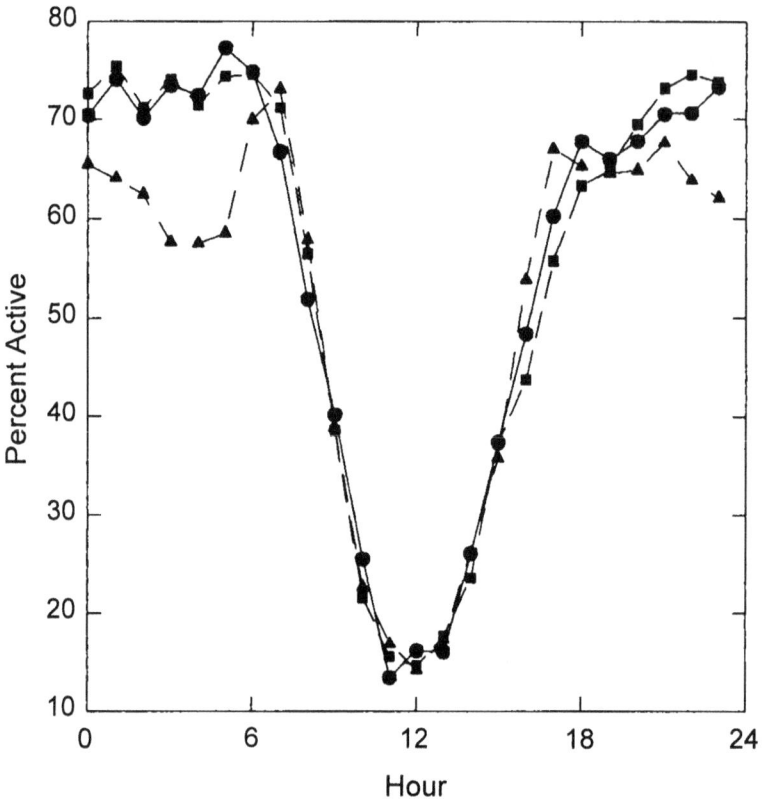

Figure 6.2.   Mean percent active of eastern grey (circles), red (triangles), and western grey (squares) kangaroos by hour for all seasons combined.

high ($R^2 = 0.99$) while those between eastern grey and reds ($R^2 = 0.90$) and western grey and reds ($R^2 = 0.90$) were lower, primarily due to the differing activity at night.

A similar overall percent activity by sex or species may be quite differently distributed over the 24-hour day. As with seasonal patterns, daily patterns were analyzed by environmental states, based on temperature and availability of green feed rather than months. Results are presented in a series of figures for each species and sex to retain the considerable amount of information on this topic (figs. 6.3 through 6.8). Mean percent activity and 0.95 percent confidence limits are plotted by hour on an international time clock (0000 h = midnight). Unlike the seasonal patterns (fig. 6.1), for which environmental states were arranged chronologically, the figures for daily activity patterns are arranged in a matrix

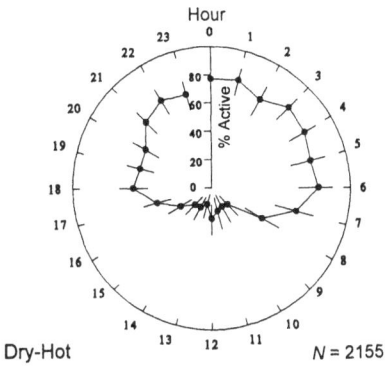

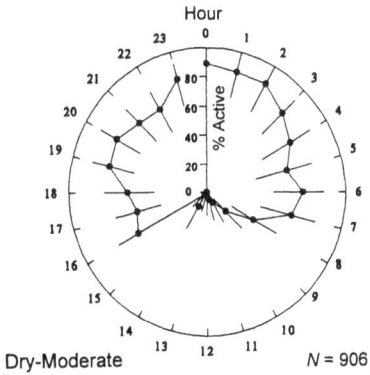

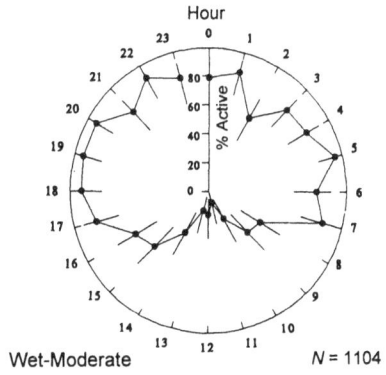

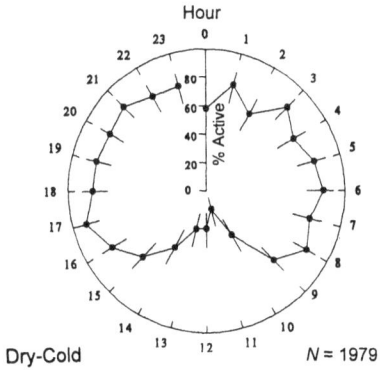

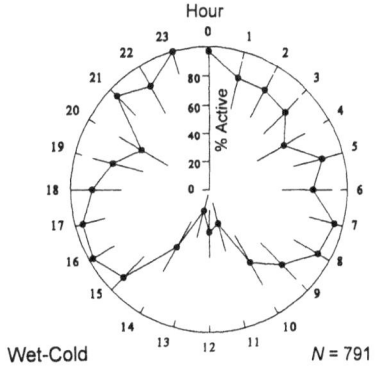

Figure 6.3. Female eastern grey kangaroo activity (mean and 95% confidence limit) by hour (international clock) for various environmental periods on Yathong Nature Reserve. Chronological sequence is left column, top to bottom, and then right column, bottom to top.

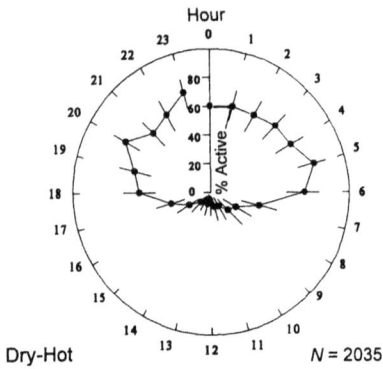

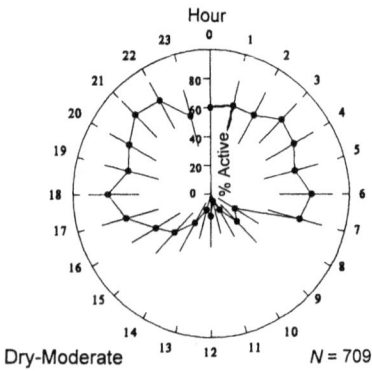

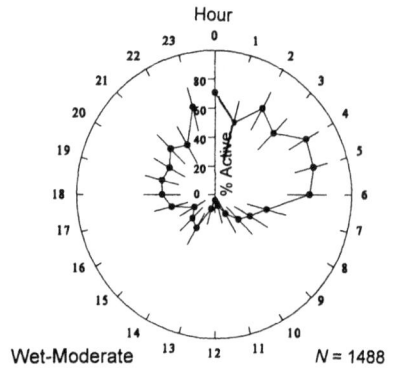

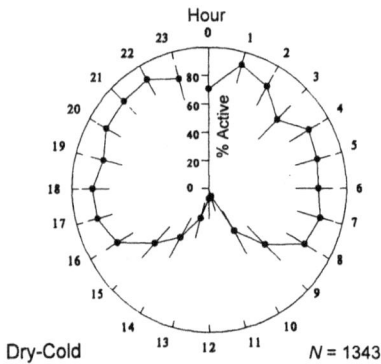

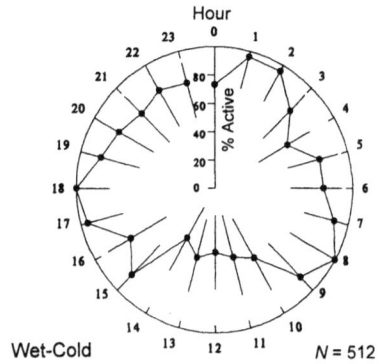

Figure 6.4. Male eastern grey activity (mean and 95% confidence limit) by hour (international clock) for various environmental periods on Yathong Nature Reserve. Chronological sequence is left column, top to bottom, and then right column, bottom to top.

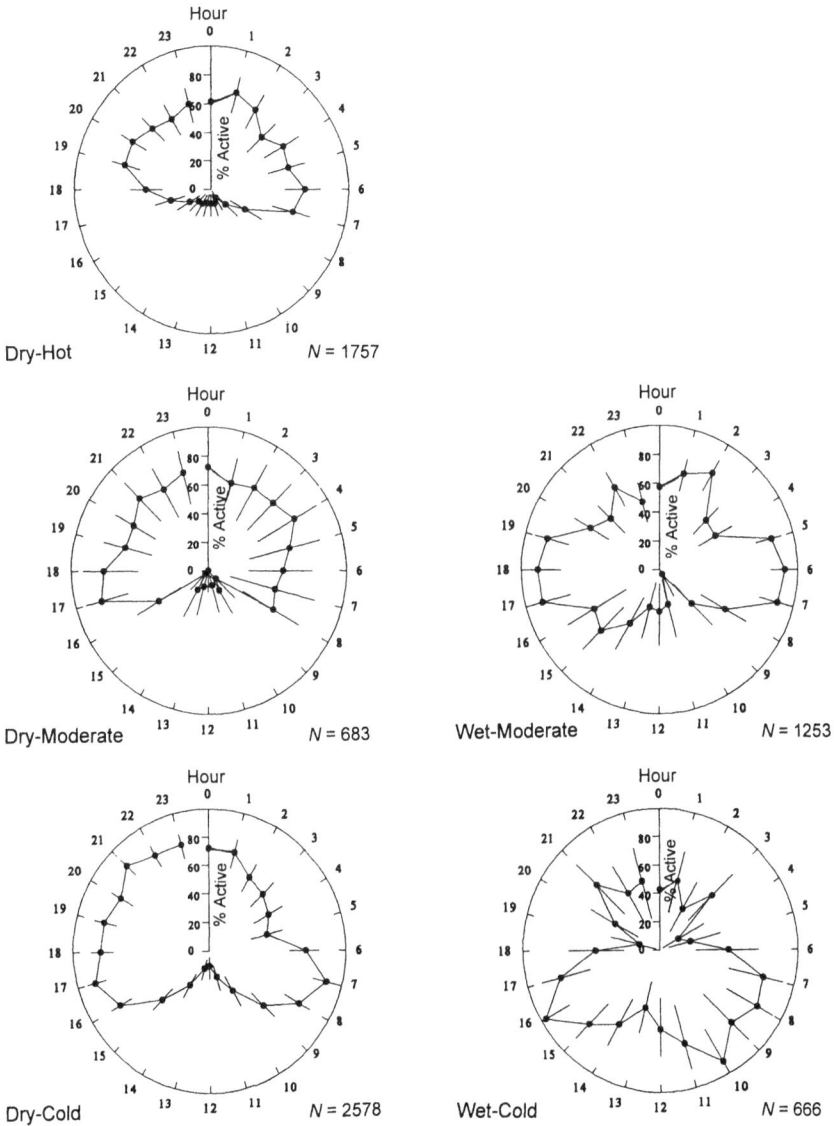

Figure 6.5. Female red kangaroo activity (mean and 95% confidence limit) by hour (international clock) for various environmental periods on Yathong Nature Reserve. Chronological sequence is left column, top to bottom, and then right column, bottom to top.

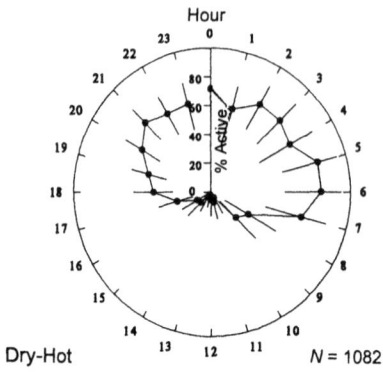

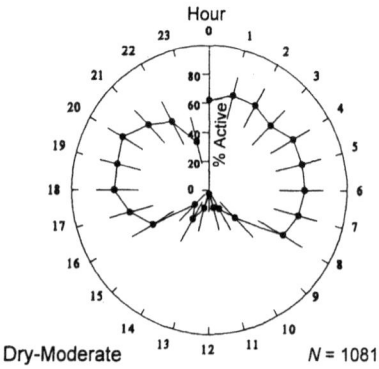

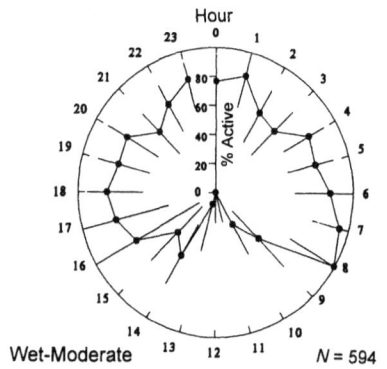

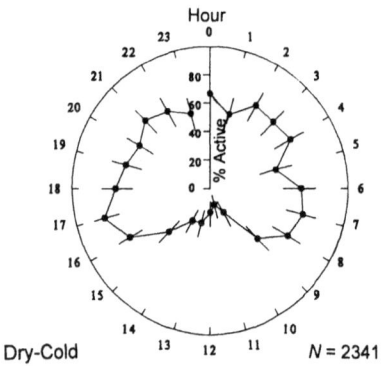

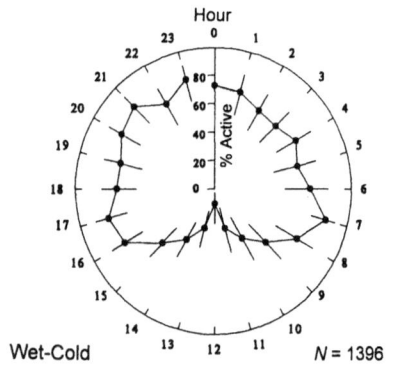

Figure 6.6. Male red kangaroo activity (mean and 95% confidence limit) by hour (international clock) for various environmental periods on Yathong Nature Reserve. Chronological sequence is left column, top to bottom, and then right column, bottom to top.

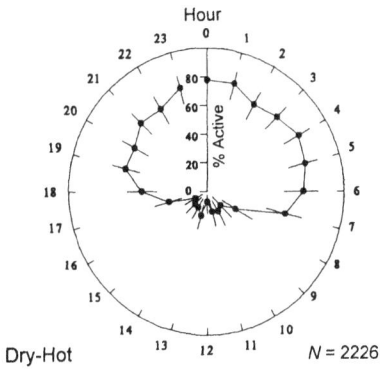

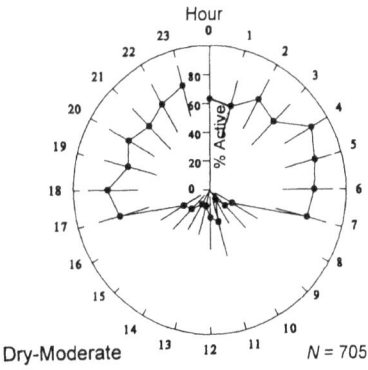

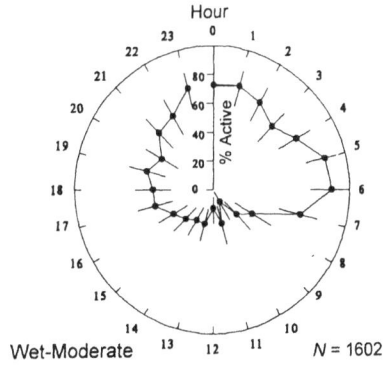

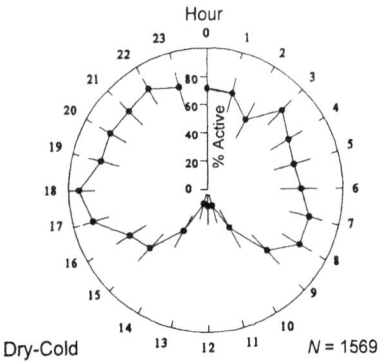

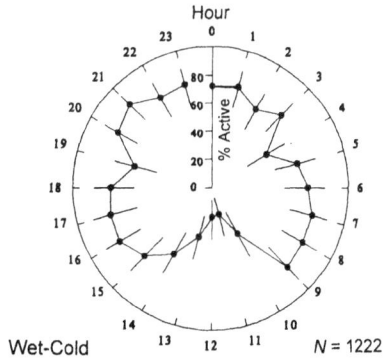

Figure 6.7. Female western grey kangaroo activity (mean and 95% confidence limit) by hour (international clock) for various environmental periods on Yathong Nature Reserve. Chronological sequence is left column, top to bottom, and then right column, bottom to top.

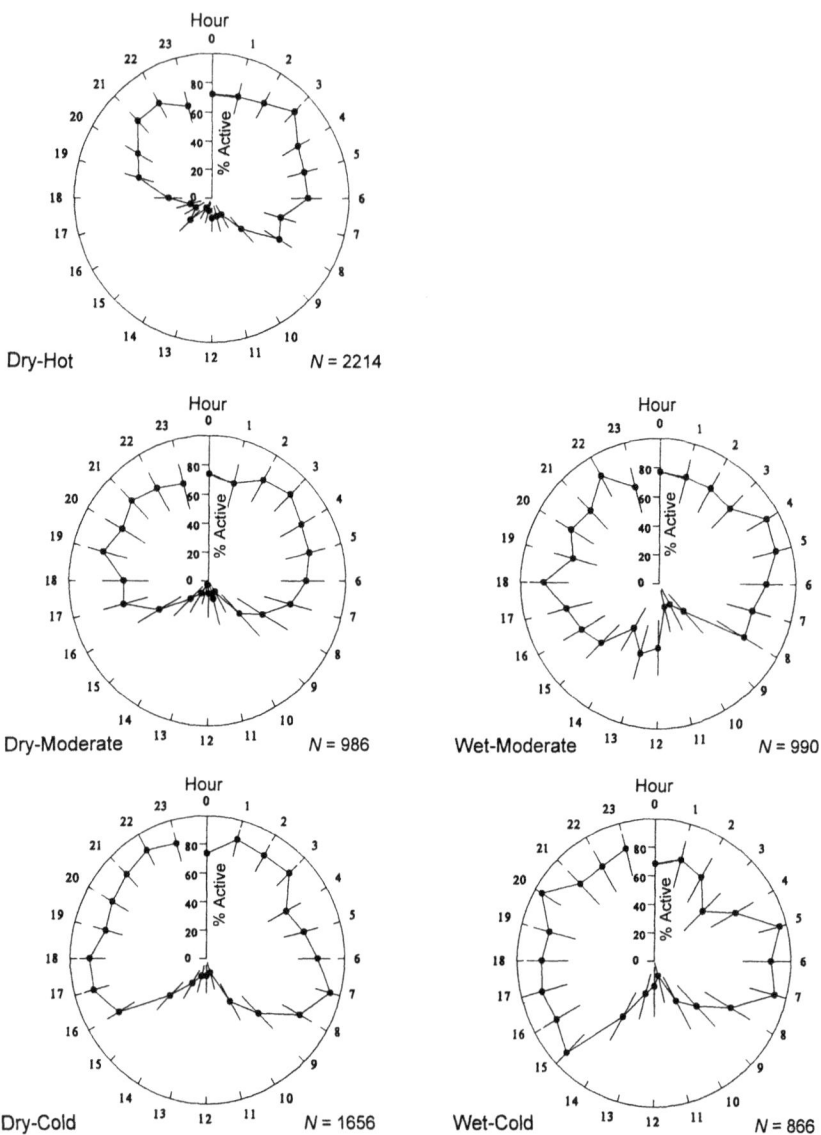

Figure 6.8. Male western grey kangaroo activity (mean and 95% confidence limit) by hour (international clock) for various environmental periods on Yathong Nature Reserve. Chronological sequence is left column, top to bottom, and then right column, bottom to top.

with temperature (from high to low) in rows, and dry or green feed in columns. This arrangement enables ready interpretation of environmental effects on activity patterns of a given species and sex category. Chronological order is represented by top to bottom of the left column, and bottom to middle of the right column (a reversed "J"). Noting differences between species and sex combinations requires comparison between figures.

Dry-Hot Environmental State

The dry-hot environmental state was the one with the lowest kangaroo activity (fig. 6.1). Green grass occurred either in isolated sites at Yathong where the last soil moisture persisted, or was not available at all. Some forbs remained green later than grasses, in agreement with the findings of Moss and Croft (1999) that forbs had a greater time lag following rains than grasses. However, the late season forbs on Yathong tended to be low-quality forage species because of secondary compounds, spines, or both. Daytime temperatures routinely exceeded 40°C and solar radiation was intense.

The most prominent pattern of the dry-hot state (figs. 6.3 through 6.8) was the low total activity and the avoidance of daytime feeding. Low daytime feeding during summer was previously reported for eastern grey kangaroos (Southwell 1981) and for red and western grey kangaroos (Priddel 1986). All activity ceased around dawn, when kangaroos moved into the shade of wooded areas to rest for the day, and activity did not resume again until dusk.

However, there were a number of differences between species and sexes during this environmental state. Eastern grey kangaroos (figs. 6.3 and 6.4) tended to cease activities an hour earlier than did red and western greys (figs. 6.5 through 6.8). Eastern greys also tended to begin the dusk feeding period more abruptly (at about 1800 h) whereas the other two species increased activities more gradually, and peaked an hour later (at about 1900 h).

Red kangaroos showed least total activity, and greys most (fig. 6.1). Eastern grey females were similar to western greys, whereas eastern grey males were more like reds. Activity of red males dropped virtually to zero at midday, whereas the other species and sexes showed activity about 10 percent of the time (figs. 6.3 through 6.8). Direct observations showed that this activity was confined to wooded areas, and consisted primarily of movements related to cooling (Dawson 1995), including occasional shifting between bedding or resting sites to follow the shade as the sun moved, and perhaps to take advantage of any breezes.

Dry-Moderate Environmental State

During this state, activity increased over that of dry-hot, but still remained lower than during other periods. Temperatures were moderate and comfortable. Kangaroos showed the greatest similarity between species and sexes in total activity (fig. 6.1) and likewise, their daily schedules were the most similar during this period (figs. 6.3 through 6.8). As compared to the dry-hot state, in the dry-moderate state, kangaroos added approximately an hour to the morning activity and began feeding about an hour earlier in the afternoon. Most species and sexes ended morning activity and began afternoon activity rather abruptly, and midday activity dropped nearly to zero. Presumably, this was because moderate temperatures allowed a single resting site to be occupied, and movement to shade for temperature regulation was less imperative. The most notable deviation from uniformity across species and sexes was shown by red females, whose pattern was skewed away from early morning, towards evening activity (fig. 6.5).

Dry-Cold Environmental State

This state saw greatly increased activity among kangaroos (fig. 6.1). Generally, an hour more of activity was added in the morning, and activity began an hour earlier in the afternoon. Food conditions were largely unchanged, so the most likely explanation for increased activity was the need for greater energy intake to offset the effects of cold temperatures, which regularly dropped below freezing at night.

Red kangaroos, particularly females, were relatively lower in activity than the other two species. As with the state previously discussed, female reds avoided late night activity when temperatures were lowest, and increased activity at 0700 h when the sun began to warm the landscape (fig. 6.5). Direct solar radiation apparently offset the thermal effects of relatively low air temperature.

Wet-Cold Environmental State

Although red kangaroos were an exception, this state had the highest kangaroo activity overall (fig. 6.1). Even for red kangaroos, relative activity was high, although not as high as during the wet-moderate state (fig. 6.1). For all species and sexes except red females, activity, mostly feeding, was high throughout most of the diel period, except midday (figs. 6.3 through 6.8). Activity declined from 0900 h (except for western grey males that ceased activity somewhat earlier) to a low at 1200 h (noon), and then increased until 1300 h. Once again, red females deviated from the general pattern. They concentrated their feeding in the daytime hours and avoided feeding at night (fig. 6.5). Lows in their activity occurred at 0500 h and 1900 h, and night activity fluctuated, suggesting alternating bouts of feeding and bedding.

Wet-Moderate Environmental State

This state had the fewest environmental constraints. High-quality green feed was abundant and widely distributed, and temperatures usually were comfortable. Not surprisingly, activity patterns showed the greatest variation between species and sexes. In a sense, each species' and sex's preferences could be expressed to the greatest latitude because of the lack of environmental constraints.

Nevertheless, some unusual results occurred, particularly the extreme difference between the sexes in eastern grey kangaroos. Eastern grey females had the highest overall activity whereas eastern grey males had the lowest (fig. 6.1). Indeed, eastern grey males showed the lowest mean percent activity (39.7%) recorded in the study (fig. 6.4). Western greys showed a wide difference between sexes as well (fig. 6.1). The differences between the sexes were surprisingly great, yet we have no plausible explanation for these results.

As usual, red females showed a different daily pattern (fig. 6.5). They continued to show some avoidance of nighttime activity and exhibited peaks of activity at 0600 h and 1800 h. Also, the results for western grey females were skewed away from activity in the early night, and towards relatively high activity in the early morning hours.

Influence of Weather on Activity Patterns

Seasonal activity is the cumulative outcome of kangaroo behavior in response to ongoing weather, moment by moment. Thus, to examine the influence of weather on activity, it is first necessary to explore the daily patterns of weather at Yathong. Weather varied greatly from day to day, yet there were overall patterns that could be expressed by mean values. Clearly, kangaroos have daily schedules of activity that are based upon the "anticipation" of weather typical of that season and time of day. In this sense, behavior is entrained towards expectation for the seasonal trends in weather. Here, we first examine the relationship of these kangaroo schedules to weather patterns, then we explore how these kangaroo activity schedules are altered by the kaleidoscope of weather on an hour-by-hour basis.

The dominant variable determining weather in summer is solar radiation. It clearly is the driving variable that influences temperature, winds, and related phenomena. By contrast, winter weather is dominated by regional air masses, as fronts move across Yathong from the south. When cloud masses move in, the sun is obscured and might not be seen for several days. Daily weather patterns, therefore, can be broken down by cold season (winter), hot season (summer), and two transitional seasons (spring, fall). However, appropriate weather and kangaroo data were not available for spring, so only the cold, fall transition, and hot seasons are examined here. The availability of green feed, an obviously important variable, is here momentarily ignored to concentrate on weather effects on kangaroo behavior.

Weather data were taken from continuously recording instruments maintained at Shearers' Quarters (chapter 3). Only data for which there were corresponding sufficient sample sizes of kangaroo activity were included in the analysis. Temperature is a useful indicator of the overall weather picture for the time state over which data were available (fig. 6.9). Figure 6.9 illustrates that the hot season can be described conveniently as January through March (Julian days 0–90), fall transitional season as April and May (Julian days 91–151), and the cold season as June and July (Julian days 152–200).

Figure 6.10 shows mean values of the recorded weather parameters for these three seasons. Note that these parameters are interrelated. Relative humidity was highly negatively correlated with temperature ($r = 0.96$, 0.96, and 0.99 for summer through winter seasons, respectively), and had no lag. Correlations of other parameters were more variable due to leads and lags that offset such relationships. At sunrise, solar radiation began to alter the weather, resulting in a typical pattern. The first parameter to respond was wind speed, which picked up rapidly, and in summer and fall, shifted in direction as well. Increases in temperature, and correlated decreases in relative humidity, lagged two hours after change in wind speed (all seasons), as established by cross correlation ($P < 0.05$). Rise in barometric pressure lagged three hours after temperature and five hours behind wind speed. Barometric pressure peaked at 0900 h, then began to fall rapidly, reaching a low at 1700–1800 h. As the sun passed its zenith (noon), weather parameters tended to hold steady (wind speed and direction) or continued the previous trend (temperature, relative humidity, barometric pressure). In all seasons, temperatures started falling and relative humidity started rising at around 1500 h, wind speed falling at 1600 h and barometric pressure rising at 1700–1800 h. Wind direction shifted gradually until after midnight.

Rainfall lagged by six to seven hours with temperature for summer, but showed no significant ($P > 0.05$) lag in fall and winter. No lag was expected in fall and winter, given that rains came with broad regional

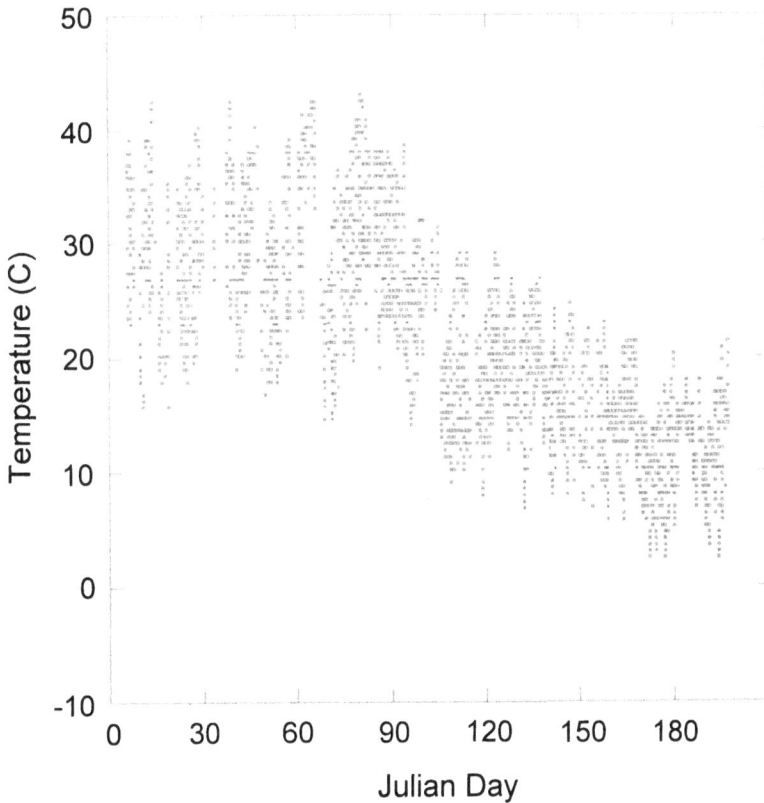

Figure 6.9.   Plot of temperature on Julian day for data taken at Yathong Shearers' Quarters between January 1 and July 19, 1986. Only data for which there are corresponding records of kangaroo activity are plotted.

storms, and occurred throughout the day and night. Still, fall and winter rain amounts were somewhat less on average in the early afternoon. During summer, there was no rain in the morning, and only traces at midday; nearly all rains began at 1500 h and lasted overnight until 0700 h. This was the typical summer thunderstorm pattern, with high temperatures occurring in the mid-afternoon, followed by continuing decline in barometric pressure, and increase in thunderclouds.

This intricate set of related, oscillatory weather parameters, with all but temperature and relative humidity showing a lag or lead, creates a complex against which kangaroo behaviorial patterns can be tested. Figure 6.11 shows mean activity patterns for the three seasons for a combined sample from all species and sexes of kangaroos. The relative decrease in activity during daytime hours is illustrated for all seasons, but

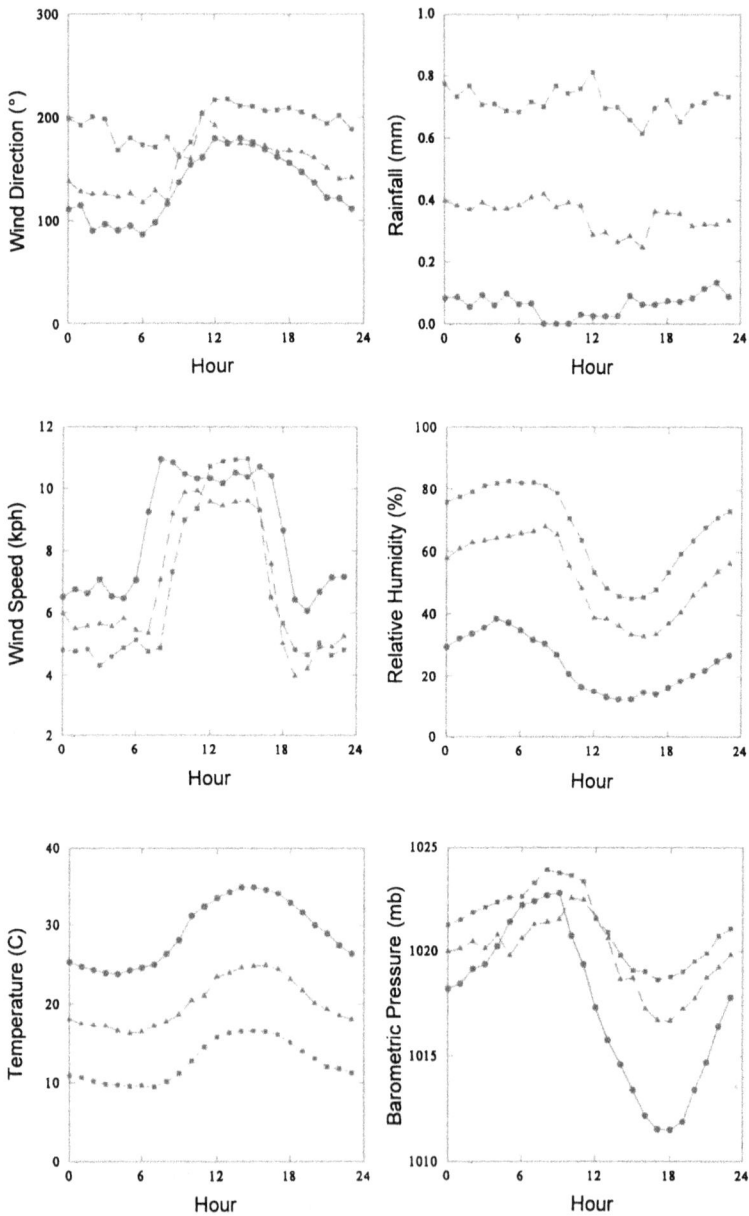

Figure 6.10. Mean values of weather variables for seasons (circles = summer, triangles = fall, and squares = winter) by hour (international clock) on Yathong Nature Reserve as taken by continuously recording instruments at Shearers' Quarters.

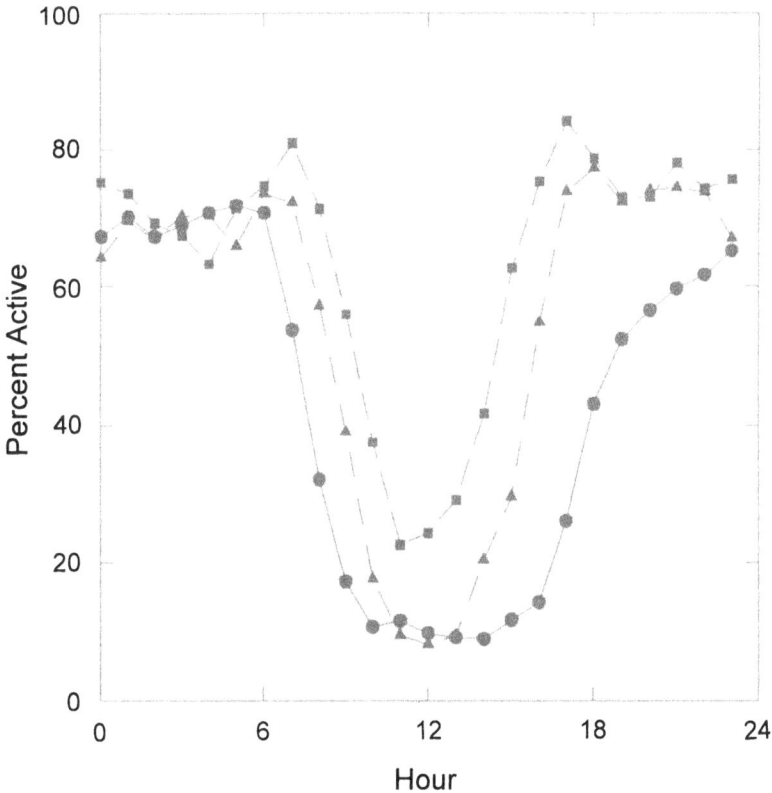

Figure 6.11.   Mean percent of time active by kangaroos (all species combined) for seasons (circles = summer, triangles = fall, and squares = winter) by hour (international clock) on Yathong Nature Reserve.

is longest for the hot summer season. Furthermore, resumption of activity is delayed into the late afternoon and evening of this season.

Nighttime activity of kangaroos tended to stabilize at around 70 percent, except during the winter season, when early morning and late afternoon activity peaks occurred at about 80 percent (fig. 6.11). The overall higher winter activity (fig. 6.1) was achieved by both extending the time of nighttime activity, and increasing daytime activity levels (fig. 6.11). The fall season showed extended nighttime feeding over that of summer, but the daytime activity dropped to the same level as for summer.

None of the weather parameters recorded showed a strong relationship with percent activity; correlation coefficients were low and $P >$

0.05. The lack of correlation could be shown by plotting mean kangaroo activity by hour (fig. 6.11) against mean weather parameters by hour (fig. 6.10) for each season; in this way a series of ovoids was obtained (fig. 6.12). Some ovoids showed very high amplitude, such as percent activity against barometric pressure in summer (independent due to large lag), while others showed little amplitude (activity independent of weather variable), such as rainfall, particularly in summer. Summer wind speed came closest to a curvilinear relationship with activity, and to a lesser degree, fall wind speed (fig. 6.12). Winter wind speed, by contrast, showed a large ovoid. Note, however, that the points for winter (in the upper left) that deviate most from the remainder of the arrays are for afternoon times (1400–1700 h). Other than these points, addressed further below, the remainder form a reasonable curvilinear relationship with percent activity, with activity being highest at low wind speeds and lowest at high wind speeds (fig. 6.12). Activity dropped markedly at wind speeds of greater than 9 km/h.

There was no lag ($P > 0.05$) in summer and fall between percent activity of kangaroos and wind speed. However, increasing wind speed should be associated with reducing the heat stress at moderate temperatures of morning, suggesting that another variable is involved, and that the kangaroos are not responding to wind speed in a cause-effect way. The obvious candidate variable was solar radiation, but unfortunately, this parameter was not measured directly. Nevertheless, the timing of solar radiation is approximately correct as shown by time of sunrise. Actual sunrise as observed in the field at Yathong was recorded on daytime systematic surveys. Sunrise in summer was at around 0600, in fall around 0630, and in winter, 0720. These times seem consistent with the times of decline in kangaroo activity in the early morning (fig. 6.11), and increase in wind speed (fig. 6.10). Solar radiation was obviously the variable driving most of the daily weather patterns, especially in the summer and fall seasons.

When wind speeds pick up anytime, kangaroos become flighty and take alarm easily. This response to wind has previously been reported for an ungulate—the tule elk (McCullough 1969)—and our experience has shown this to be a common response among large, openland herbivores. McCullough speculated that its cause was a decreased probability of detecting predators under these conditions. Harden (1985) reported that peak activity of dingoes occurred an hour or two after dawn, and this too may be related to wind improving their chances of hunting success. However, the increase in activity of dingoes after dawn over other times of day was small, and caution should be exercised in relating it to

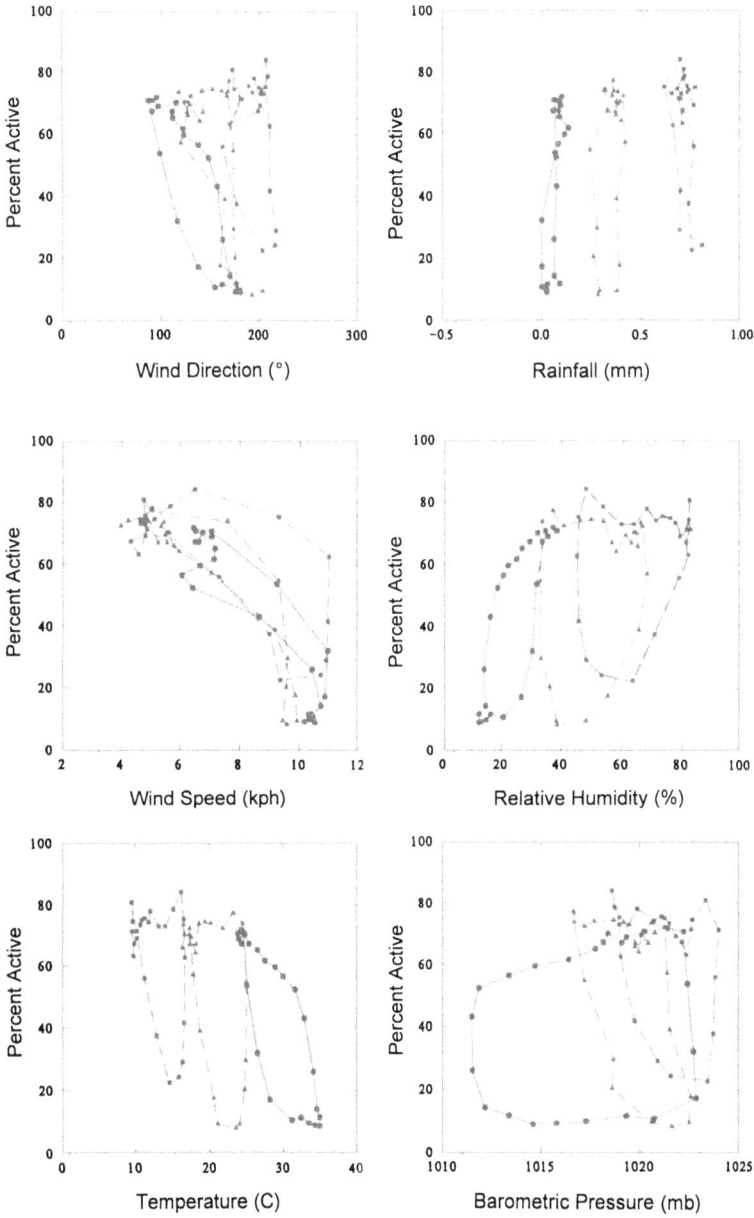

Figure 6.12. Relationship between kangaroo (all species combined) activity and weather parameters for seasons (circles = summer, triangles = fall, and squares = winter) on Yathong Nature Reserve. Data are taken from figures 6.10 and 6.11, thus each point represents an hourly mean value.

Table 6.3. Significant ($P = 0.01$) relationships of weather parameters and kangaroo activity by time of day for different environmental states. Lack of significance is indicated by n.s. Signs indicate direction of correlation of the weather variable. Weather variables are RH = relative humidity, TEMP = temperature (°C), WSPED = wind speed, WDIR = wind direction, and BP = barometric pressure. Times of day are late night = 0000 to 0700, morning = 0700−1100, midday = 1100−1400, afternoon = 1400−1700, and early night = 1700−0000.

	Late Night	*Morning*	*Midday*	*Afternoon*	*Early Night*
Dry-Hot	n.s.	+ WSPED + BP − TEMP − RH	+ WSPED − WDIR	+ RAIN − TEMP	+ WSPED + BP
Dry-Cold	n.s.	+ RH + TEMP − WSPED − BP	+ RH − WSPED − BP	+ RH − WSPED − BP	+ TEMP − WSPED
Wet-Cold	n.s.	n.s.	+ RH + TEMP − RAIN	+ RAIN − WDIR	+ TEMP
Wet-Mod	n.s.	n.s.	n.s.	n.s.	+ TEMP

kangaroo behavior. In summer, winds pick up in the morning at a time when kangaroos must curtail activity in any event due to heat stress. Although this factor may contribute to their flightiness, it would not account for their general response to winds, including times when heat stress is not a problem.

After removing the major seasonal effects of weather on activity, examination of residuals showed that the same variables that were important overall tended to be important for short-term deviations from the typical values as well. However, the influence of weather variables changed with time of day, so the data were sorted into five diel periods (table 6.3). There were no significant effects for any environmental state for late night (table 6.3). For other times of day during cold states, the pattern showed increased activity as temperatures and relative humidity increased, and decreased activity when wind speeds and barometric pressure were high. These kangaroo responses would be consistent with avoidance of cold, stressful weather conditions.

The same weather variables were important during the dry-hot environmental state, but the signs of the variables were reversed. Thus, activity increased as wind speed and barometric pressure increased, but decreased as temperatures increased and relative humidity decreased (table 6.3). These responses were consistent with avoiding heat stress under summer conditions. Rainfall had a negative influence at midday during the wet-cold environmental state and a positive influence in the afternoon during the dry-hot state (table 6.3). Again, these results appear consistent with behavior to decrease cold stress during cold seasons and to decrease heat stress during hot seasons. Highest significance was achieved when the interaction between temperature and wind speed was included in the equations, and this was true for all species and sexes treated separately. Thus, although the species and sexes differed in their degree of response, the pattern of response was consistent.

Wind direction had a negative effect on activity at two times of day by environmental states (table 6.3). We do not believe this represents a direct influence of wind direction per se, but rather, that it reflects the covariance of wind direction and other weather parameters. The possibility of spurious correlations must also be considered. We feel confident in accepting the biological importance of weather variables effecting consistent patterns, and not making too much of isolated instances of statistical significance.

These results show that during the extreme seasons of cold and hot, kangaroos alter their activity patterns to moderate the effects of these extremes. During moderate states, they are seldom stressed by weather conditions, and do not usually alter their entrained schedules, with the possible exception of increasing activity in the early night if temperatures are warm. Finally, there is a consistent time of feeding activity during late night that is independent of weather conditions.

We did not find that red kangaroos had constant activity amounts across seasons, as reported by Priddel (1986) for Kinchega National Park and Watson and Dawson (1993) at Fowler's Gap. Although they varied in degree, red kangaroos showed seasonal shifts in percent of time active that paralleled the patterns of grey kangaroos (fig. 6.1, table 6.1). Cold season activity for eastern grey kangaroos in this study matched the high percent activity (71 percent) reported by Southwell (1981) and exceeded it in August, the coldest month (table 6.1). Activity of eastern grey kangaroos was less during the warm season, dropping below the 63 percent reported by Southwell (1981). Perhaps the greater extremes shown at Yathong are due to its more extreme conditions, compared to eastern Australia where Southwell worked. Our results also differed in many details from those of Clarke et al. (1989) at Wallaby Creek in northeastern New South Wales, although overall patterns were similar. Eastern grey kangaroos, particularly the females, generally showed greater activity during cold states than the other two species.

Overall, our results for western grey females agreed reasonably well with those of Arnold et al. (1988) from Western Australia. Detailed comparisons are difficult given the totally different methods employed by the two studies; nonetheless, the diel patterns and responses to weather variables were similar.

Discussion of Activity Patterns

A number of factors influence kangaroo activity patterns. There is the need to obtain food, water, and shelter, and these vary in distribution, abundance, and quality through the seasons. Physiological requirements must be met and homeostasis maintained by metabolic or behavioral means. Finally, there is the need to avoid enemies and interact with other kangaroos in social encounters. Social encounters, based on direct observations, constitute only a small fraction of the total activity time and cannot be examined quantitatively; instead, these are discussed qualitatively in chapter 7. By far, the majority of time active was spent feeding, as found by Clarke et al. (1989) for eastern grey kangaroos in eastern Australia. The remainder of time was spent primarily in moving between feeding and resting places and within feeding patches. Movement constitutes no more than five percent of the active time, as also previously reported by Watson and Dawson (1993).

Except for the newly emerged pouch young, natural predators are not much threat to kangaroos on Yathong today (see chapter 11). However, they continue to contribute indirectly, in that residual behavior may reflect the evolutionary consequences of a long prior history of selection by the presence of predators. Thus, activity included time spent in vigilance, movements due to alarm, and selection of habitats for feeding and resting based upon their concealment and predator escape values.

No doubt, response to predators is reinforced by human disturbance. For example, some poaching occurred along Belford Road during our study. This was evidenced by blood found on the road where professional shooters had stopped their pickup trucks with kangaroo carcasses racked on the back, as well as the suspicious disappearance of a radio-collared animal (chapter 3). However, poaching was not common. Because the rest of Yathong Nature Reserve is not open to the public, the resident caretaker's and research workers' vehicles caused most of the human disturbance. Although these contacts represented no threat, kangaroos often showed alarm behavior and moved away from the road. Given the small number of people and vast area of the reserve, the sum total of these impacts was small, and probably quantitatively trivial. All things considered, it seems that at least 98 percent of the activity time recorded for kangaroos was directed to feeding and movement in an undisturbed state. The remaining two percent would include social activity between kangaroos and a negligible percent attributed to all other things, including response to enemies.

Seasonal Patterns

Highest kangaroo activity occurred during cold seasons when an average of 65 percent of the time was spent active (fig. 6.1). This agrees with results previously reported by Southwell (1981), Arnold et al. (1988), and Clarke et al. (1989). In the coldest month (August), kangaroos were active about 70 percent of the time. Lowest activity (just over 50 percent), by contrast, occurred during hot seasons. These differences by season probably are due to two factors. First is the higher metabolic requirement imposed upon kangaroos by cold weather in winter, which also is associated with more frequent rainfall (fig. 6.10). Wet and cold conditions often persisted for several days because of the regional extent of winter storms, subjecting kangaroos to wetting and exposure. Two of our radio-collared animals, a red male and an eastern grey male, died under such conditions. Extreme low temperatures (in the range of $-10\ °C$) occur during clear nights in open habitats. Nonetheless, kangaroos can moderate these extremes by small shifts in feeding time away from early morning hours when the temperatures are lowest (0700 h, fig. 6.10), and by remaining in wooded habitats. Although temperatures are much more moderate during winter rainy periods (typically $10°C$ low, $16°C$ high), the combination of relative cold and wetting demands more energy to maintain homeothermy than does cold alone. After rains, kangaroos often are observed to move to open areas to dry out. Roads are favorite sites—in fact, wet kangaroos are even reluctant to leave roads upon the approach of a vehicle.

The second constraining factor is summer heat. In contrast to the cold and wet of winter, the problem confronting kangaroos in summer is excessive heat buildup. The most obvious manifestation of heat avoidance is the almost total absence of daytime activity (fig. 6.11) and the selection of shaded habitats for protection from solar radiation (evidenced by radio telemetry locations and direct observations). Direct solar radiation seemed to be the principal factor constraining kangaroo activity. Parker and Robbins (1984) and Parker and Gillingham (1990) have shown that deer are relatively susceptible to heat buildup, and kangaroos—comparable in size to deer—are subject to the same problem (Dawson 1995). The more arid-adapted red kangaroos accumulate stored heat that subsequently must be dissipated, whereas the more mesic eastern grey kangaroos maintain their body temperature, but are more dependent on water (McCarron and Dawson 1989). Brown and Dawson (1977) found that red kangaroos, in order to be able to absorb more heat, lowered their body temperatures by 2 to 4°C prior to the hot time of day. Denny (1982) reported that red kangaroos could, on a temporary basis, lose water amounting to two to four percent of their body weight daily; amazingly, they are unaffected by water loss of up to 20 percent of total body mass (Dawson 1995).

Viewing solar radiation as the most important variable would explain why daily kangaroo activity dropped to as low a level during spring as in summer, although for fewer hours (fig. 6.11). Even in winter, some days have high solar radiation, and kangaroo activity drops to a low at midday (fig. 6.11). Nevertheless, the mean activity values at midday in winter are substantially higher than in summer, because overall radiation is less intense and many more cloudy days occur during this season. Wind speed is high during winter sunny days, and kangaroos appear to rest in microsites providing shade and exposure to breezes.

Under cool conditions, kangaroos lie flat on the ground in open, splayed postures. Under the hottest conditions, kangaroos seek shade, remain upright, and hunch over on their legs with their tails curled underneath their bodies (Russell and Harrop 1976; McCarron and Dawson 1989; Dawson 1995). This posture minimizes radiant energy input and exposes the body to airflows. Dawson and Denny (1969) found that even small trees could reduce solar radiation by about 80 percent. Caughley (1964) and Russell (1971) reported restlessness during midday heat, and upright posture and other motions associated with cooling are responsible for our recording of about 10 percent activity (fig. 6.11) during the hottest times of day (fig. 6.10). Other activities at hot times are shifts between sites. However, these movements are sporadic and confined to brief shifts to other sites in close proximity. Dawson et al. (1974) demonstrated a high heat buildup associated with exercise, and kangaroos avoid such activities at midday, unless forced by disturbance. Movement between sites (as opposed to shifting about within a site) would likely constitute less than 1 percent of the 10 percent activity observed. Thus, under hot midday conditions, recorded activity should not be taken as feeding or movements. Rather, it is activity primarily carried out in one or several resting sites, and directed mainly at reducing heat load. Even panting, by itself, if the head were in the proper position, could result in continuous shifts in the radio-collar tip switch. Note that during other seasons, when movements for cooling are not necessary, midday activity drops closer to zero (figs. 6.3 through 6.8).

During summer, the lowered energetic cost of homeothermy and the need to avoid a buildup in body heat reduces food intake requirements. Even during early night, following a long period without feeding, kangaroos are less active during summer than at other seasons, and during late night when temperatures are lowest (fig. 6.10), they are no more active than at other seasons (fig. 6.11). Thus, the inability to feed due to daytime heat is not compensated by increased feeding at night when conditions are more moderate. The summer season results reported here were obtained when dry feed was prevalent and food quality was low. Kangaroos could not have compensated, therefore, for shortened feeding times by selecting for high-quality forage. This fact further supports the interpretation that lower feed requirements are responsible for overall decline in activity in the summer season.

In contrast to food intake, water intake increases in summertime. Bentley (1960) and Dawson et al. (1974) pointed out that during summer, kangaroos depend upon evaporative cooling from panting and licking the forearms at midday, resulting in higher water loss. Kangaroos also depend upon sweating when running, but not when resting (McCarron and Dawson 1989). Furthermore, dry feed reduces substantially the amount of water obtained from food intake (Green 1989). When plant water content drops below 55 to 65 percent, kangaroos become dependent on free water. If water is unavailable, food consumption declines. Although kangaroos do obtain metabolic water from oxidation of dry feed, and also minimize water loss by recycling nitrogen (Hume 1982) and concentrating urine and feces (Barker 1987; Dawson 1995; chapter 9), in the heat of summer, they usually become dependent on free water.

During this study, green feed was widely available through winter and spring of 1985 (fig. 2.9). Checking for tracks in the mud at the perimeter of water in the earthen tanks showed no evidence of kangaroo visitation during these seasons. Drying up of tanks was common in summer, and by late September, a few kangaroos (mainly eastern greys) began to visit those tanks that still contained water. However, most kangaroos did not use the tanks, and remained widely scattered. A heavy rain on October 12 caused a major greenup, and free water use by kangaroos ceased. Green feed persisted until mid-December, when kangaroos once again began to visit tanks.

As vegetation dried out, the first to come to water were eastern grey kangaroos, followed by western grey and red kangaroos at about the same time. At Fowler's Gap, Dawson (1995) found that red kangaroos visited water every few days. At Yathong, all species visited water tanks, typically once per day, and this behavior may simply reflect Yathong's greater availability of water. Visits usually were during nighttime hours. However, if undisturbed, western greys and reds would sometimes remain in the vicinity of tanks well into daylight. Eastern greys usually came in relatively late at night (usually not before midnight) and left well before dawn.

Thereafter, kangaroos, as well as emus (which, in contrast to kangaroos, visit during daytime), feral goats, and pigs, made heavy use of tanks until rains in July 1986 resulted in a small greenup. Many tanks on Yathong dried up during this period. During this time, all radio-collared kangaroos except two had tanks with water within their home ranges. The two exceptions, surprisingly, were both eastern grey females. We strongly suspect that these two individuals, often together in a social group and with overlapping home ranges, periodically made undetected visits to a neighboring tank. Also, they ranged on the lower portions of Merrimerawa Ridge, where woods were dense and temperatures more moderate than the flatlands occupied by most kangaroos in the study.

Availability of green feed was a third important factor influencing kangaroo seasonal activity patterns. For moderate seasons, when temperatures and other weather parameters were similar, the activity patterns varied markedly between dry feed and green feed environmental states (fig. 6.1). Moderate weather with dry feed gave the most uniform response between kangaroo sexes and species, whereas moderate weather with green feed gave the greatest differences between species and sexes. We cannot explain why, under the latter conditions, western grey females and eastern grey males had so much lower amounts of activity than the other species and sexes (fig. 6.1). Nevertheless, the lack of constraint by either food or weather obviously allowed behavior to vary widely.

A less-pronounced influence of green feed availability was seen in the cold environmental state (fig. 6.1). Although some shifts went counter to the rest (i.e., red females), overall, the array of activity values was elevated by about five percent during the green feed versus the dry feed periods (fig. 6.1). This may reflect the time it takes to collect a meal. The relative abundance of a uniformly distributed standing crop of dry grass takes relatively little time to collect. Green feed, by contrast, although of high quality, is rather sparsely spread at ground level, and patchily distributed by site. Bite sizes are small on new blades, compared to the concentrated clumps of dried grasses. These characteristics of the food make green feed relatively more time-consuming to gather than dry feed, particularly early and late in the green period, when distribution is patchy. At these times, most red and western grey kangaroos were concentrated in a few favorable feeding sites, and most of the remaining area was completely unused (see chapter 9).

During the study, the combination of hot weather and green feed did not occur at a time when a radio-collared sample of kangaroos was available. However, we would expect that green feed in hot weather would result in increased activity at night, when temperatures were more moderate, and therefore a greater amount of time would be spent active during this period. Because of temperature constraints, we would not expect much change in daytime activity, although the high water content in the available food would decrease dependence on free water.

Differences between Species and Sexes

Figures 6.3 through 6.8 show a number of differences in activity patterns between species and sexes. Indeed, the differences between sexes nearly equaled those between species (table 6.1). For most species of large mammals, it would be reasonable to assume that the differences between the sexes would be substantially smaller than the differences between species, independent of sex. However, in this study, the differences between sexes were about as great as those between species. For example, eastern grey sexes were different at the beginning (dry-hot) and end (wet-moderate) of the study, with the differences at the end being the greatest between the sexes for any species by environmental state combination. For the middle states (dry-moderate, dry-cold, and wet-cold), the sexes were not significantly different, and for the two former, they were virtually the same. Western grey kangaroos were not significantly different for the first four environmental states, but were different in the last state. Red kangaroos were the most similar between the sexes of the three species, only showing significant difference in the dry-cold state.

Eastern grey kangaroos were less tolerant of heat than the other two species. This is consistent with their major distribution being in eastern Australia, where mountains, high rainfall, and moderate to cool climates prevail. In the hot periods, they occupied habitats with greater density of large trees and a more closed canopy, and generally, better shading. They also ceased morning activity about an hour earlier than the other two species (figs. 6.3 through 6.8). Under drought conditions, according to local observers, they are the first kangaroo species to die. Both western grey and red kangaroos are more resistant to drought, although according to Robertson (1986), western grey kangaroos begin dying before red kangaroos.

As the vegetation dries out, eastern grey kangaroos are the first to come to tanks to drink free water. In a series of high rainfall years, their range expands westward into the arid zone, moving along river courses (Poole 1975; Shepherd 1982), and during drought years, their range contracts (Denny 1975). Yathong is near the western limit where woodlands are sufficiently dense to constitute continuously distributed habitat (as opposed to stringer woodlands along stream courses). Even on Yathong, wooded habitats are mainly on the eastern side of the Reserve in association with the ridge systems.

Red kangaroos, by contrast, are much more susceptible to cold, particularly during wet, rainy conditions. This is especially true of females. Male reds seem less subject to cold than females, probably because of their much greater body mass. The red kangaroo body conformation appears the most angular of the three species, and if corrected for size, they probably have the highest surface-to-volume ratio. Yet, even male red kangaroos seem less able to deal with cold weather than either sex of the two species of grey kangaroo.

Female red kangaroos avoid nighttime activity when temperatures are coldest (fig. 6.5), and that avoidance is most pronounced when it is wet as well. When it is wet, activity is lower even during periods of moderate temperature, and red kangaroos were the only ones seen to move beneath trees from open areas when it started raining. The coat of the red kangaroo has a fine, soft, underfur-like texture with high reflectance that reduces heat penetration (Dawson and Brown 1970; Hume et al. 1989), but lacks the long, shiny guard hairs of the eastern and western grey kangaroos that are better at shedding water. When the pelage of red kangaroos gets soaked in heavy rain, it becomes dark gray or black in appearance, and they look bedraggled. Eastern grey kangaroos look "furry" year-round, whereas red kangaroos appear to have a thin summer pelage even in winter. Western grey kangaroos look slick in summer and furry in winter. They seem to be the most weather-resistant of the three species and are most likely to be out actively feeding in winter when it is windy or rainy. Neither eastern nor western grey kangaroos look soaked when exposed to rain. In fact, they are invariably seen with raindrops glistening on their guard hairs.

Long rain-shedding guard hairs are an obvious adaptation in eastern grey kangaroos that live in the cooler mountains in a higher rainfall zone and western grey kangaroos that live in the southern zone with cold winters and high winter rainfall. The red kangaroo's stronghold, by contrast, is the dry interior of the continent, where even winter temperatures are moderate and most rainfall occurs during summer at the hottest season. Over most of their range, the easily wetted pelage of the red kangaroo is not a disadvantage. However, at Yathong, near the southern and eastern limits of their range, a wettable pelage is a disadvantage, and subjects the animal to cold stress and exposure. Distribution of the red kangaroo was previously reported to be limited to the 250 mm isohyte (Short et al. 1983; Horton 1984), although Newsome (1965) reported the limit at the 375 mm isohyte. The variation, we believe, is due to trends in rainfall over years that shift the occupied distribution away from the mean isohyte. There is anecdotal evidence from station managers neighboring Yathong that the red kangaroo range may contract during series of high rainfall years and expand during low rainfall years. Some station owners commented to us about how much more common red kangaroos were now than 10 years ago, when they were rare; what one saw then was nearly all "brown stinkers," the local term for western greys.

If our hypothesis that red kangaroos are subject to cold stress due to wetting is correct, then temperature and rainfall would govern their southern and eastern distribution. Still, areas with both high rainfall and moderate temperatures could be occupied by red kangaroos. This may explain the occurrence of red kangaroos in a 400 mm isohyte zone in one part of Western Australia (Short et al. 1983).

Just as red kangaroos are limited by cold, wet climates, eastern grey kangaroos are limited by hot, dry climates. Of the three species, the western grey kangaroo seemed best adapted to the variation of climate at Yathong. It can withstand both cold, wet and hot, dry conditions. It is most likely to be observed feeding in openlands, regardless of season or weather. It is a tough survivor. From all indications, its numbers on Yathong are more stable over changing climatic patterns than either eastern grey or red kangaroos.

7

SOCIAL GROUPING

Just as kangaroos live in known places with a spatial structure, they live in known social structures based upon established relationships. Group formation is an obvious manifestation of social organization. In large groups, social relationships are more readily discerned because the individuals, due to close proximity, interact more frequently. Kangaroos at Yathong seldom form large groups, and groupings that do occur are often temporary. Nevertheless, there is every indication that kangaroos are organized into relationships rather than existing in a social vacuum. Social relationships are not continuous, as with social group-forming species, but are instead sporadic as individuals or small groups encounter each other in addition to evidence of each other's past presence in the form of odors from droppings, urine, and scent marks. These relationships occur not only within species, but between species of kangaroos at Yathong.

In this chapter, we explore the size of groups, including mixed-species groups, and we consider the variables that influence group dynamics. In the next chapter, we consider the behavior of individuals and social interactions of kangaroo society.

Species Abundance

In the systematic surveys, a total of 5513 groups was observed in the combined daylight and night spotlight samples, of which 53.8% were red, 6.5% eastern grey, and 37.9% western grey kangaroos. Mixed-species groups constituted 1.8% of the total number of groups (table 7.1). The total number of individuals observed was 11,273, of which 56.4% were red, 6.9% eastern grey, and 36.7% western grey kangaroos. This is consistent with the discussion on populations in chapter 4, showing that on Yathong Nature Reserve, red kangaroos were the most prevalent species, western grey kangaroos somewhat less numerous, and eastern greys relatively few during this study.

Group Size

On Yathong, the social units of kangaroos are small. The mean group size overall was 2.04 individuals per group (2.01 if mixed-species groups are excluded). All three species are characterized by small group sizes. The largest group of red kangaroos was 14 (two males, eight females, four young-at-foot), for eastern grey, 8 (five females and three young-at-foot), and for western grey, 13 (two males, undetermined female and young composition) individuals. Mean group sizes per species were 2.06 for red, 2.08 for eastern grey, and 1.91 for western grey kangaroos. No tests for differences were applied because such tests would be biologically meaningless in the absence of sex and age distributions of groups within species.

These results are at odds with frequent reports of large aggregations (sometimes called "mobs") of kangaroos, particularly eastern greys in eastern New South Wales (Kaufmann 1975; Taylor 1982; Southwell 1984; Jarman and Southwell 1986; Jarman and Coulson 1989). Group size apparently varies in stability and persistence across the geographic range, with much larger group sizes in eastern Australia where habitat is better and population size greater. Other studies in the arid zones have found results more similar to ours. Dawson (1995) reported groups of one to four eastern grey kangaroos at Fowler's Gap, near the western end of the range. Our mean group size for red kangaroos, 2.06, was lower than the 2.4 reported by Caughley (1964), the 2.6 reported for Fowler's Gap in far western New South Wales by Russell (1979), and 2.6 at Kinchega National Park by Johnson (1983a). Johnson (1983a) also reported mean group sizes of 1.54 for red males and 1.71 for red females at Kinchega—again, lower than the 1.91 for sexes combined at Yathong.

Much of the difference between these results may be due to habitat differences, but population density may be an important variable. Many authors have reported a correlation between population density and mean group size (Caughley 1964; Russell 1979; Taylor 1982; Johnson

Table 7.1.   Statistics on groups and individuals by species observed in systematic counts. Note that daytime and night spotlight data have been combined.

Species	No. Groups	Percent of Groups	No. Individuals	Individuals as Percent of Total	$\bar{x}$ Group Size
Eastern grey	361	6.5	752	6.7	2.08
Red	2965	53.8	6122	54.3	2.06
Western grey	2089	37.9	3990	35.4	1.91
Eastern grey + Red	10	0.2	36	0.3	3.60
Eastern grey + Western grey	22	0.4	80	0.7	3.64
Red + Western grey	65	1.2	287	2.5	4.42
Eastern grey + Red + Western grey	1	0.0	6	0.1	6.00
Total	5513	100.0	11,273	100.0	2.04

1983a; Coulson 1993b; Dawson 1995). As discussed in chapter 4, populations at Yathong were lower than in many of the areas studied by others. Thus, Coulson (1993b) reported mean group size of western grey kangaroos at Hattah-Kulkyne National Park of 2.0 to 4.61 (for different habitat types) as compared to our 1.91 for all habitats combined; respective, comparable densities were 18 to 58/km^2 and 4.9/km^2. A notable exception was the low group size reported for western grey kangaroos at Kinchega by Johnson (1983a), even though the population there is higher than at Yathong. Probably the most complete data set for the relationship of group size to population was that of Southwell (1984) for eastern grey kangaroos. His regression predicted a mean group size of 2.62 for Yathong as compared to the observed 2.08, a bit lower than expected but well within the 95% confidence limits of the regression.

On Yathong, "mobs" were temporary aggregations at concentrated food sources, usually where "green pick" (green grass) was first available at the beginning or end of a rainfall-initiated vegetative growth period. Usually, only red and western grey kangaroos gathered in temporary mobs, while eastern greys at Yathong remained dispersed. Concentrated food patches were usually located in openlands, which eastern greys generally avoided, and they appeared skittish when found there. Within these temporary aggregations, the interindividual space was small within social units, and larger between social units. Social units joined and left such aggregations on their own schedules, as also reported by Jarman and Southwell (1986) and Jarman and Coulson (1989), although there was a certain amount of synchronization due to similar activity patterns, particularly during hot months (chapter 6). Resting sites were not shared in common by mobs. Similarly, when large aggregations were disturbed by vehicles or other intrusions, they took flight as separate social units, each scattering on its own. Red kangaroos were more likely to stay together longer before splitting into separate social units, owing to their common dependence upon distance rather than cover to retreat from disturbances. Still, if one continued to track retreating groups, they would ultimately fractionate into small units of one or a few individuals before resuming activity or moving to resting sites.

Thus, the distribution of food rather than the need for social bonds prompted the formation of these aggregations. They were never seen during periods when food distribution was more uniform, whether dry (low quality) or green (high quality). If one had only direct observational evidence, the erroneous impression that kangaroos were moving in nomadic mobs, pursuing green patches of food that follow rainstorms would seem the most parsimonious way to account for mobs of kangaroos. However, as pointed out in chapter 5, radio-collared individuals did not abandon their individual home ranges to move to nearby green feed produced by rainfall from summer thunderstorms. Surely, kangaroos could see and hear the violent storms so overwhelmingly obvious to the human observer. Some authors have speculated that kangaroos may be able to smell the presence of green feed at great distances. Nevertheless, such cues did not draw kangaroos from their usual home ranges. The two cases of a shift in home range (chapter 5) involved neither movement to green feed from dry feed nor from low to high quality. Curiously, these movements did not occur when food sources were concentrated. One kangaroo moved during a period of widespread availability of green feed (Scarlett O'Hara), whereas the other moved during a uniform dry period (a western grey male). Neither, however, joined a kangaroo aggregation by the move.

Feeding mobs, consequently, must be composed of individuals whose home ranges overlap a concentrated food patch. These aggregations can be quite large (20 to 30 individuals occurring as several nearby social groups is common), illustrating a large overlap of kangaroo home ranges and a high degree of shared area. This, plus the lack of observation of defended areas by any species or sex, clearly shows that kangaroos are not territorial. Individuals do defend their individual space, both intraspecifically and interspecifically, but this is not linked to either a specific geographic area or specific resource, except in the immediate and temporary sense; that is, the defense of individual space. Behavior related to feeding aggregations at concentrated food sources is discussed further in chapter 9.

One circumstance, other than feeding, that produces temporary aggregations is disturbance. When fleeing a vehicle, kangaroos sometimes encounter other social groups of kangaroos, and they frequently join together in one fleeing group. Given the amount of tree cover on Yathong, such groups usually do not go far or last long before moving out of sight. They quickly break apart once the need to flee is past. In more open areas, such aggregations may be more persistent. At Willandra National Park, for instance, where the uplands lack tree cover, at times it seemed as if every kangaroo in the countryside was running in one aggregation or another in response to our vehicle. It seems reasonable to presume that under such habitat conditions, frequent disturbance in pre-European times by coursing predators (such as dingoes) may have selected for more permanent and socially integrated aggregations. However, extrapolation of the results from Yathong to different situations should be done with caution.

Group Composition

Juveniles, individuals independent of their mothers but not yet adult in size, were only rarely identified in this study. It seemed that young animals essentially remained young-at-foot with their mothers until they were adult size. Juveniles seem to show rapid growth rates and the juvenile phase was short. Although a high mortality rate would also explain the low number of juveniles recorded, we would have detected such mortality by the presence of carcasses, readily indicated by carrion-feeding birds. Thus, smaller individuals ranging without their mothers (except in a few cases where dependent young-at-foot were temporarily separated from their mothers) were categorized as females or males along with fully mature animals.

Young-at-foot were dependent upon their mothers, and their associations, movements, and so forth were determined by their mothers. Otherwise, females and males were free to choose their associates. Kangaroos of both sexes defend individual space, and enforce dominance relationships. We saw no aggressive behavior, however, that could be construed as preventing association of other individuals that chose to remain in proximity and maintain a social group. Therefore, the observed distribution of males and females as solitary or associates of others of the same or opposite sexes were considered as expression of choices among alternatives by individual kangaroos.

Males of all species were most likely to be solitary (49.4%), followed by being associated with one or more females (39.3%) (table 7.2). Note that for purposes of this analysis, a "group" may be composed of a

Table 7.2.   Type of groups formed by adult male and female kangaroos from systematic counts, exclusive of mixed species groups.

Type of Group	Eastern Grey N	Eastern Grey %	Red N	Red %	Western Grey N	Western Grey %	Combined Species N	Combined Species %
Single male	79	60.3	352	44.4	451	55.4	885	49.4
Multimale	10	7.6	112	14.1	70	8.6	202	11.3
Male and female [1]	42	32.1	329	41.5	293	36.0	703	39.3
Total	131	100.0	793	100.0	814	100.0	1790	100.0
Single female	142	57.0	1491	61.6	953	64.6	2588	61.4
Multifemale	65	26.1	599	24.8	228	15.5	923	21.9
Female and male [1]	42	16.9	329	13.6	293	19.9	703	16.7
Total	249	100.0	2419	100.0	1474	100.0	4214	100.0

[1] Note that this category is repeated to derive separate percentages for mixed-sex groups.

single individual. Relatively few kangaroo groups occurred as multimale groups (11.3%). Group associations of male eastern grey and western grey kangaroos were similar (chi-square = 1.11, df = 2, P = 0.58), whereas males of both species of greys were significantly different from red males (eastern grey chi-square = 12. 45, df = 2, P < 0.002; western grey chi-square = 23.83, df = 2, P < 0.001). Male red kangaroos were less likely to occur as solitary individuals, and more likely to occur with other males or with females in social groups.

For all three species, there was a highly significant difference (P < 0.001) in group type between the sexes (red chi-square = 261.27, eastern grey chi-square = 26.03, western grey chi-square = 78.63, df = 2). Females were more likely to be alone (61.4% for all species) or in multi-female groups (21.9% for all species), and less likely to be associated with a male (16.7% for all species) (table 7.2). Red and eastern grey females were similar in group type distribution (chi-square = 2.59, df = 2, P = 0.27), whereas western grey females were significantly more likely to be solitary or with a male (P < 0.001; eastern grey chi-square = 15.55, red chi-square = 62.31, df = 2) (table 7.2).

The basic social unit among females is the adult female and her dependent young. At times, several females form associations, but these are labile over time. Stuart-Dick and Higginbottom (1989) found such female associations were often mother and grown daughter, and Dawson (1995) suggested that female associations in kangaroos are based on matrilines. Direct observations of radio-collared females (particularly Scarlett O'Hara and eastern grey females number 1860 and 1921) showed that groups formed and broke up periodically. These two eastern grey females were the most commonly associated pair of adult kangaroos in this study, and their home ranges were nearly the same (fig. 5.1). Yet from both radio location and direct observation, they were known to regularly separate. Grant (1973) reported similar associations between pairs of eastern grey females.

Mixed-sex groups formed when males joined females, as one would expect in these polygynous species. However, mixed-sex groups were not exclusively related to breeding opportunities, as evidenced by their presence year-round and during times when no sexual behavior was observed in mixed-sex groups.

Composition of mixed-sex groups varied among species (table 7.3). Red and western grey kangaroos were nearly significant (chi-square = 7.00, df = 3, P = 0.07), and both had a much higher number of groups consisting of just one male and one female. The next most common mixed-sex group was one male and multiple females (table 7.3). Multi-male mixed groups were least common, and occurred in similar frequencies with a single female or multiple female groups. Presumably, the high frequency of single male groups in red and western grey kangaroos was due to the relatively high number of mixed groups related to breeding activity, and the active exclusion of rival males by the dominant male.

Eastern grey kangaroos were significantly different from both red (chi-square = 11.53, df = 3, P = 0.009) and western grey kangaroos (chi-square = 24.88, df = 3, P < 0.001), primarily in the greater frequency of one male–multifemale groups (table 7.3). Yet even in this species, in which sexual activity in mixed-sex groups was observed less frequently, one-male groups predominated. This would suggest that breeding activity may have been involved; mixed-sex groups were spread across seasons more evenly in eastern greys than in red and western grey kangaroos. Poole (1983a) reported that eastern grey kangaroos

Table 7.3.   Composition of mixed-sex groups by species of kangaroo.

Composition		Eastern Grey		Red		Western Grey	
Males	Females	N	%	N	%	N	%
1	1	12	28.6	179	54.4	186	63.5
1	>1	17	40.5	89	27.1	69	23.5
>1	1	4	9.5	29	8.8	22	7.5
>1	>1	9	21.4	32	9.7	16	5.5
	Total	42	100.0	329	100.0	293	100.0

bred throughout the year over a broad geographic range, and this would favor association of the sexes throughout the seasons.

Seasonal Changes in Groups

Kangaroo groups change over time due to seasonal climatic changes, reproductive activities, and rainfall (and associated green feed). Kangaroos show less seasonal predictability in groupings than is typical of temperate-zone ungulates because, other than temperature, climatic factors are less predictable over time.

Group size showed considerable change over the period of this study and was not the same in July of 1985 and 1986, the one month observed in both years (fig. 7.1). At the beginning of the study, both red and west-

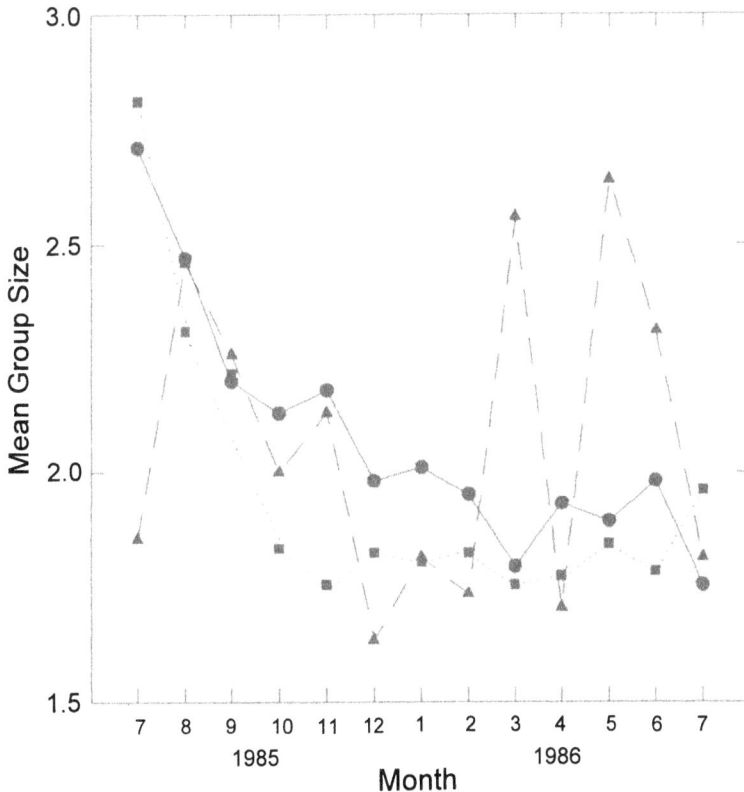

Figure 7.1. Mean group size by month of eastern grey (triangles), red (circles), and western grey (squares) kangaroos observed in combined daytime and night spotlight systematic surveys.

ern grey kangaroos had high mean group sizes that declined and re-mained lower thereafter. Mean group size of eastern grey kangaroos varied greatly over time, partly due to lower sample sizes, but also apparently because of less consistency.

Figures 7.2 to 7.4 show these shifts in group types across time for the three kangaroo species. Group types of eastern grey kangaroo showed considerable variation over time (fig. 7.2) and this accounted for the high variance in mean group sizes in this species. Regression analysis showed that mean group size was significantly correlated with groups of males and females with young ($r = 0.63$, $P = 0.02$), and nearly significant for females with young ($r = 0.54$, $P = 0.06$). Stepwise regression, however, gave a model ($R^2 = 0.80$, $P = 0.002$) including females and young as a positive variable, and single males and females both as negative variables. These results suggest that the frequency of single animals of either sex is important in reducing mean group size in this species.

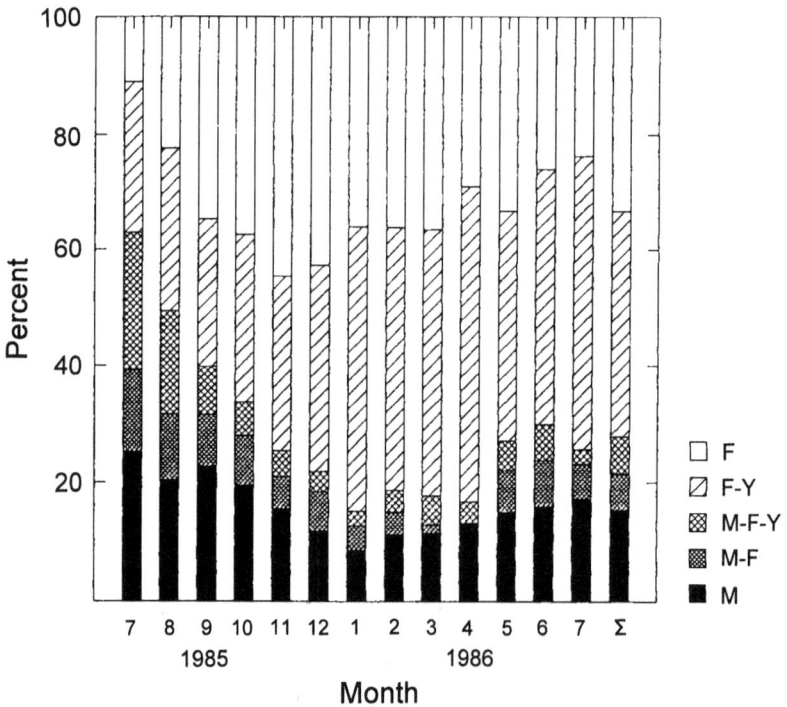

Figure 7.2.   Percent of group types of eastern grey kangaroos observed in combined daytime and night spotlight systematic surveys. Group types are single males (solid bars), males with females (small cross-hatched bars), males with females and young (large cross-hatched bars), females with young (slashed bars), and single females (open bars).

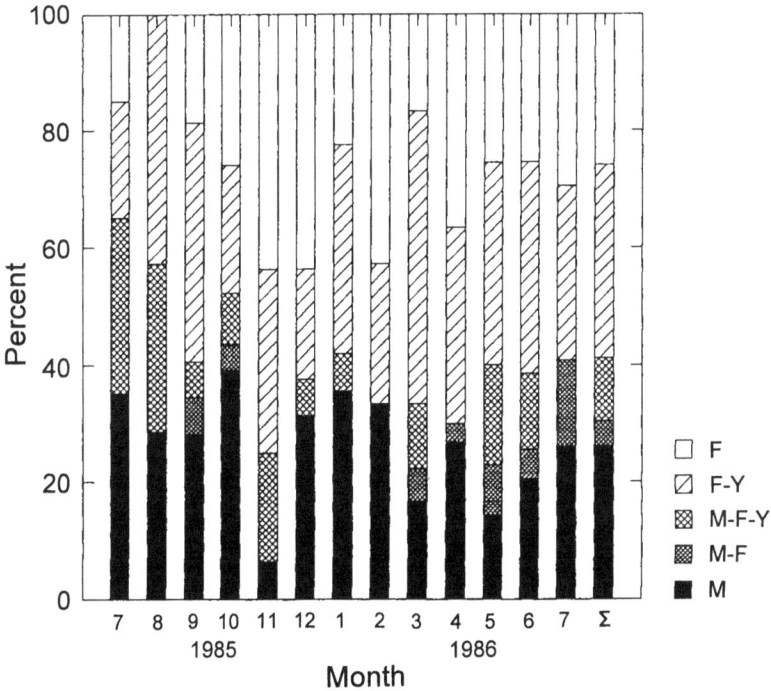

Figure 7.3. Percent of group types of red kangaroos observed in combined daytime and night spotlight systematic surveys. Group types are single males (solid bars), males with females (small cross-hatched bars), males with females and young (large cross-hatched bars), females with young (slashed bars), and single females (open bars).

For red kangaroos, shifts in group size over time were strongly related to shifts in group types (fig. 7.3). Single males and males in association with females or females with young-at-foot were high early in the study, declined during the middle, and increased somewhat towards the end of the study. The percent of females with young-at-foot, or very large pouch young, was the inverse of the male pattern (fig. 7.3). Pouch young were included with young-at-foot because only very large pouch young were recorded, so the distance at which the kangaroos were observed would not seriously bias the number. Furthermore, there is a fairly lengthy period over which the young are sometimes in the pouch, and sometimes at foot.

Regression analysis showed that mean group size in red kangaroos was positively related to male-female-young combined groups ($r = 0.89$, $P < 0.001$) and negatively with female-young groups ($r = -0.81$, $P = 0.001$). Inclusion of these variables in a stepwise regres-

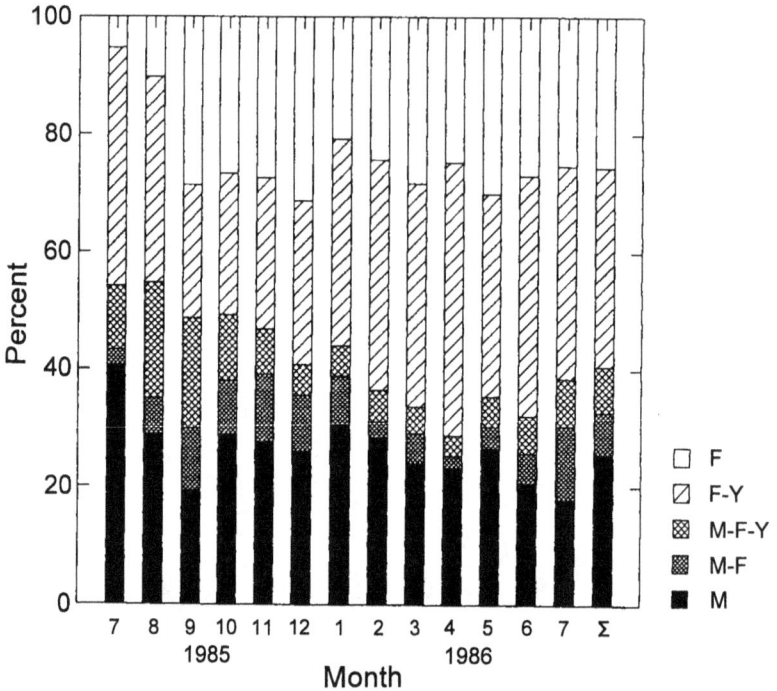

Figure 7.4. Percent of group types of western grey kangaroos observed in combined daytime and night spotlight systematic surveys. Group types are single males (solid bars), males with females (small cross-hatched bars), males with females and young (large cross-hatched bars), females with young (slashed bars), and single females (open bars).

sion ($R^2 = 0.84$) accounted for 84 percent of the variance in mean group size.

Western grey kangaroos (fig. 7.4) showed a decline in male groups over time, and an increase in female groups. Groups of males with females or females with young were high early in the study, lower in the middle, and somewhat increased towards the end. Males ($r = 0.72$, $P = 0.006$), females with young ($r = -0.76$, $P = 0.003$) and single females ($r = -0.78$, $P = 0.002$) were all negatively correlated with mean group size. Stepwise regression gave a model ($R^2 = 0.69$, $P = 0.003$) including females and young and single females as negative variables. The importance of the occurrence of single individuals on mean group size in western grey kangaroos would be expected given their prevalence (table 7.2); this resulted in western greys having the lowest overall group size of the three species (table 7.1).

Mixed-Species Groups

Mixed-species groups of kangaroos occurred regularly on Yathong Nature Reserve (table 7.1). Of 5513 total groups observed in the combined daytime and night spotlight counts, 98 (1.8%) were mixed-species groups (table 7.1). Mean group sizes of mixed-species groups were approximately the sum of the mean group sizes of the separate species (table 7.1). This suggests that mixed groups form by the joining of a social group from each species. Because mixed-species groups are larger than single species groups (table 7.1), the number of individuals in mixed-species groups (409) is higher (3.6%) than the percentage by groups (1.8%).

All combinations of species were observed (table 7.1), including one group of all three species. Another group of all three species was observed outside of the systematic surveys as well. Red and western grey mixed groups were most common and red and eastern grey least common. This was expected in view of the greater abundance of the red and western grey kangaroos and their more similar use of habitat (chapter 10). Yet, in relation to the numerical abundance of reds and eastern greys (table 7.1), eastern greys associated with western greys far more frequently, and with reds far less frequently, than expected by chance. This result agrees with Coulson's (1999) finding that the two species of grey kangaroos readily combined in Grampians National Park in western Victoria.

Although some mixed-species groups were observed to persist over several days, most were more fleeting. For the most part, mixed-species groups were opportunistic associations formed while feeding in openland areas. As previously noted, they usually broke up when alarmed, or when they moved into cover to rest. Although our evidence is weak, we suspect that persistence of mixed groups was longer in winter when temperatures were cool and feeding groups, if undisturbed, were more inclined to rest on the fringes of adjacent cover.

Seasonally, mixed-species groups showed a small peak early in the study, a major peak in January and February, and perhaps another rise in June (fig. 7.5). When mixed-species groups were few, they were composed predominantly of red and western grey kangaroos. Eastern grey kangaroo mixed-species groups occurred mainly at times when red and western grey mixed groups were also most common (fig. 7.5). This result suggests that whatever factors favored mixed-species group formation influenced all three species similarly.

Comparing numbers of mixed-species groups (fig. 7.5) with avail-

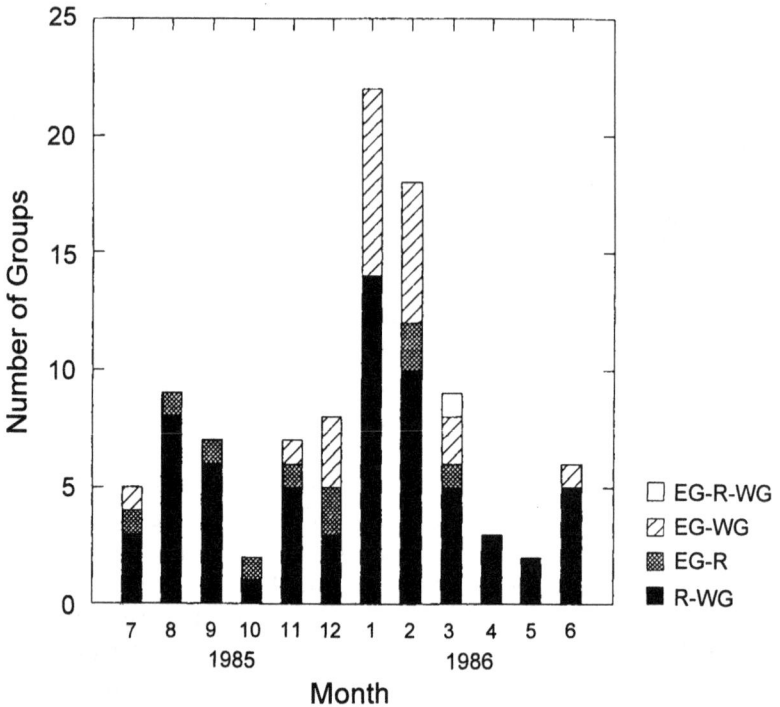

Figure 7.5.   Number of mixed-species groups observed in systematic surveys (daytime and night spotlight data combined). Cross-hatched bars = eastern grey + red; cross-slashed bars = eastern grey + western grey; solid bars = red + western grey; and open bars = eastern grey + red + western grey.

ability of green feed (fig. 2.9) showed that most mixed-species groups formed during periods when green feed was declining rapidly. Conversely, during periods of abundant green feed, few mixed-species groups were observed. When food quality was high, kangaroos were widely scattered, and came into interspecific contact infrequently (chapter 5). Concentration of kangaroos at restricted feeding areas at the beginning and end of green feed periods seemed to be responsible for bringing the different species into contact. The peaks of mixed-species groups match almost exactly the time, from our field notes, when kangaroos were feeding in concentrated green feed areas. The feeding behavior of kangaroos is discussed further in chapter 9.

The common occurrence of mixed-species groups, including during breeding periods, raises the question of possible hybridization between species. Hybrids have been produced in captive populations, but those between red and grey kangaroos are invariably sterile (Gray 1972). Hybrids of this cross have 18 chromosomes, intermediate between red (20) and grey (16) kangaroos. Hybridization between the two species of greys is somewhat more complicated. These two are closely related (sibling species) and have the same chromosome number. In fact, they are sufficiently similar that they were not recognized as distinct species until 1967 (Kirsch and Poole 1967, 1972). Hybrids produced in captivity are usually from matings of western grey males and eastern grey females. The reverse combination always failed in Poole and Catling's (1974) experiments, but Kirsch (1984) produced a successful cross. In Poole and Catling's (1974) studies, hybrid males were always sterile, but hybrid females mated to western grey males produced live young. F2 males again were sterile, but females were fertile, and may have produced a fertile male in the F3 generation (Poole and Catling 1974). According to Poole and Catling (1974), no grey kangaroo hybrids are known in the wild.

Throughout this study, we observed large numbers of grey kangaroos, and we became adept at distinguishing the two species at a glance based on their appearance and behavior. Nevertheless, in three instances we encountered individual grey kangaroos that we could not place to species, even though they were examined at relatively close range with binoculars. These individuals (perhaps the same individual observed more than once) showed characteristics intermediate between the eastern and western greys (in coat texture and color patterns, particularly on the face and ears), and we believe they were hybrids. Because we encountered only three such cases in a total of over 5000 observations of grey kangaroos, hybridization, if present at Yathong, is rare.

Discussion of Social Groupings

Our results on social groupings are in substantial agreement with published reports for the same species of kangaroos (see review of Jarman and Coulson 1989). However, some differences are worth noting. Eastern grey kangaroos at Yathong, near the western edge of the continuous distribution of the species, showed grouping behavior different from those in richer habitat further east in the mesic, mountainous areas of Australia. Group size was not nearly as large, nor did socially integrated social units occur as reported by researchers in other areas (Jaremovic 1984; Jarman and Southwell 1986; Jarman 1987; Jarman and Coulson 1989; Jaremovic and Croft 1991). At Yathong, eastern grey kangaroos have group dynamics more similar to those of red and western grey kangaroos.

In contrast to eastern greys, red and western grey kangaroos at Yathong have social dynamics similar to those reported for other areas of their respective ranges. But then, Yathong deviates less from the typical habitat of these species than it does from the moist, cold, heavily forested stronghold of the eastern grey kangaroo in the Great Dividing Range of eastern Australia.

For all three species, group sizes were somewhat smaller at Yathong than in other areas studied. This may be due to the relatively low density of kangaroos at Yathong (chapter 4). Numerous authors have reported a correlation between group size and density in these species. Nevertheless, by choice, kangaroos distribute themselves more individually over Yathong, whereas they could have chosen to clump locally and form much larger groups. This shows that habitat quality and patchiness do play a role in shaping social grouping and organization beyond the influence of population density. Some of the studies reporting a correlation of group size with population density use local density as the independent variable rather than regional density. This would indicate high habitat patchiness that favored clumping of kangaroos in good habitat and avoidance of poor habitat. Thus, the correlation may be a reflection of the influence of habitat quality rather than that of population density per se. At Yathong, habitat quality in the intensive study area was not so strongly patchy, and the distribution of kangaroos was more uniform.

We believe that most kangaroos on Yathong recognize conspecifics with overlapping ranges as individuals. Yet they choose to not form larger social units. This would suggest that the large, socially integrated groups of eastern grey kangaroos in mesic habitat may be forced to aggregate due to high density and habitat patchiness, and are not the social organization of choice.

Group size at Yathong changes largely due to males joining female groups. The search for reproductive opportunities is no doubt the prevalent motivation for males to join females. Still, if it were so simple, one would expect much stronger seasonality than was observed in this study (figs. 7.2 to 7.4), particularly in the most seasonally breeding species, the western grey kangaroo. The unpredictability of rainfall at Yathong, in both summer and winter, probably mitigates against a highly predictable breeding period with a constant timing between years. Consequently, males, even among the seasonally breeding western grey, would be required to monitor female groups more frequently for the presence of estrous females.

Mixed-species groups were relatively common at Yathong, particularly in January and February 1986 (fig. 7.5). In our experience, such groups are quite uncommon among ungulates in North America. For example, in a study of seven coexisting species of ungulates on the National Bison Range in Montana, Y. McCullough (1980) observed over 9000 groups without a single mixed-species group. Mixed-species groups are more common in plains ungulates in Africa. Wildebeest (*Connochaetes taurinus*) commonly associate with zebra (*Equus burchelli*) (Talbot and Talbot 1963; Sinclair and Norton-Griffiths 1979) and Thompson's (*Gazella thomsoni*) and Grant's (*G. granti*) gazelles (Fitzgibbon 1990) on the east African plains. As with kangaroos, these species do not form integrated social units, and disaggregate frequently (Talbot and Talbot 1963).

On open plains, which have no effective cover, such aggregations are usually considered a means of avoiding predation (Sinclair 1985; Fitzgibbon 1990). Such an argument has been put forward to account for mixed-species groups in kangaroos (Coulson 1999). Arguments usually are based on the observation that individuals in larger groups spend less time in vigilance and more time feeding. Coulson (1999) presents data that this argument is supported by behavior of eastern and western grey kangaroos. However, if mixed-species groups were formed to avoid predation, then conspecifics should join to form larger groups as well. Yet, in January and February, 1986, when mixed groups were most frequent (fig. 7.5), the kangaroos at Yathong showed no tendency to form larger single-species groups (fig. 7.1). This would suggest that predation was not a factor in the formation of mixed-species kangaroo groups. We further discuss the role of predation in group formation in kangaroos in chapter 12.

Mixed-species groups were most common at a time when food quality was poor overall, but remained high in localized patches. We think that congregation in concentrated food patches accounted for the peak in mixed-species groups shown in figure 7.5. Nevertheless, not all mixed-species groups were associated with such feeding sites, because mixed-species groups occurred at a low frequency throughout the study. Their frequency during dispersed food periods was of similar magnitude to that of larger conspecific groups, and it may have been simply that independent social groups occasionally joined for brief periods, regardless of whether one or several species were involved.

8

BEHAVIOR

It is easily understandable why the first Europeans to reach Australia were amazed by the kangaroo's peculiar habits, because the only previously known bipedal mammals were small desert-dwelling rodents. To see, for the first time, a group of deer-sized animals saltating across the countryside is a unique experience even for the modern biologist with the benefit of mental preparation.

Early researchers reported kangaroos as uninteresting behaviorally; this can probably be traced back to the prevalent notion of the time that they were primitive, had small brains, and acted mainly from "instinct." More recent work by behaviorists has shown kangaroos to be complex in repertoire and social interactions. Coulson (1997b) reported that eastern and western grey kangaroos shared most of the same repertoire (45 of 46 stereotypic acts), and our observations suggest that with minor differences, red kangaroos share a similar repertoire. In this respect, they are comparable to ungulates in other parts of the world. True, interaction rates between individuals may be infrequent, but this is due to low population densities, small group sizes, and protracted breeding seasons. The impression that ungulates were more complex and interesting socially arose from the studies conducted during rutting seasons when interaction rates were high and fitness benefits great; ungulates at other seasons occurring in dispersed, small social units are equally subdued, with a low frequency of interactions. Furthermore, kangaroos living at high densities, such as the eastern grey population at Wallaby Creek studied by Peter Jarman and his students, show high interaction rates and complex social structure—the equal of ungulates under comparable circumstances.

In this chapter, we report our behavior observations, which are mainly of interest as comparative results to those reported elsewhere under different environmental conditions. Because interaction rates were low, and breeding periods sporadic, we did not establish regularly scheduled, systematic observational methods. Our behavioral work occurred upon opportunity and otherwise, as time permitted. Consequently, our behavioral results are more narrative and less quantitative in character.

Individual Behavior

Posture

For kangaroos in the upright posture, the powerful tail serves as a prop that, with the hind legs, forms a three-point stance. At rest, the long distal limbs of the hind legs rest on the ground, as does much of the tail, forming a stable base from which the upper body can assume many positions.

The upright posture varies among the three species, with the eastern and western grey kangaroos being more similar, and the red the most different. This traces to differences in the basic morphological structure of the three species. In the red kangaroo, the hindlegs are placed relatively farther back on the body, and a longer upper body is associated with a stronger pectoral girdle and better developed forelegs, which function more like arms than legs. The red kangaroo body is more angular than that of the other two species, which are more sleek in conformation. The red kangaroo tail is more flexible, with a broad distal section lying flat on the ground in the three-point stance, whereas mainly the tip of the tail is on the ground in the two grey species. The grey kangaroos have less-developed pectoral girdles and less well-developed forelegs, although the eastern grey is more like the red in being generally a more robust and heavy-bodied animal than the western grey. At total relaxation in the upright posture, the grey kangaroos have a nearly vertical orientation, and the small forequarters present a slope-shouldered appearance, tapering outward from the neck downward. This appearance is enhanced by their tendency to let the forelegs hang downward in an unflexed position.

All three species can rise from the upright three-point stance to much greater height on their toes and tail tip, which they do in aggressive encounters. However, in alert situations, red kangaroos are most likely to raise the trunk to the vertical, and often lean over backwards in a slightly overbalanced position. Most striking in appearance in this posture are red kangaroo males, flexing their powerful shoulders and forelegs and looking for all the world like steroid-pumped body builders (fig. 1.4). We wonder if this posture evolved to intimidate rival males and potential enemies.

Locomotion

Windsor and Dagg (1971) have described kangaroo gaits. When feeding, kangaroos use a five-point "pentapedal" posture utilizing all four legs and the tail. To move a short distance (about 15 cm), the hind legs move alternately while the body weight rests on the forelegs and tail. To move a longer distance (about 30 cm), both hind legs are brought forward simultaneously while the animal balances on the forelegs and tail. Several of these simultaneous "hitches" are performed in succession to move between feeding patches that are several meters apart. If feeding patches are three or more meters apart, one or a series of bipedal hops are used to move between patches. Thus, pentapedal locomotion is used to deal with patchiness on a finer scale, and bipedal locomotion is used to deal with patchiness on a coarser scale, as well as for long-distance movements.

Bipedal locomotion (Dawson 1977, 1995) involves hopping on both hindlegs with the forelegs held next to the arched body, and the tail extended behind as a balance organ. The tail ordinarily does not touch the ground. Speed of hopping is determined more by the height and length of each bound than by the number of bounds per unit of time. Sustained high speed of kangaroos clocked by speedometer on Yathong was about 40 kph, although they are capable of greater speeds for short distances. Dawson and Taylor (1973) have shown that bipedal locomotion is energetically more efficient than quadrupedal locomotion at speeds above 17 kph. This gain apparently comes from the elasticity of leg ligaments, which function much like a pogo stick spring. At low speed, the vertical component of trajectory is sufficiently costly that bipedal locomotion is less efficient than quadrupedal locomotion.

Because of differences in body conformation, the three species of kangaroos differ in their hopping behavior. Western grey kangaroos have a short stride, relatively high trajectory, and rapid frequency of hopping. They hop in a bent-over position and seem to bounce along like a ping pong ball. They are apparently the slowest of the three species, and are more inclined to head for woody cover when pressed. The red kangaroo has a longer stride and lower hop frequency, but a similar trajectory to the western grey. They hold the head more erect, particularly at higher speeds. The eastern grey has a long stride, low trajectory, and relatively high hop frequency. They hop with the body in a much more upright position with the head held high. This locomotion appears energetically more costly than that of the other two species and seems to be geared for high speed over short distances, which would be adaptive in an animal that uses heavy cover to avoid predation. The gait of eastern greys appears less amenable to slow speeds, and this species seems to pause more frequently between a series of hops during a slow retreat, as if the range in lower speeds were limited. As with western grey kangaroos, eastern greys seldom engage in races with vehicles, but quickly head for wooded areas.

One disadvantage of bipedal locomotion is its relative instability. In observing kangaroos, we were struck by the number of times they tripped (and sometimes fell) when their legs struck a tree limb, bush, or old fence wire. Ungulates almost never trip. Obviously, bipedal locomotion is best adapted to open, level environments where there are few such objects over which to trip. The thickest cover is used by western grey kangaroos, which have the shortest legs and are the least clumsy of the three species. Still, whereas alarmed ungulates, such as deer, escape by running into cover, kangaroos tend to escape by running behind cover, using it as a visual screen.

In contrast to hopping, the pentapedal gait, despite its awkward appearance, is quite stable. Kangaroos can lower themselves to and rise up from recumbent resting positions with surprising agility. They can crawl under or through stranded-wire fences, which they prefer to do, although they can also hop over. They can escape from capture with remarkable quickness and agility for an animal that appears so ungainly. When tackled, they must be held down firmly or they will slip through the handler's grasp.

Grooming

All grooming we observed was self-grooming. Grooming of other individuals, which has been reported frequently by other researchers, must occur but at a low frequency at Yathong for us never to have seen it. Certainly mothers must groom their young-at-foot, but we never actually observed such behavior. Once we observed a female grooming herself, and 15 minutes later a large pouch young jumped out of her pouch and began grooming itself. They then bounded off together into the trees. Perhaps the low level of mutual grooming is related to the small size of social groups at Yathong.

Grooming is done mainly with the forepaws and by licking with the tongue. Kangaroos also have a double toe on the rear foot, the two claws of which are used as a comb (Russell 1970b). It is curious that we never once observed such grooming behavior in our study. We have no explanation, for if it were common behavior, we would expect to have seen it, given the amount of time we spent in observation.

All species show similar grooming behavior with bouts lasting a few seconds to several minutes. Short bouts usually involve grooming of one or several body parts by scratching and licking. All parts of the body can be reached by the tongue except the upper back, neck, and head, and even these parts can be reached by the forepaws. Scratching is done with the claws using an inward flexing stroke. Licking, by contrast, is done with an outward motion of the head.

Occasionally, long, thorough grooming bouts were observed. For example, a female eastern grey kangaroo groomed for 90 minutes, systematically going over her body. She started with the stomach and pouch, moved to the back and sides, worked up to the ears, scratching only, and then to the hind legs.

Response to Insects

Outback Australia is blessed with an abundance of insects. In the summer, they are the bane of humans. One inhales them, and they crawl into the eyes and nose. Their crushed bodies smear paper during data recording. At times in summer, while spotlighting at night, the observer had to place the light on the roof of the vehicle to be able to write without being blinded by fluttering moth bodies. Flies emerge when it begins to warm up in August and September, and by October or November, they blacken all living things with their numbers. Being covered by flies is a constant until fall when cold weather arrives in April or May. Brushing flies away from the face is known locally as the "Aussie salute." Strange as it may sound, by the end of fly season, one becomes so inured to flies, one hardly notices their disappearance. Kangaroos are similarly plagued by flies, and they constantly shake their heads and flap their ears. The red kangaroo seems to have the advantage in this behavior because of its large ears. The eastern and western greys have noticeably smaller ears. The flies do not bite, so the cost to kangaroos involves only the energy to chase the flies off of their faces and heads. Flies are primarily a terrible annoyance, with little direct cost to either kangaroos or humans.

At times there are large outbreaks of mosquitoes, and for a few weeks they add to the torment. Also, large tabanid flies sometimes abound, and kangaroos have been seen to shake their heads periodically while looking intently, and swiping with their forepaws. Although none were seen at close enough range for positive identification, we believe large biting flies were involved. Speare et al. (1989) reported kangaroos shaking their heads in response to biting black flies (*Austrosimulium*) as well as inflamed eyelids and impaired vision in severely afflicted animals.

Alarm Behavior

When alerted, kangaroos usually rise up on their hindlegs and tails and stare at the source of disturbance. If satisfied that the intruder is harmless, they will return to their previous activity, usually feeding. If nervous, they usually bound away, stopping at a distance, or in the case of grey kangaroos, near concealment cover, to look back. If taken by surprise, kangaroos simply bound away at high speed.

Aggregations of kangaroos often form at concentrated feeding sites, and these typically break up as kangaroos flee from disturbance. In a conspecific aggregation, the group usually remains together until they are some distance from the disturbance. In an interspecific aggregation, they often go their separate ways immediately as each species utilizes its usual escape means: eastern greys heading for large trees, western greys heading for mallee for more scrubby cover, and reds putting distance between themselves and the disturbance in open country.

Alarm behavior is most pronounced under windy conditions and is less likely under quiet, balmy conditions. During strong winds, kangaroos take alarm at great distances, and if one is driving a vehicle at these times, it seems as if every kangaroo and emu in the countryside is fleeing in the distance. On calm days, sometimes kangaroos will remain on the side of the road and will allow the vehicle to pass within a few meters without fleeing. The generally less alarmable behavior that kangaroos show at night may be partly attributable to the usually lower winds at that time (fig. 6.10).

Among the three species, the eastern grey was clearly the most skittish. Coulson (1999) also reported that eastern greys were less at ease in open cover. This showed up in simple ways, such as the difficulty of doing direct observations of this species and the difficulty of approaching them closely enough to capture for radio collaring. For example, in the major capture effort for this study (chapter 3), the last seven animals caught were eastern greys, which were being pursued exclusively because the samples of the other two species had long since been obtained.

Quantitative expression of alarm behavior was obtained from systematic surveys (chapter 3) in which kangaroos observed were coded for alarm by categories as calm, alert, hop, and bound, in increasing intensity of response. The three species differed significantly in alarm scores (chi-square = 138.14, df = 6, $P < 0.001$; results of this and following tests in this section were the same when tested by log-linear models). Eastern greys showed less alert (i.e., watching the observer) responses and much greater hopping (i.e., retreating at a normal speed) than the other two species. Eastern greys were significantly different in alarm behavior from both red (chi-square = 136.84, df = 3, $P < 0.001$) and western grey kangaroos (chi-square = 109.05, df = 3, $P < 0.001$). Red and western grey kangaroos did not differ in alarm behavior (chi-square = 4.75, df = 3, $P = 0.19$).

Alarm behavior of all three species (table 8.1) differed between daylight and night spotlight counts (eastern grey chi-square = 19.97, df = 3, $P = 0.002$; red chi-square = 68.64, df = 3, $P < 0.001$; western grey chi-square = 65.72, df = 3, $P < 0.001$). The major difference at night for all three species was a much lower frequency of bounding away at full speed, and for eastern greys, a greater frequency of showing alert behavior. Overall, eastern greys showed substantially less alert and more escape behavior than the other two species (fig. 8.1). These results show

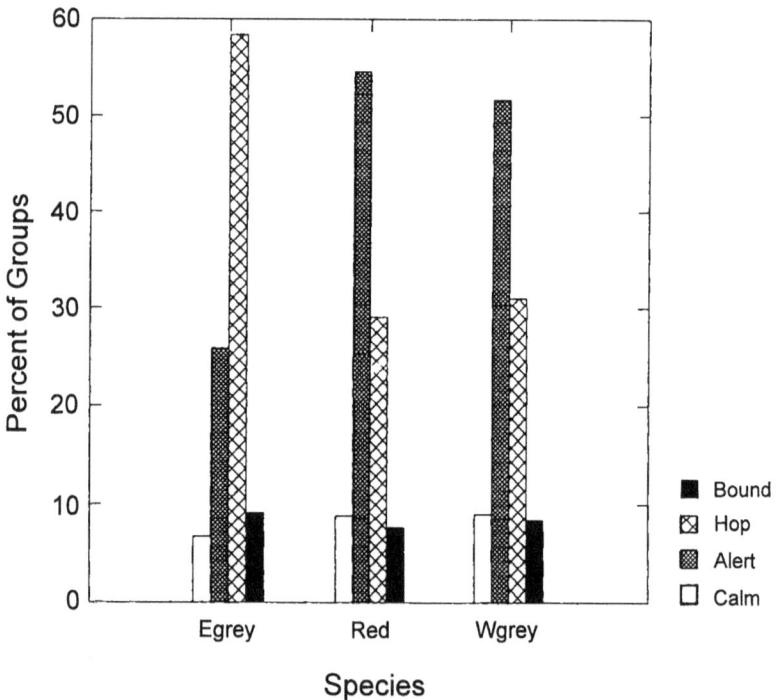

Figure 8.1.   Percent of groups observed in systematic surveys showing alarm behavior of increasing alarm by species. Alarm categories are calm (open bars), alert (small cross-hatched bars), hop (large cross-hatched bars), and bound (black bars).

Table 8.1.   Percent of kangaroo groups showing different alarm behaviors by species for daytime and night spotlight systematic surveys.

Species	Time	N	Alarm Behavior			
			Calm	Alert	Hop	Bound
Eastern grey	Day	161	5.0	20.5	58.4	16.1
Eastern grey	Night	200	8.0	30.0	58.5	3.5
Red	Day	1851	7.7	53.4	27.9	10.5
Red	Night	1113	10.5	56.0	31.0	2.5
Western grey	Day	1411	7.0	50.7	31.0	11.3
Western grey	Night	677	13.0	53.5	31.5	2.1

that the eastern grey kangaroo was most readily alarmed, being much less likely to remain calm or alert and much more likely to hop or bound away, especially during daytime hours.

One would expect that greater avoidance would result in lowered numbers of eastern grey kangaroos being observed in the daytime than in spotlight counts, which proved to be true. Fewer eastern grey kangaroos were seen in daylight (161) than night spotlight systematic surveys (200; ratio of 0.81) with equal sampling effort (table 8.1). The comparable figures for western grey kangaroos were 1411 and 677 (ratio of 2.08), and for red kangaroos, 1851 and 113 (ratio of 1.66) (table 8.1). Thus, a lower number of eastern greys than the other two species were seen in daylight, despite the fact that overall activity patterns of the two species of greys were quite similar (fig. 6.2). Alarm behavior accounted for part of this difference, although the habitats used and their concealment values also were important (chapter 10).

Alarm behavior varied by group composition (table 8.2) for red (chi-square = 099.28, df = 12, $P < 0.001$) and western grey kangaroos (chi-square = 44.77, df = 12, $P < 0.001$) but not for eastern greys (chi-square = 10.14, df = 12, $P = 0.60$). This result was partly due to a substantially smaller sample size for eastern greys ($N = 341$) than for reds ($N = 2821$) or western greys ($N = 1954$); but still, the high tendency to flee in eastern greys seemed to occur independently of group composition. Again, red and western grey kangaroos were similar in that the major group accounting for the differences was all-male groups, which were less likely to remain alert, and more likely to flee from the scene than other groups, including mixed-sex groups (table 8.2).

Though we made every attempt to standardize systematic survey methods, it was not feasible to maintain a constant vehicle speed. From the sample sizes given in table 8.3, it can be seen that 20 kph was the

Table 8.2. Percent of kangaroo groups showing differing alarm behaviors by species and group type. Note that data from daylight and night spotlight systematic surveys have been combined.

			Alarm Behavior			
Species	*Group Type*	*N*	*Calm*	*Alert*	*Hop*	*Bound*
Eastern grey	Male	86	7.0	25.6	64.0	3.5
Eastern grey	Female	86	5.8	31.4	51.2	11.6
Eastern grey	Male + Female	14	7.1	14.3	78.6	0.0
Eastern grey	Female + Young	108	6.5	25.9	62.0	6.5
Eastern grey	Male + Female + Young	47	2.9	31.4	57.1	8.6
Red	Male	434	9.7	40.8	40.6	9.0
Red	Female	938	10.9	62.4	22.6	4.2
Red	Male + Female	176	6.3	54.5	30.7	8.5
Red	Female + Young	1100	5.6	59.2	29.3	5.9
Red	Male + Female + Young	173	8.1	47.4	35.3	9.3
Western grey	Male	498	10.4	45.0	38.0	6.4
Western grey	Female	502	9.6	55.6	29.1	5.8
Western grey	Male + Female	140	8.6	55.0	22.9	13.6
Western grey	Female + Young	659	5.0	56.3	31.1	7.6
Western grey	Male + Female + Young	155	9.7	49.7	32.9	7.7

Table 8.3. Percent of kangaroo groups showing different alarm behaviors by speed of observer's vehicle. Note that data from daytime and nighttime systematic surveys have been combined.

			Alarm Behavior			
Species	*Speed (kph)*	*N*	*Calm*	*Alert*	*Hop*	*Bound*
Eastern grey	5	23	21.7	26.1	52.2	0.0
Eastern grey	10	20	20.0	15.0	50.0	15.0
Eastern grey	20	277	4.7	28.5	58.5	8.3
Eastern grey	30	38	2.6	13.2	68.4	15.8
Eastern grey	40	3	—	—	—	—
Red	5	141	24.1	41.1	27.7	7.1
Red	10	229	25.3	41.5	22.7	10.5
Red	20	2014	7.6	55.5	29.3	7.6
Red	30	563	2.7	61.1	30.7	5.5
Red	40	18	5.6	27.8	44.4	22.2
Western grey	5	101	32.7	33.7	24.8	8.9
Western grey	10	130	16.9	41.5	30.8	10.8
Western grey	20	1451	7.8	53.5	31.0	7.7
Western grey	30	391	3.8	53.2	33.5	9.5
Western grey	40	15	26.7	40.0	26.7	6.7

modal speed driven. This was the desired speed, but slower speeds were sometimes imposed by road or observation conditions, or when new groups were sighted just after recording a previous group. Higher speeds were driven in large, open areas with excellent visibility and lower density of kangaroos.

Vehicle speed influenced alarm behavior of all three species (table 8.3; eastern grey chi-square = 31.99, df = 12, P = 0.001; red chi-square = 171.39, df = 12, $P < 0.001$; western grey chi-square 106.39, df = 12, $P < 0.001$). For all three species, an important difference was that speeds above 10 kph resulted in a decline in calm responses and an increase in stronger reactions. Also, the greater likelihood of eastern greys to hop away and less likelihood of remaining calm or alert than the other species is clearly shown in table 8.3.

Another variable we identified as influencing alarm behavior was the distance of the kangaroo group from the observer vehicle. For all species, distances at which kangaroos retreated were greater in daytime surveys than in nighttime surveys (all cases, $P < 0.001$; Kolmogorov-Smirnov z = 9.63 for reds, 2.94 for eastern greys, and 9.08 for western greys). This difference persisted even when distances in the daytime counts were truncated to equal the shorter distances over which animals could be observed in night spotlight counts.

During daylight, there were significant differences in the alarm distances between eastern grey and both red and western grey kangaroos, with eastern grey kangaroos taking alarm at greater distances (table 8.4). There was no difference between red and western grey kangaroos in the daytime (table 8.4). At night, the reverse was true—there was no difference in alarm distributions between eastern greys and the other two species, whereas a significant difference existed between red and western greys, with the former showing greater alarm distances (table 8.4). Thus, during the day, eastern grey kangaroos showed the greatest tendency to take alarm at longer distances, whereas the alarm behavior of red and

Table 8.4. Kolmogorov-Smirnov tests for differences in distances of alarmed kangaroos by species and daytime versus nighttime systematic surveys.

	Day		Night	
Species Pairs	*K-S z*	*P*	*K-S z*	*P*
Eastern grey–Red	1.488	0.024	0.798	0.547
Eastern grey–Western grey	1.493	0.023	1.000	0.270
Red–Western grey	0.902	0.391	2.324	<0.001

western grey kangaroos was similar. At night, western greys were least alarmable, while red and eastern grey kangaroos took alarm at greater distances.

Social Behavior

The behavioral repertoire of kangaroos has been well documented by a series of authors (e.g., Russell 1970a, 1970b, 1984; Ganslosser 1980, 1989, 1995; Coulson 1989). Although we mention some stereotyped acts here, our intention was not to study behavior per se, but rather, how behavior related to the ecology of the three species.

Aggressive Behavior

We often observed aggressive behavior in kangaroos, this being the most common overt social interaction between adults. Aggression occurred primarily between animals of the same sex, usually among conspecifics, but sometimes interspecifically. Though displacement of subordinate individuals by more dominant ones was occasionally seen, it probably was more prevalent, but unrecognized, as encounters were often subtle, the subordinate animal moving as the dominant one approached. An overt threat was given only if a subordinate (or equal) individual resisted displacement. If the disparity in size between the contestants was great, a threat was usually sufficient to move the smaller one away. A few hops over 3 or 4 m distance usually ended the encounter.

Aggressive threats began with a direct stare from the pentapedal stance, with the aggressor leaning forward and lifting the head, and rising partially to release the forelegs for combat. In the most abbreviated form of threat, the nose was thrust outward and upward towards the opponent, from a distance of several meters. This head snap was abrupt and often repeated several times. Sometimes western grey kangaroos extended the head, nose out, from the pentapedal position and waved the head from side to side in snakelike fashion. This seemed a lower level threat than head snapping. Threat in the most pronounced form, rearing, was accompanied by slashes of the forepaws in shadow fighting, while the opponent was well out of reach. These movements appeared to be intention movements.

Sometimes the recipient of such threats twisted toward the aggressor, rose partially upward, and raised the forearms with claws exposed briefly, before hopping off. If the recipient did not concede, but held its ground, the encounter would escalate into a full-fledged fight.

Fights were characterized by the contestants facing each other directly, with full eye contact, and rising to their full heights in the three-point stance. In this posture, they were balanced on their hind toes and tail tip, with their bodies raised as high as physically possible. This posture seemed directed at illustrating their size to the opponent, for the stance is rather unstable, and the contestants drop down somewhat before making actual contact. Presumably, size would be correlated with strength in most cases, and height would convey advantage in fighting position as well. Barnard and Burk (1979) have noted that even unfamiliar opponents can assess each other by size and behavior, probably giving larger individuals the advantage. It is also of benefit to smaller individuals to avoid a fight (and consequent risk of injury) in which a loss is a forgone conclusion. Rearing to maximum height would allow such an assessment.

Kangaroo fighting consisted of striking with the forepaws, biting, wrestling to force the opponent off balance, and kicking with the hind feet (Dawson 1995). At Yathong, hind foot kicking usually involved only one foot, but at times, an animal balanced on its tail to use both feet. Bouts typically lasted a few minutes, with rest breaks following strenuous fighting. The shortest bouts were clashes of a second or two whereas the longest bout observed was 11 minutes. Sometimes the contestants fought for a few minutes and then returned to feeding, only to resume fighting again after 10 to 20 minutes. Croft and Snaith (1991) have described play-fighting in kangaroos, and a few of the fights we observed could be considered play fights. However, most fights we observed seemed of an intensity and seriousness to be considered as true fights. Certainly, the potential for serious injury was present, and western grey kangaroos were particularly vicious in intent.

The three species differed in their mode of fighting. Red kangaroos were basically wrestlers. They used their powerful forearms and pectoral girdles to engage the opponent in a direct physical contest of strength. Like sumo wrestling, the object of the contest was to force the opponent backwards or off balance. Slashing with the forepaws was used, but it is the scratching, rather than cutting, of such blows that is important, because the claws in this species are dull. Usually such blows were attempted prior to engagement, and were not resorted to much during wrestling. Similarly, hind foot kicks and biting were less commonly used by red kangaroos than by the other two species. Sharman and Calaby (1964) reported that males "clucked" with their tongues and jaws while fighting, but we could not detect such calls at the distances from which we observed fights. They also reported "quivering" of the upper body and forearms, and licking a glandular area on the chest, neither of which was apparent to us.

In fights between red kangaroos, there was usually a clearly discernible winner and loser. The loser broke off the match and moved away, much as most ungulates do. Unlike ungulates, however, the winner seldom pursued the loser, and the long postfight chases common in ungulates were not seen in kangaroos. Another difference was that kangaroos were seldom seen engaging in mock or ritualized fights common in ungulates. In fact, the only mock fighting that might be construed to contribute to development of fighting ability was between mother and young (see below), as previously reported by Croft (1981).

In contrast to reds, the western grey kangaroo was a slasher. It has formidable weapons. Its claws, both fore and rear, are razor-sharp, and it is also capable of delivering serious lacerations by biting. One western male kangaroo, captured for radio-collaring, delivered a nasty bite to the senior author's finger when the holders lost control of its head. Had the animal had a fraction more leeway, it may well have bitten the finger clean off.

Among western grey kangaroos, a high percentage of adults of both sexes had split ears (fig. 1.3) or large scars on the head and face. Perhaps 20 percent of the population showed obvious effects of combat with conspecifics. Such dangerous weapons explained the relative reluctance among the radio-collaring catching crew to go after western greys. Whereas the largest red kangaroo males, standing 2 m tall, would engender enthusiasm for the chase, a small western grey female, standing less than 1 m tall and weighing less than 30 kg, would prompt a serious discussion of being too far away, or that there were old fences, uneven ground, or too many tree branches. Note that although some authors (Lee and Ward 1989) list western grey females as larger, in our population they were clearly smaller than both red and eastern grey females. Given their small size, quickness, and agility, female western greys were even more dangerous to catchers than males because it was more difficult to restrain their limbs and head while bringing them under control.

The formidable weapons of the western grey kangaroo were accompanied by a pugnacious personality. Over twice as many aggressive encounters were observed in western grey than in red kangaroos, even though the latter were observed 1.5 times more frequently (table 7.1). Furthermore, the sexes in western greys showed approximately equal frequency of aggression, whereas aggression among red kangaroos was almost entirely between males. Ganslosser (1989) reported that female kangaroos hardly fight at all. However, this certainly was not the case for western grey females at Yathong, although female red kangaroos at Yathong were remarkably docile, and rarely showed aggression.

Because of their weapons, fighting behavior of western grey kangaroos was entirely different from that of red kangaroos. Wrestling was virtually absent. Instead, slashing with the forepaws was the major element of a fight. Their objective appeared to be to deliver cuts to the opponent while avoiding receiving cuts. The head and ears were the most vulnerable parts, and they were held as far back away from the opponent as possible. Thus, the combatants reared up to their full height in the three-point stance and threw their heads and necks back. Typically, the muzzle was pointed upward, or backward away from the opponent. Meanwhile, the forepaws alternately slashed out at the opponent with great rapidity—finesse was not involved—with each contestant making as many slashes as possible.

Usually, fights by western grey kangaroos consisted of alternating stages in which brief flurries of slashing were interrupted by standoffs in which the opponents stared at each other from just beyond reach, presumably gauging their opponent. Reengagement was initiated by contestants suddenly hopping forward while slashing, somewhat reminiscent of cockfights. Hindfoot blows were few. Feints with a hind foot elicited strong evasive response, suggesting that kicks were quite dangerous, but seldom delivered successfully. Also, since the contestants were usually reared to their full height, they may have risked losing balance by delivering a hind foot kick. Similarly, bites were not common in fights among western greys, because the head was kept as far away from the opponent as possible. Although we did not observe an instance, it seems likely that an opponent exposed by a slip or other mishap would be subjected to bites and hindfoot kicks.

During fighting, cough-roaring was repeated regularly (about every four to five seconds). This aggressive sound differs from the "cough" used in alarm situations, in that the latter is usually given in bursts, short in duration and more from the mouth, whereas the former is more drawn out, and more from the throat.

The dangerousness of weapons in western grey kangaroos also appeared to influence the combination of opponents and outcomes of fights. A greater disparity in size of opponents was seen in western greys than in red kangaroos. Red kangaroo opponents tended to be relatively closely paired in size, whereas mismatches in size of western greys were more common. Also in contrast to reds, western grey kangaroo fights often ended in draws. Even when there was a large disparity in size, the lesser contestant seldom was clearly defeated. Even when one combatant broke off the engagement, it would resume if pressed by the opponent. Although the lesser contestant usually broke off the encounter, it did not concede defeat, and returned aggression with aggression if the other contestant persisted. This suggests that, because of their serious weapons, western greys are better off settling for immediate advantage than pursuing a clear victory, and risk receiving a serious wound. Possession of such dangerous weapons seems to raise the risks while simultaneously leveling the playing field. Large body size is a lesser advantage than in the other two species, which use wrestling as the predominant fighting technique.

Many fewer aggressive encounters were observed among eastern grey kangaroos because they were less often observed (table 7.1), more likely to be found in heavy cover (see chapter 10), and were more alarmable (table 8.1). However, the fights observed showed that eastern grey kangaroos were more similar to red than western grey kangaroos in their fighting behavior. Wrestling was the principal mode of fighting. The eastern grey kangaroo had much less sharp claws than western greys, although not so dull as those of red kangaroos. In fighting, they did not slash as frequently as western greys. Hind foot kicks were both attempted and successfully delivered more frequently than in the other two species, and the thump of their delivery to the opponent's body could be heard from a considerable distance.

The only interspecific aggression observed occurred when a western grey male threatened an eastern grey female. The eastern grey female retreated and the brief encounter ended.

Grass-Pulling

Western greys also showed a male aggressive behavioral act that has been described as grass-pulling (Kaufman 1974c). Grass-pulling has been reported for red kangaroos (Croft 1981; Russell 1985) and eastern greys (Coulson 1989), but it was not observed in either of those species in this study; Coulson (1997b) reported grass-pulling and rubbing in both grey species. Western grey males at Yathong often scratch up dry forbs or bunch grasses with their forepaws in mock fighting motions. They start this behavior in the pentapedal position, rubbing chin, neck, chest and stomach on the dried clump of vegetation. They then begin slashing at the vegetation with the claws on their forepaws, and as a collection of dried plant material is gathered they roll it up on their stomachs and chests while rearing up in the three-point stance. Sometimes they flail the debris around in the air and make various waving motions with their forearms, sometimes jumping into the air to strike the debris. In one instance, a male made movements like plucking the strings of an imaginary fiddle.

The significance of grass-pulling was not clear. It was not performed in the presence of females, and it was never observed in a consort-pair male. It may have a scent-marking function, as suggested by the rubbing that preceded the scraping. Russell (1985) reported a glandular area on the sternum that shows pigmentation (Nicholls and Rienits 1971) and the shredded vegetation may be marked by this gland during grass-pulling. However, sites examined after the scraping gave off no odor detectable to us, nor signs of an exudate, urine or feces. We did not observe extrusions of the penis during this act. No other kangaroo in association with the grass-puller showed a behavioral response to the act itself, or to the site later on. Usually, it was performed while other males were in the vicinity, but not close to the site. The nearest males were usually 50 m or more away. Although this behavior was usually seen at times when fighting between males was occurring, it had no obvious connection to the fighting, either in time or place. Scraping sites were not concentrated in areas of consort pairs.

It was our impression that grass-pulling was associated primarily with dominance interactions among bachelor males. Some very large males were seen performing the behavior, but only when associated with subordinate males. It was almost as if they were putting on a machismo display to "psych" themselves up and intimidate associated males. In some ways, the slashing at the end of the act mimics fighting. But if development of fighting skills were the purpose, one would expect males to attack more realistic targets, such as tree branches that would be at the right level, and present resistance. A better understanding of the significance of this behavior must await more detailed study.

Sexual Behavior

Few observations of sexual behavior in eastern grey kangaroos were obtained, so this discussion pertains mainly to the red and western grey kangaroos. In those two species, the most obvious manifestation of sexual activity was the frequent occurrence of consort pairs (Croft 1981; Jarman and Southwell 1986), consisting of a single male and female and, if she had one, her young-at-foot. Consort pairs were seen often enough among eastern grey kangaroos to believe that the same breeding system applies to that species as well. Often another single female, and sometimes several males, were nearby, usually 50 to 100 m distance from the consort pair.

In ungulates, this system of breeding is usually referred to as a "tending bond," and nearby males are referred to as bachelor (or sometimes satellite) males. Clearly, these species of kangaroos were not territorial, nor was a harem defended, two other breeding systems common in ungulates. The tending bond breeding system has sometimes been viewed as serial monogamy in that males form pairs with females, and remain in those pair bonds until copulation is achieved. However, given the relatively short period before copulation (several days), the term "monogamy" seems misdirected. Shortly after copulation, the pair bond dissolves and the male moves on in search of another female coming into estrus with which to form a pair bond. The process continues until no more females are in estrus, and the breeding period ends.

In all respects, the red and western grey kangaroos conformed to the typical tending bond system as seen in ungulates. The tending male remained in close proximity to the female, usually 2 m away and typically behind or slightly to the rear and off to the side. It is as if the male were linked to the female by an invisible tether. If she took a step forward, he took a step forward. If she shifted direction, he shifted direction, and so on in lockstep.

The female being tended took the initiative in normal movements and largely ignored the male. Periodically, the male approached the female. Such approaches were made very slowly and tentatively while in the pentapedal posture with the neck and head extended parallel to the ground and the nose angled slightly upwards. These approaches were the antithesis of aggressive approaches, which apparently minimized the likelihood of the female being alarmed by the much larger male. If the female suddenly looked back at the male, or began to hop away, the male stopped, and did not approach for the moment.

Successful approaches were initially ignored by the female who typically continued to feed. The male smelled and licked the female's tail, back and shoulders, grasped her tail, and gently pawed or scratched her shoulders. The retractable penis was extruded from the cloaca during this stage of the approach. If the female did not move away, the male worked his way towards the front of the female, gradually assuming a more erect body posture. As the male came around the side of the female, she usually rose up and there was a nose-to-nose contact between the pair. The male pawed the female's neck or shoulders and palpated his penis. The female smelled the penis. The female usually then turned away and resumed normal feeding activity. However, if she was receptive, she hunched down and accepted mounting. This sequence is similar to that reported by Coulson (1997b), except for flehmen. Coulson (1989) reported that males elicit urine from females and then show flehmen, a posture associated with testing the urine for indication of estrus. We did not observe this behavior in our study, but we may have missed it because of the distances from which we observed.

Only three copulations were observed in this study, two by red and one by western grey kangaroos. Few copulations have been reported from the wild. The female was crouched close to the ground and was covered by the male who was hunched over her and held her with his forearms just behind her pectoral girdle with his hindlegs to the outside of hers. Essentially, the female was contained by his body to the rear, and her forward movement was restricted by his grasp about the middle of her body, as reported previously by Sharman and Calaby (1964).

Pelvic thrusts of males were slow, almost rocking motions. Such rocking motions would occur for 10 to 40 seconds, followed by quiet periods of 2 to 5 minutes. This alternation characterized the entire copulatory period. According to Poole and Pilton (1964), multiple ejaculations occur during a copulation. Sometimes the male would nuzzle the female's head, ears, neck and shoulders. If the female stirred, the male would tighten his grip, clearly preventing interruption of the copulation by the female.

Copulation was terminated when the male released the female, either because she struggled or he backed off. Little postcopulatory behavior was observed. In one case, the male briefly sniffed the female's rump, but essentially, both members of the pair simply resumed normal feeding.

Copulation was long and drawn out. One red kangaroo copulation observed in total lasted 25 minutes, and the two others observed in progress continued for about 10 minutes. Sharman and Calaby (1964) reported that copulations by red kangaroos in captivity lasted about 30 minutes on average. Copulations in captive western grey kangaroos reported by Poole and Pilton (1964) lasted about 50 minutes, and those of eastern grey (Tyndale-Biscoe and Renfree 1987) 40 to 50 minutes. Coulson (1997b) reported on four eastern grey and three western grey copulations in captivity that lasted up to 55 minutes. This is in marked contrast to ungulates in which copulation consists of mounting of the female by the male and a single, powerful ejaculatory thrust.

The dominance relationship among males plays an important role in determining reproductive success of males (Croft 1981, 1989; Jarman and Southwell 1986; Ganslosser 1989, 1995; Dawson 1995). If a subordinate male approached a consort pair, the consort male would turn and drive him away. On a few occasions, a larger male was observed to arrive and displace a smaller male from a consort pair. We do not know if dominance relationships in male kangaroos are stable and linear, or determined encounter by encounter. Too few known individuals were available, and the lack of consistent social grouping made it difficult to observe combinations of males to address this question. However, outcomes of male-male encounters were related to size because the larger individual (if difference could be determined visually) invariably prevailed, as previously reported by Russell (1970a) for red and Grant (1973) for eastern grey kangaroos. Grant (1973) also reported stable but separate hierarchies for males and females, but this was in a coastal population with a much more predictable environment than at Yathong. At Yathong, low population density, dispersion of individuals, large home range sizes, frequent movements, and low social tendencies make it unlikely that stable dominance hierarchies are common, a conclusion previously reached by Caughley (1964).

Mother-Young Interactions

No observations of birth were made in this study. Sharman and Calaby (1964) reported in detail on red kangaroo reproductive biology in a captive population, and Lee and Ward (1989) and Dawson (1995) have reviewed kangaroo life histories. Sexual maturity in females occurs at 15 to 20 months of age, and in males at two years. The estrus cycle is 34 to 35 days, and gestation is 35 days.

At birth, the small young, weighing less than a gram (Poole and Pilton 1964), makes its own way into the pouch, and attaches to one of the four teats. Usually, a single young is born, but twins, although rare, do occur (van Oorschot and Cooper 1989; Arnold et al. 1991). Lactation is induced by suckling, because lactation could be induced in virgin or unmated females by placing a foster young on a teat. Postpartum estrus usually occurs within two days after the young reaches the pouch (Sharman 1965).

Red females may have up to three generations of offspring simultaneously: a young-at-foot suckling from an elongated teat, a young in the pouch attached to a second teat, and a blastula in arrested development in the uterus. No estrus occurs for about eight months while the pouch young is suckled. If the pouch young is lost, development of the blastula proceeds. Thus, delays in reproduction are minimized. This system is believed to foster rapid population growth in an uncertain environment, where natural events, particularly drought, cause offspring mortality (Newsome 1975; Low 1978).

The young in the pouch first protrudes its head at 150 days, starts leaving the pouch for short periods at around 190 days, and leaves permanently at about 235 days. The female continues to suckle the young-at-foot for another 130 days with weaning occurring at about one year of age. The mother enforces leaving the pouch when she prevents the young-at-foot from reentering (Sharman and Calaby 1964). The pouch retracts and birth and attachment of the next young occurs. These authors also reported "squeals" by young and "clucking" by mothers not heard in this study.

Timing of reproductive events at Yathong was measured by the occurrence of large pouch young. The presence of readily identified pouch young observed over time in the systematic surveys is shown in figure 8.2. The pouch young schedules of the three kangaroo species were essentially the same, and were significantly correlated (red and eastern grey $r = 0.629$, $P = 0.03$; red and western grey $r = 0.89$, $P < 0.001$; eastern grey and western grey $r = 0.76$, $P = 0.004$). Pouch young

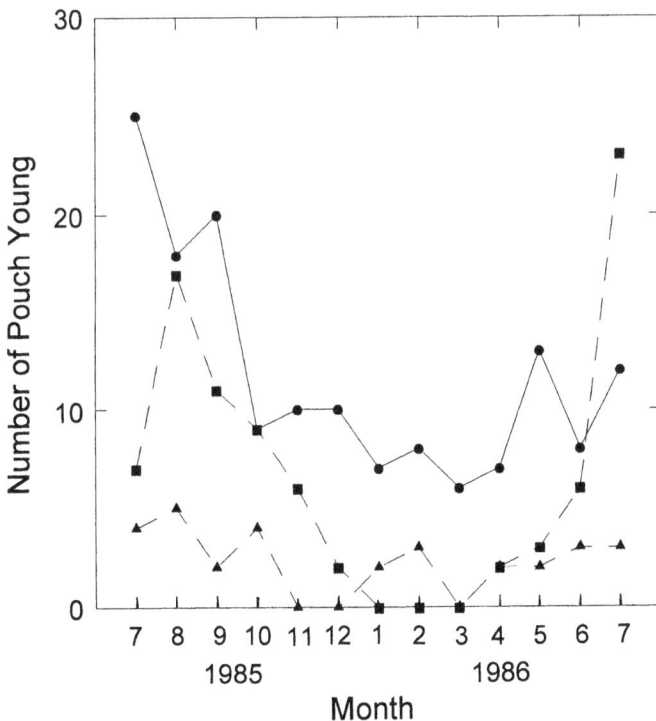

Figure 8.2. Number of pouch young observed in systematic surveys by month for eastern grey (triangles), red (circles), and western grey (squares) kangaroos.

numbers peaked in August or September 1985, and declined thereafter, until rising again in July 1986. These peaks probably reflected both the growth stage of the offspring as well as the greater time spent in the pouch due to cold weather. In eastern and western greys, large pouch young declined almost to zero at other seasons, whereas in red kangaroos, large pouch young occurred in low numbers throughout the year (fig. 8.2). This reflects greater aseasonality of breeding in the red kangaroo.

Numbers of young-at-foot observed in the systematic surveys are given in figure 8.3. In red and western grey kangaroos, numbers of young-at-foot were low in July 1985, but increased thereafter to a high in July 1986. Generally, there was an inverse relationship between large pouch young and young-at-foot that reflects the population turnover in generations (i.e., the previous young-at-foot disperse from their mothers

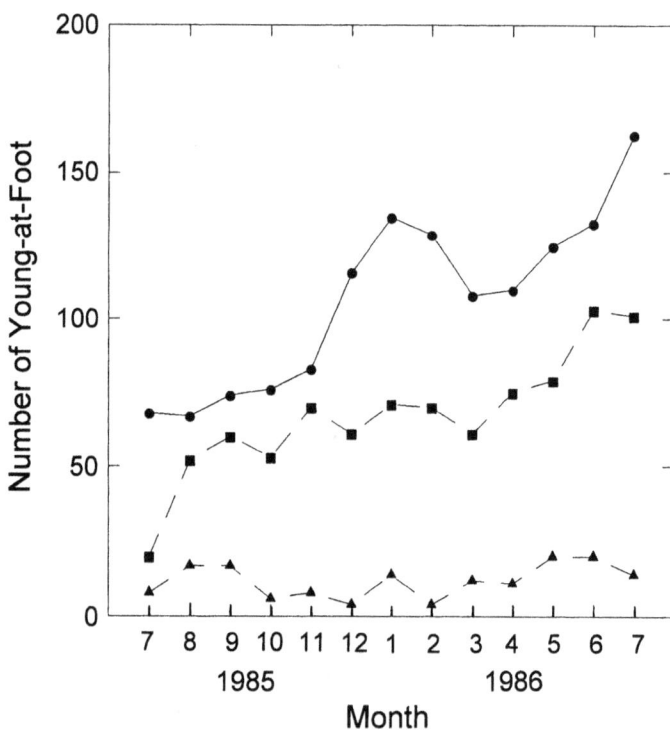

Figure 8.3.   Number of young-at-foot observed in systematic surveys by month for eastern grey (triangles), red (circles), and western grey (squares) kangaroos.

as the next generation of pouch young matures). This relationship was strongest in red kangaroos. Cross-correlation showed a significant negative correlation ($P < 0.05$) for zero and minus one-month lags (i.e., when pouch young were highest, young-at-foot were lowest) and the highest positive correlation (but not significant) at a lag of six months. Western grey kangaroos showed a similarly smooth cross-correlation plot between pouch young and young-at-foot, but cross-correlations were not significant. However, the highest negative lag was at minus four months and positive at six months. These results suggest that the seasonally breeding western grey young-at-foot disperse prior to maturation of pouch young (and thus are not with the mothers at the emergence of pouch young), whereas the aseasonally breeding red kangaroo shows dispersal of young-at-foot around the time that pouch young are emerging. Sample sizes for eastern grey kangaroos were too small to give a meaningful test.

It was not possible to observe nursing by pouch young under the conditions of this study. However, nursing by young-at-foot was observed on occasion. The mother sat upright and the offspring hunched facing the pouch and nursed. Uninterrupted bouts ranged from about seven to nine minutes in duration. No particular signals were detected prior to these bouts, and the nursing was terminated by the female who simply turned away and resumed feeding.

Leaving the pouch occurred gradually as the young animal spent a greater proportion of time outside the pouch. According to Russell (1989), the young initiates leaving the pouch, or falls out accidentally. As young grew large, it was common to observe them leaning from the pouch to feed actively on green feed while the mother fed. When it was very cold, very large young could be seen in the pouch.

Usually, the young-at-foot remains within 3 or 4 m of the mother, but at times would go as far as 20 m to interact with other kangaroos, usually by nose-to-nose contact, following which, the young returned to the mother's vicinity. At times, mother and young become separated. On one occasion, while walking in a mallee thicket, we alarmed six western grey kangaroos, which hopped away quickly. A very small young-at-foot, which was either unaware that the mother had fled or which had hidden, jumped up at our approach, and ran towards us, stopping at about 10 m away. The young-at-foot stood in the upright alarm posture while issuing a series of isolation "coughs" (Baker and Croft 1993; Coulson 1997b), and after about a minute, bounded off rapidly in the direction taken by the others.

At times, mothers show behavior towards their young that can only be described as affectionate. The mother leans back in the three-point stance, and the young faces her, sitting virtually in her lap. Then the young and mother cuff each other about the head and ears in play-fighting, as described by Croft (1981) and Coulson (1997b). The young fights with youthful vigor, landing blows, whereas the mother responds slowly and gently. Periodically, the mother embraces her young in a tender, furry hug that prompts the human observer to think "aaawh." Sometimes these mother-young bouts go on for five minutes or more before the kangaroos return to being animals and the observers return to being scientists.

Breeding Seasons

Red kangaroos have a postpartum estrus (Sharman 1965) and thus, breeding activity also reflects the season of birth. Insufficient data were obtained for eastern grey kangaroos, but western grey and red kangaroos at Yathong were similar in their breeding seasons. Aggressive (by both sexes) and sexual behavior (including tending bonds, sexual approaches, grass-pulling, and copulations) were associated in time, and they, in turn, were related to the formation of social groups containing at least one male and one female (fig. 8.4). There was a major peak in mixed-sex groups and fighting and sexual behavior in the spring months, and a smaller peak in late summer (January and February, fig. 8.4). The spring increase in breeding activity began in June, peaked in September, and declined until December, at which time there was a small resurgence. No sexual behavior was observed from March through May, and in December (fig. 8.4). Still, the time extent over which the major peak occurred and the fact that sexual activity was observed in nine out of twelve months shows that breeding in kangaroos is not confined to a specific period. It is also notable that red and western grey kangaroos showed such a similar periodicity.

Fights and sexual behavior were also correlated with the observed numbers of large pouch young ($r = 0.72$, $P = 0.009$) (fig. 8.2); furthermore, observed pouch young correlated with the percent of mixed-sex groups in red and western grey kangaroos (red $r = 0.75$, $P = 0.005$; western grey $r = 0.69$, $P = 0.012$). Lagging of red kangaroos showed little difference, but in western greys, lagging percent of mixed groups by one month resulted in a much higher correlation (cross correlation, $r = 0.91$, $P < 0.001$).

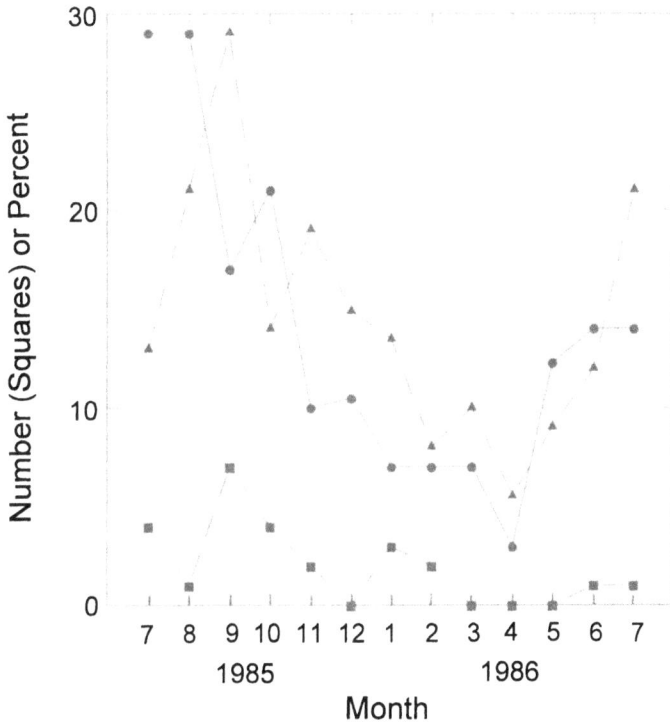

Figure 8.4.   Relationship of number of fights and sexual acts (squares) to percent of groups that are mixed sex for red (circles) and western grey (triangles) kangaroos by month.

Discussion of Behavior

Behavior of kangaroos at Yathong is strongly influenced by low population density in a harsh environment. Small group sizes, dispersed distribution, and low interaction rate are characteristic of this population. Most interactions are related to breeding activity of adults, and mothers and young. Western grey kangaroos were particularly pugnacious, and red kangaroos (especially females) most docile.

It seems likely that large dominant males performed most of the breeding, as has been reported for kangaroos (Croft 1981; Jarman and Southwell 1986; Stuart-Dick and Higginbottom 1989; Dawson 1995). Sequential tending bonds are the characteristic breeding system in ungulates when resource distribution is unpredictable over time and social structure of females consists of solitary individuals or small groups. Unpredictable resource distribution disfavors territorial systems, because neither the time nor place where animals will congregate at the time of breeding is consistent. Dispersion of females as singles or in small groups eliminates the possibility of harem formation, the breeding system in the most polygynous species of large mammals. Such systems depend, at minimum, upon sufficient aggregation of females to form the basis of a female group that can be defended by a male against rival males.

Formation of tending bonds between kangaroo pairs is a system that maximizes copulations by dominant males, given the circumstances at Yathong. Large males remain in a tending bond with a given female until copulation is achieved, then move on to form tending bonds with other estrous females. Subordinate males have reliably sorted out the estrus status of unbred females, with the larger males in tending bonds with the females nearest to estrus, the smaller males with those females farther from estrus, and no males with females showing no sign of estrus. Dominant males, by searching for tending bonds held by the largest males they can displace without a really serious fight, can maximize their copulatory success and avoid having to sort through females to determine their reproductive status. Within a short time, the female in the consort pair of the displaced male will be coming into estrus, and the short period maximizes the number of sequential tending bonds that can be completed in a breeding period.

In his review of macropod courtship behavior, Ganslosser (1995) noted that kangaroos show behavior that contributes to dominant males achieving most copulations. Estrous females roam broadly and use conspicuous signals that are likely to attract the attention of males. The long consort pairing (several days) and long copulatory act also give notice to rival males of the receptive state of the female, and the rewards of a challenge, if successful.

In captive grey kangaroos, Poole and Pilton (1964) reported that males formed pairs with females several days before estrus, and that agrees with our observations at Yathong. However, for red kangaroos in captivity, males formed pairs with females only a few hours before estrus. We think this short pair bonding was a consequence of captive conditions where only a single male had access to the female, and the female was restricted in movement. We know that tending bonds of red kangaroos were much longer in the wild, although probably not so long as those of the grey kangaroos. The longer tending period for red kangaroos in the wild was further supported by the low number of observed copulations in relationship to the high number of tending bonds we observed. If behavior for animals held in captivity, as reported by Sharman and Calaby (1964), applied to ones in the wild, then a two-hour tending bond and half-hour copulation should give a time ratio of tending bonds to copulations of 4:1 in the wild. Tending bonds were observed so commonly during breeding periods that not all tending bonds observed were individually recorded. Yet the ratio of tending bonds to copulations observed in red kangaroos at Yathong were of the order of 50:1. We believe that smaller males began guarding females approaching estrus at least several days (and probably longer) in advance of copulation, and a majority were displaced by large males prior to female receptivity.

In situations where the breeding season is short, a number of females come into estrus simultaneously, and the time needed to complete a copulation and form the next tending bond will limit reproductive success of dominant males. Breeding will be spread more widely among males, as smaller males are able to copulate with females when dominant males are occupied in other tending bonds. However, as pointed out by Jarman and Southwell (1986), with a protracted breeding season (such as at Yathong, see below), a smaller proportion of females is in estrus simultaneously, and dominant males will be able to achieve a longer string of serial pairings. Thus, dominant males will achieve a greater proportion of the copulations at the expense of breeding success of smaller males.

Dominance struggles among males were particularly obvious in western grey kangaroos. As previously noted, this species shows a much higher level of aggression than the other two. Even three-way fights and round-robin contests were seen, despite the smaller group size shown by this species.

The results on reproduction show that July, August, and September were a time of important life history phenomena in kangaroos at Yathong. Previous young-at-foot were nearing independence from their mothers and approaching sexual maturity. Pouch young were growing large and were about to become young-at-foot with the birth of a new offspring. Mixed-sex groups were forming, and fights and sexual behavior accompanied the attempts of males to breed.

The question arises about the causality of the peak in breeding activity. Is the peak facultative, simply a reflection of kangaroo response to locally favorable conditions for breeding, or is there a breeding season that is consistent from year to year? Indeed, a third possibility must be considered: that the peak was a manifestation of the 1982 drought "resetting" females of all species to the same schedule, and this peak was simply the third cycle of reproduction following the end of the drought. These three hypotheses can be summarized as follows:

1. The breeding peaks observed were a historical artifact of the 1982 drought, and a reasonably consistent interval of the production of young. The breeding of females was synchronized by the rains that broke the drought in early 1983, and because of the few reproductive cycles since then, synchronization is still apparent in this observed peak. Over time, synchronization should disappear.
2. Breeding peaks are caused by locally prevalent plant growth conditions that favor successful reproduction.
3. Breeding peaks occur in more or less predictable seasons that are consistent between years. This implies that peaks, to some extent, are genetically determined due to past selection because that season was most favorable for survival of offspring.

While logically reasonable, the drought synchronization hypothesis was not well supported by the evidence. There was little doubt that the 1982 drought resulted in some synchronization of kangaroos on Yathong. It was severe and resulted in significant adult mortality. Undoubtedly, juvenile mortality was extreme, as was anestrus (Newsome 1964), and temporary male sterility (Newsome 1973) was likely. The drought was broken by heavy rains in February, March, and April of 1983. Given that a complete cycle of reproduction (particularly in the two species of grey kangaroo) takes about one year, it is conceivable that the peak we observed was simply the third cycle since the drought was broken, and synchronization might easily persist over that interval. Studies of captive, penned red (Sharman and Calaby, 1964) and the two species of grey kangaroos (Poole and Catling 1974) demonstrated that females can show estrus year-round. Thus, the reproductive physiology of kangaroos would allow this hypothesis. Despite the logical possibility of this hypothesis, we believe the evidence rejects it. First, a year-long cycle should have resulted in the third peak occurring in March or April, when, in fact, this period was a low point in reproductive behavior (fig. 8.4). Second, the red kangaroo has an inherently higher reproductive rate and shorter interval between offspring (Shepherd 1987; Lee and Ward 1989) (table 8.5) than either species of grey kangaroo. Given the relatively high rainfalls in years since the drought (fig. 2.7), the peak in red kangaroo reproduction would have occurred at shorter intervals, and thus, should have deviated in time from that of grey kangaroos. Although the peak for the red kangaroo we observed may have been about

Table 8.5. Reproductive parameters of red and grey kangaroos. Data from Sharman and Calaby (1964), Poole and Pilton (1964), and Poole and Catling (1974).

	Red	Eastern Grey	Western Grey
Age at sexual maturity—male	24 mo.	48 mo.	31 mo.
Age at sexual maturity—female	15–20 mo.	18 mo.	14 mo.
Estrous cycle	24–35 days	45 days	35 days
Mean time to internal estrus if pouch young survives	—	345 days	307 days
Return to estrus from loss of pouch young	—	10 days	8 days
Gestation period	33 days	37–38 days	29–31 days
Postpartum estrus	Yes	No	No
Embryonic diapause present	Yes	Sometimes	No
Mean no. of estrous periods before conception	1.1	1.9	1.5
Age at which young first leaves pouch	190 days	—	—
Age at which young leaves pouch permanently	235 days	312 days	—
Age at weaning	365 days	540 days	—

one month ahead of that of the western grey (fig. 8.4), they were largely synchronized. Also, given the shorter cycle of red kangaroos, one would expect the synchronization to have dissipated, and the peak to be less pronounced. Figure 8.4 shows that this was not the case. Finally, Shepherd's (1987) data from Kinchega National Park showed a surge of reproduction in red kangaroos following the rains in February–April 1983, and peaks thereafter that could reflect subsequent synchronization (fig. 8.5). However, western grey kangaroos did not show a surge in reproduction in February–April 1983, and did not respond reproductively until late 1983 (fig. 8.5). It seems likely therefore, that whatever synchronizing effects the 1982 drought may have had at Yathong, such effects were no longer apparent in 1985–86.

There is substantial support for the second hypothesis of a direct reproductive response to locally favorable conditions for reproduction. The breeding peak (fig. 8.4) was generally associated in time with both rainfall and availability of green feed (fig. 2.9). However, correlations, although suggestive, were not statistically significant (rainfall $r = -0.43$, $P = 0.16$, green feed $r = -0.47$, $P = 0.12$). Cross-correlation analysis showed no lag ($P > 0.05$) between rainfall or green feed and sexual behavior. If the breeding season at Yathong was related to the highest probability of rainfall, based upon long-term records for Cobar (chapter 2), it should have occurred in December, January, and February. Yet, no breeding behavior was seen in December and relatively little in Janu-

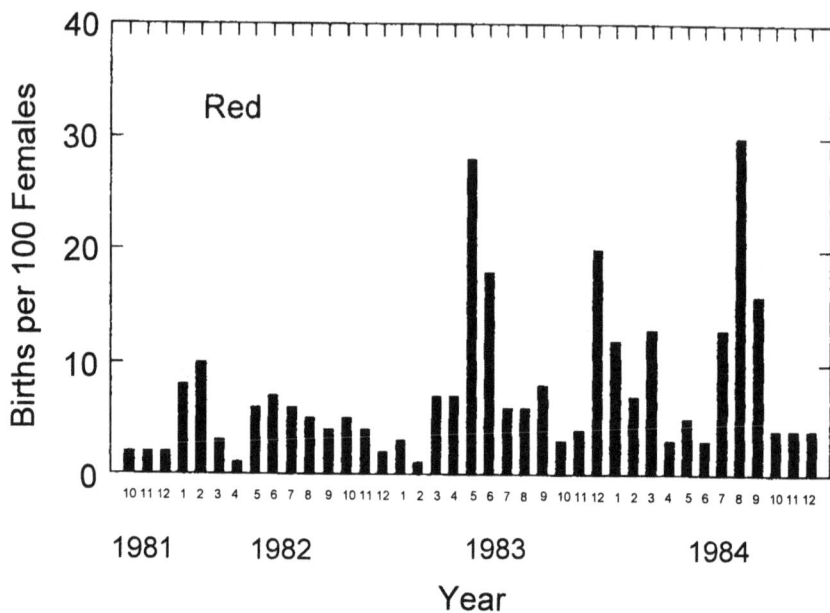

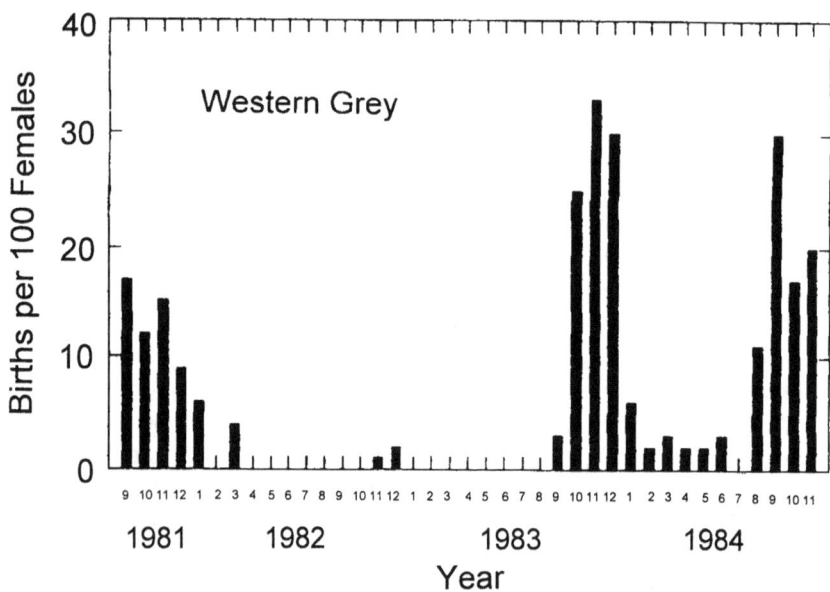

Figure 8.5. Reproductive activity of red (above) and western grey kangaroos over time at Kinchega National Park. Data from Shepherd (1987).

ary and February. It could be argued that breeding should not occur in either winter or summer (the extreme seasons), and spring should be the period with most favorable conditions—in terms of both temperature and rainfall. Perhaps the association of green feed, rainfall, and a seasonal breeding season during this study was coincidental. Yet, the resurgence of breeding in January and February is inimical to this coincidence, and favorable to a direct relationship of breeding to presence of green feed.

There is also considerable evidence of a breeding season at a fairly consistent period between years. This is especially true for the two species of grey kangaroos, but, to some extent, for the red kangaroos as well. Based upon large collections from the Mt. Hope area (about 70 km east of Yathong), Poole and Pilton (1964) found peaks of reproduction in the summer over three reproductive periods (fig. 8.6). Given that similar numbers of animals were observed over each of the three years, the 1961–62 breeding period was substantially lower in number of reproductive females than those of the following two years. Poole and Catling (1974) noted that data from penned animals held near Canberra paralleled those from the Mt. Hope collections of wild animals during the breeding season.

In Kinchega National Park in western New South Wales, Shepherd

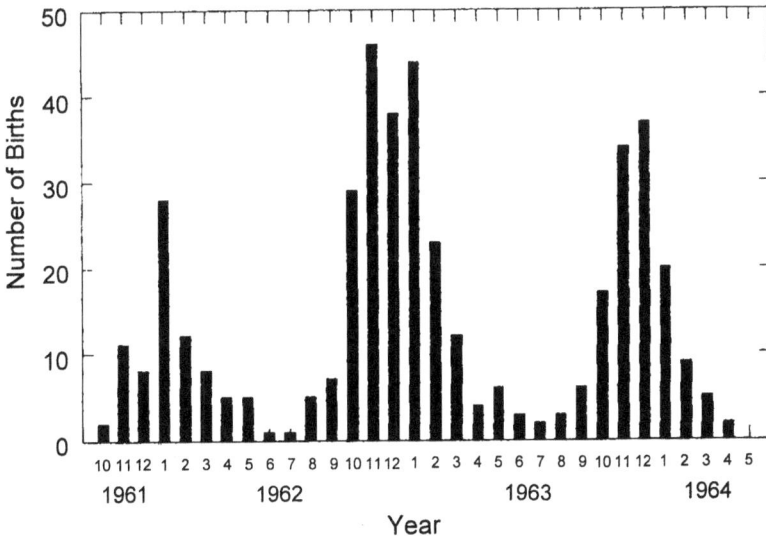

Figure 8.6. Reproduction of western grey kangaroos over time near Mt. Hope. Data from Poole and Pilton (1964).

(1987) found that western grey kangaroos showed strong seasonal breeding over four years, although breeding was nearly absent in the drought year of 1982 (fig. 8.5). The results of Poole and Pilton (1964) (fig. 8.6), Shepherd (1987) (fig. 8.5), Norbury et al. (1988) for northern Victoria, and Arnold et al. (1991) for Western Australia strongly suggest that western grey kangaroos have seasonal breeding periods.

How, then, can the apparent evidence for both hypotheses 2 and 3 be made compatible? First, seasonal breeding can be shifted forward or backward a month or so by local conditions in most temperate-zone ungulates (e.g., McCullough 1969, 1979), and the same may be true of kangaroos. Note that in Shepherd's (1987) data, the peak occurred somewhat earlier in 1981 and 1984 than it did in 1983 (fig. 8.5). And, of course, both Poole and Pilton's (1964) and Shepherd's (1987) data indicate that breeding was less prevalent in some years due to local conditions, although this may have been due more to mortality of pouch young rather than the number of females breeding. Anestrus in the nonbreeding season (Poole and Pilton 1964) and failure to conceive at first mating in the breeding season give grey kangaroo females considerable physiological control over when reproduction will occur.

Note that the timing of breeding of western grey kangaroos at Yathong (fig. 8.4) was more similar to that at Kinchega National Park (fig. 8.5) than at Mt. Hope (fig. 8.6), although the latter location was much nearer to Yathong. Although Mt. Hope is within 70 km, the area east of Yathong shows a marked increase in rainfall (fig. 2.1). Also, Mt. Hope is on a plateau at higher elevation and, in general, has more woodland habitats than Yathong. Thus, different timings of the breeding season may be present in the two areas, with Yathong being more typical of populations much farther west. This variation in timing between years may have been sufficient to shift the observed peak at Yathong much earlier, in response to the favorable rainfall in 1985.

The concurrent peak in breeding in both red and grey kangaroos at Yathong (fig. 8.4) might be entirely due to the favorable local conditions. Shepherd's (1987) data from Kinchega National Park (fig. 8.5) show red kangaroo reproduction as largely linked to rainfall, and although peaks occurred, substantial numbers of females reproduced at other times as well, as also observed at Yathong. Similarly, although red females showed estrus all year long in Sharman and Calaby's (1964) colony, most females were in estrus during a peak season and fewest during a low season.

Embryonic diapause in red kangaroos allows rapid response of reproduction to favorable environmental conditions. Some eastern grey females show embryonic diapause as well (Poole and Catling 1974), and this may allow greater flexibility in times of breeding, depending upon local conditions. Western grey kangaroos, in contrast, do not show embryonic diapause (Poole and Catling 1974), and they dropped to virtually zero reproduction between breeding seasons at Yathong (fig. 8.4), Kinchega (fig. 8.5), and Mt. Hope (fig. 8.6).

Although eastern grey kangaroos seemed to have a less specific breeding season, the data from Yathong were few; based upon obvious pouch young (fig. 8.2), their reproduction also seemed to drop to near zero, more similar to western grey than red kangaroos, but still intermediate overall between those two species.

9

FEEDING ECOLOGY

The feeding ecology of kangaroos should be viewed from the perspective of their digestive capabilities. Like ruminants, kangaroos depend upon microbial fermentation for digestion (McIntosh 1966; Hume 1978, 1982; Dellow 1982; Dellow et al. 1983). Hume (1984) has reported in detail on the kangaroo digestive system. Kangaroos have a tubular forestomach that allows them to pass food more rapidly, and consequently, to consume and process more poor quality forage than ruminants of comparable size. Kangaroos also have an enlarged caecum and colon that allow greater hindgut fermentation than ruminants.

Much like desert-adapted ruminants, the evolution of kangaroos' ability to survive on poor forage includes recycling in saliva of phosphates, sulphur, and urea (i.e., nitrogen) (Forbes and Tribe 1970; Beal 1987, 1989). Presumably, as with ruminants, microbial digestion allows detoxification of secondary compounds, although this topic is less well studied in kangaroos. A notable exception is the remarkable ability of western grey kangaroos to detoxify sodium fluoroacetate (Oliver et al. 1977; King et al. 1978), a deadly poison, also called Compound 1080, used on rodents and canids. It is a colorless, odorless, tasteless compound to which the target organisms cannot develop an aversion. It occurs naturally in leguminous shrubs of the genera *Gastrolobium* and *Oxylobium* in Western Australia, where the western grey kangaroo evolved resistance to it. Western greys from other parts of the range have resistance as well, which bespeaks their speciation in the isolation of southwestern Australia (Oliver et al. 1979). Moreover, neither the red kangaroo (King et al. 1978) nor the eastern grey (Oliver et al. 1979) shows resistance to Compound 1080.

This resistance in western grey kangaroos is a part of their overall capacity to deal with poor quality forages. As noted by Prince (1976), the digestive efficiency of western greys is greater than red kangaroos. Thus, although they are less capable of gathering high-quality green forage (Short 1985, 1987), they are better at digesting low-quality forage than the more selective red kangaroo. Prince (1976) assumed that eastern and western grey kangaroos were similar in their digestive capacities; but from our observations, we believe that western greys are better able to utilize poor-quality forage than eastern greys.

In contrast to ruminants that chew a cud (by regurgitating and masticating a bolus of vegetation previously swallowed, largely unchewed), kangaroos do not chew a cud. However, as reported by Barker et al. (1963), kangaroos do regurgitate. We, too, observed this behavior at Yathong, and we agree that these regurgitations are not chewed, as in ruminants, but usually are swallowed after desultory chewing. Prior to swallowing, kangaroos appear to grind forage, when first cropped, to a fine particle size. Why then do kangaroos regurgitate? Ian Hume (personal communication) suggested that regurgitation may stimulate salivary flow to maintain an appropriate pH in the foregut. However, further work is needed to more thoroughly address this question.

Overall, kangaroos are highly specialized to deal with great fluctuations in food quality associated with rainfall and drought patterns in Australia. Their efficiency exceeds that of ruminants of comparable size, and this fact helps to account for their success in the face of competition from exotic herbivores such as sheep, goats, and European rabbits.

Feeding at Yathong

Two statements sum up kangaroo feeding ecology at Yathong. First, kangaroos eat grass and palatable species of forbs (fig. 9.1). Numerous authors have reported that, by preference, kangaroos eat green grass and green palatable forbs (Griffiths and Barker 1966; Griffiths et al. 1974; Ellis et al. 1977; Barker 1987; Edwards et al. 1996; Moss and Croft 1999). Moss and Croft (1999) found that red kangaroos lost body condition following disappearance of green feed with a three-month lag at Fowler's Gap. Second, variation in rainfall results in dramatic fluctuations in availability of grasses and forbs across time and space. The effectiveness of rainfall, in turn, is modified by temperature (which slows or speeds drying and plant physiological processes) and fire

Figure 9.1.   Adult female red kangaroo feeding on new green grass.

(which consumes dry matter and thereby eliminates mulch and its moisture-conserving influence).

When conditions are favorable, there is a bountiful—indeed, a superabundant—supply of high-quality food. When this occurs, kangaroos are enveloped in a carpet of deep green vegetation that blankets the landscape from horizon to horizon. Then, it is as if nature has run amok with splendor. Paper daisies flush the scene with white and then gold, and flowers of Patterson's curse flood the low-lying swales with blue, looking from a distance like water. At these times, kangaroo feeding does not even make a dent in the biomass. Levels of consumption are so small relative to supply that conventional methods of measuring herbivore removal of plant material fail, because consumption is less than the error in the method.

Favorable conditions do not last long—feast is quickly followed by famine. Soon the landscape dries and kangaroos settle like liquid into the swales where the last remnants of green feed persist. Henceforth consumption is easily measured, as all of the remaining greenery in these patches is eaten to the ground. And then it is gone. Color disappears from the landscape, as it turns harsh and gray. Kangaroos have no choice but to eat the dry remnants of nature's previous splurge. We think the dried vegetation at Yathong is probably better than that at Fowler's

Gap, where Moss and Croft (1999) found red kangaroos to lose condition, and the two species of grey kangaroo probably do better than reds on this feed. Still, there is little doubt that compared to green feed, dried feed is substantially inferior in quality and digestibility.

Abundant at first, the standing dry residue is gradually reduced. It is consumed by kangaroos, rabbits, goats, and veritable armies of ants and termites. It is also blown away by the winds. If a fire starts in the dry grasslands, it spreads quickly and is virtually unstoppable. It will roar through the dry herbaceous tinder until the fuel in its path is gone — or until the rains come. Some of the bunch grasses, predominantly *Stipa* at Yathong, cure in a state that burns like tinder. This is their way of preventing the establishment of woody plants in their domain. Woody fuels, too, are so dry and burn so well that they are consumed completely by fire, as if in a furnace. Wooden fence posts simply disappear. Dead trees or branches on the ground burn so completely they leave a "silhouette" of white ash in the exact shape of their previous form, down to the smallest twigs. Often, no trace of charcoal is left to indicate their previous carbon structure.

During drought, kangaroos literally are "kept hopping" trying to maintain an adequate diet. Kangaroo mortality begins with the very old and the juveniles (Dawson 1995). Then, if the rains still do not come, as happened in 1982, the dry herbaceous layer disappears, the ground is bare, and any woody stems within reach are eaten (fig. 2.8). All kangaroos are in jeopardy. During this study, the rains came in time. No mortality occurred other than that due to old age or chance.

Yet clearly, kangaroos shifted their food habits and feeding areas to obtain the best foods available. As Barker (1987) noted, diets vary depending on what is available. In times of shortage, both red and western grey kangaroos ate substantial quantities of shrubs at Kinchega National Park (Barker 1987). On Yathong, however, the low-growing shrubs typical of Kinchega are rare, and shrubs of any kind are uncommon. Of 7500 points in the randomly placed step-point vegetation surveys (chapter 3), only 18 (0.24%) fell within shrub canopy cover. What little browsing we found by examination of shrubs was attributable to feral goats. Of the numerous direct observations of feeding by kangaroos, only three cases of feeding on woody plants were recorded. On November 20, 1985, a western grey was observed nibbling on the foliage of a belah tree. This tasting was desultory, and the animal quickly moved off to feed on grasses. On January 7, 1986, a red kangaroo was observed eating cypress-pine, and on January 10, 1986, a western grey was observed feeding on cypress-pine as well. These latter cases involved a moderate intake; but overall, use of woody species was so rare that total intake of woody species would have been little more than a trace.

Feeding Site Characteristics

From the systematic surveys in which feeding site characteristics were recorded for each kangaroo observed, it was apparent that there were relative differences in site selection by the three species (figs. 9.2 to 9.4). All three species chose sites with green plant growth whenever available, but red kangaroos selected a higher proportion of sites with low green growth (fig. 9.3). By comparison, both western and eastern grey kangaroos occupied more tall dry standing grass sites that contained green growth at the ground level (figs. 1.2 and 1.3). Use of such sites was most prevalent in eastern grey kangaroos (fig. 9.2), with western

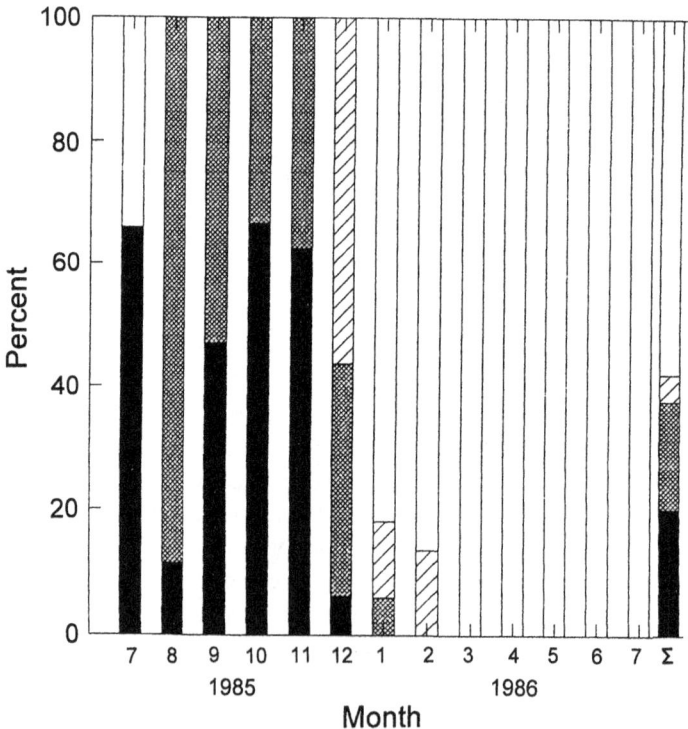

Figure 9.2. Feeding sites of eastern grey kangaroos observed in systematic surveys; low green forage (solid bars), green forage under tall dry grass (cross-hatched bars), green forage in isolated concentration areas (slashed bars), and dry forage (open bars). The Σ column on the right indicates the overall average.

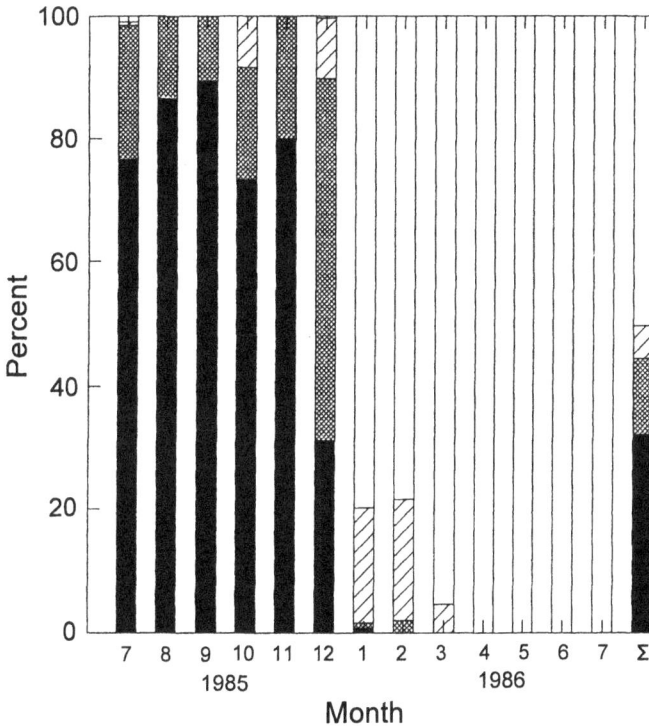

Figure 9.3. Feeding sites of red kangaroos observed in systematic surveys; low green forage (solid bars), green forage under tall dry grass (cross-hatched bars), green forage in isolated concentration areas (slashed bars), and dry forage (open bars). The Σ column on the right indicates the overall average.

greys (fig. 9.4) being intermediate between easterns and reds. Kangaroos observed feeding in these sites could easily be differentiated as feeding on the ground-level green growth, in which case their heads were down and bodies bent over, or on the dry tall grass, in which case their bodies were more erect and their heads in clear view.

When kangaroos consumed dry feed, it was predominantly the seed stalks of *Stipa*, which were clearly visible protruding from their mouths as they chewed (fig. 1.3). Jarman (1994) reported feeding on seed heads by all three species as well.

Although red kangaroos were at times seen eating dry seed heads during dry feed periods, they were never seen eating dry feed when green feed was available. Western grey kangaroos, by contrast, ate considerable amounts of dry *Stipa* seed heads, even when green feed was

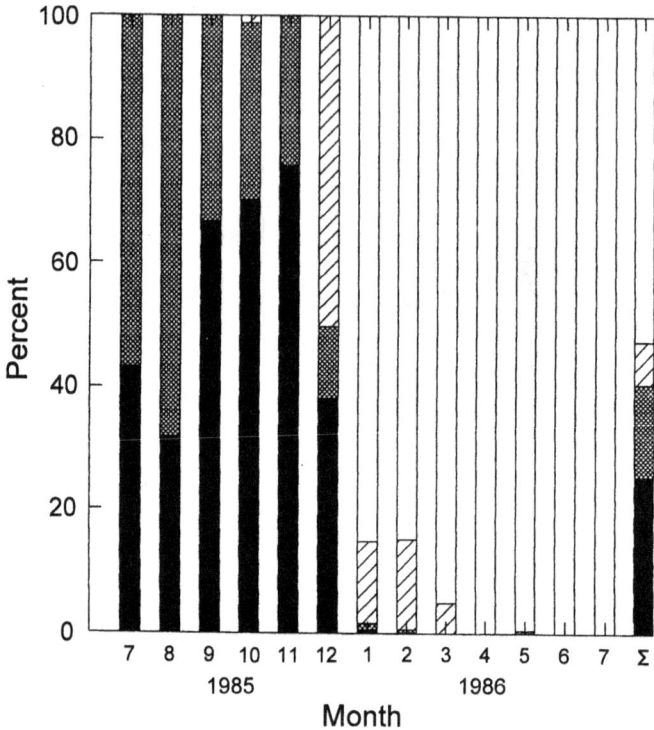

Figure 9.4. Feeding sites of western grey kangaroos observed in systematic surveys; low green forage (solid bars), green forage under tall dry grass (cross-hatched bars), green forage in isolated concentration areas (slashed bars), and dry forage (open bars). The Σ column on the right indicates the overall average.

readily available. Obviously, western greys consumed dry feed in considerable amounts by choice. Because of their alarmable nature, we obtained too few direct observations of feeding by eastern grey kangaroos to positively characterize their use of dry feed when green feed was available; our impression was that they were similar to red kangaroos in selection of green feed as reported for eastern New South Wales by Jarman and Phillips (1989). However, more of the green feed taken at Yathong was at ground level in stands of dry grass, whereas Jarman and Phillips reported eastern greys moving to burned areas to feed. Southwell and Jarman (1987) likewise reported congregations of eastern greys on burned areas.

Vegetation was assessed by step-point methods on randomly selected areas in burned and unburned areas in the grassland and bimble box vegetation types (table 9.1). In both types, burned areas had significantly

Table 9.1. Percent vegetative cover by layer obtained for burned and unburned vegetation types and two kangaroo feeding concentration areas (both in the grassland vegetation types), as determined by the step-point method.

	Grassland		Bimble box		Pine-box	Pine Assoc.	Hordeum Conc. Area	Thistle Conc. Area
	Burned	Unburned	Burned	Unburned	Unburned	Burned	Unburned	Burned
No. transects [1]	38	8	6	6	4	7	5	6
Herbaceous layer								
Green pick	16.6	13.3	3.7	1.8	6.5	16.1	34.4	15.0
Green bunch	1.6	14.5	0.5	0.8	2.5	0.9	4.6	0.8
Dry bunch	1.6	10.2	1.3	5.7	13.5	3.2	11.4	1.0
Forb	25.6	23.5	6.3	8.2	11.0	28.4	15.6	21.8
Litter	9.1	14.5	29.8	55.8	25.8	14.0	19.0	16.3
Bare ground	45.5	24.0	58.4	27.7	40.7	37.4	15.0	45.1
Shrub layer	0.5	0.2	0.2	0.3	0.0	0.5	4.6	0.2
Tree layer	<0.1	<0.1	25.3	25.6	33.0	8.4	<0.1	<0.1

[1] 100 step points/transect.

different herbaceous cover than unburned areas (grassland chi-square = 535.07, df = 5, $P < 0.001$; bimble box chi-square = 134.61, df = 5, $P < 0.001$). In both types, green pick was higher in burned areas, and green bunch grasses higher in unburned areas. Dry bunch grasses were sparse in burned areas because they were largely consumed by fire, as was litter (table 9.1). Bare ground, thus, was much higher in burned than unburned areas.

Step-point transects revealed a high proportion of herbaceous cover comprised of forbs in all vegetation types (table 9.1). Usually, forbs exceeded new green grasses (except in a *Hordeum* concentration area), and often exceeded the combined green pick and green bunch grass (table 9.1). Although we speak of kangaroos seeking green grasses, it is apparent that forbs, especially blue crowfoot, comprise a large part of the kangaroo diet.

Green growth of plants occurred first in areas without a dry grass stand (predominantly burned areas), but without shade from standing vegetation and litter to insulate the soil from the sun's rays, such green feed dried up earliest in the season (chapter 2). The importance of litter in modifying soil moisture conditions, and consequent plant growth, has been documented previously in Mediterranean climates by Heady (1956).

Because most of the burned area at Yathong resulted from backfires purposefully set along roads, the contrast between like sites on opposite sides of roads was readily apparent. Total plant production on the burned side, although not measured quantitatively, was obviously much less by visual comparison. Green growth was relatively short and wispy in burned areas because of the soil moisture regime. Production was approximately twice as great under the dry grass stands as on the areas exposed by burning, and green growth persisted here much later in the season.

Seasonal Changes in Feeding Sites

Sequentially, during the first half of this study at Yathong, there was a surge of plant growth in July, that dried in October, and then large production of green growth thereafter in 1985 (fig. 2.9). During August, the two species of grey kangaroos shifted away from burned areas to feed on green feed under tall grass in unburned areas (figs. 9.5 and 9.7). Red kangaroos continued to favor burned areas through this period (fig. 9.6), and the relative preference of reds for feeding sites in burned areas continued throughout the study (fig. 9.6). Overall, 80.7% of observations

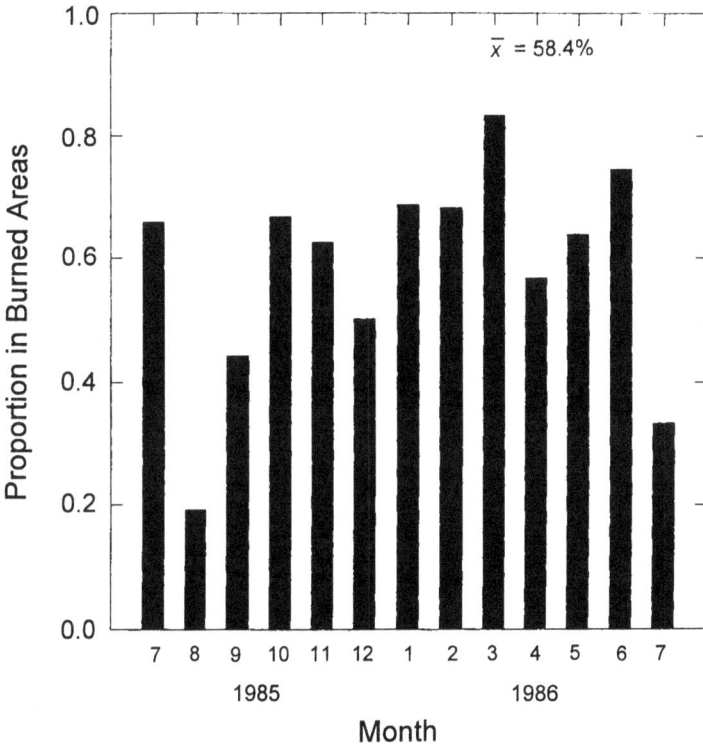

Figure 9.5. Proportion of observations of eastern grey kangaroos in systematic surveys that occurred in burned areas (December 1984 burn) as opposed to unburned areas that contained a tall, dry grass stand.

(N = 2965) of red kangaroos were in burned areas even though only 72% of the systematic survey routes sampled was burned. Eastern grey kangaroos were much more variable in their use of burned areas than red kangaroos (fig. 9.5), partly due to the grey's much smaller sample size (N = 361). Generally, eastern greys fed more heavily in burned areas when green feed was available and used unburned areas more heavily during the dry periods. Overall, 58.4% of observed eastern grey kangaroos were in burned areas, or less than the area burned. Western grey kangaroos used burned areas 68.8% of the time, intermediate between the other two species and, due to their large sample size (N = 2089), the monthly variance was less (fig. 9.7).

As green feed became less available in burned areas in December 1985, kangaroos of all three species began to shift to unburned areas where green feed persisted under the tall dry grass stands (figs. 9.2 to 9.4). By January 1986, green feed was available in only a few sites where soil moisture persisted longer, such as swales or shallow drainages. By April 1986, all green feed was gone, and all kangaroos were eating dry feed, mainly *Stipa* seed heads. Dry feed conditions continued unbroken through mid-July 1986. All feeding sites, therefore, were dry (figs. 9.2 to 9.4). Eastern grey kangaroos did not use the same feeding concentration sites as red and western grey kangaroos, which tended to be in open areas. Eastern greys confined their activities more to creek drainages, which were lined with strands of bimble box trees. Some green feed did occur in the dry creek bottoms, and this was consumed by eastern grey kangaroos at times when the other two species were concentrating in open sites.

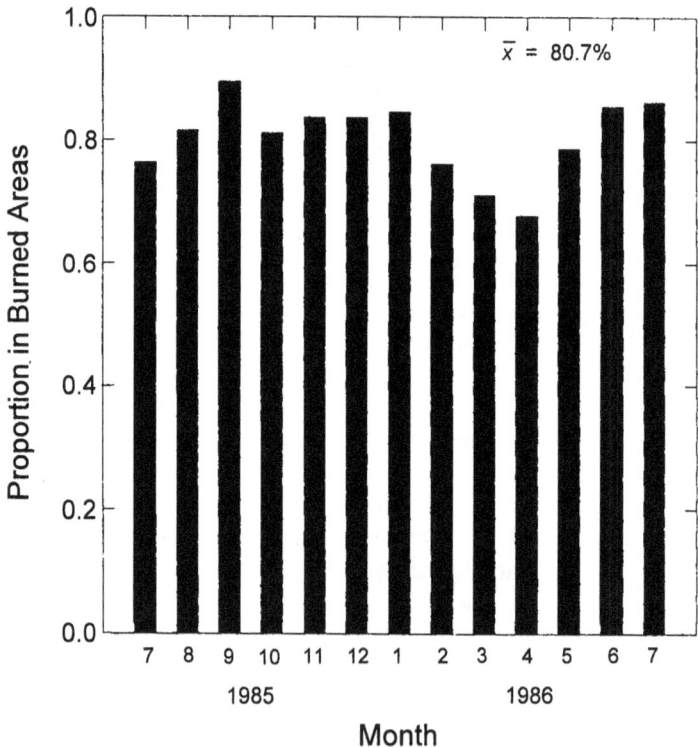

Figure 9.6.   Proportion of observations of red kangaroos in systematic surveys that occurred in burned areas (December 1984 burn) as opposed to unburned areas that contained a tall, dry grass stand.

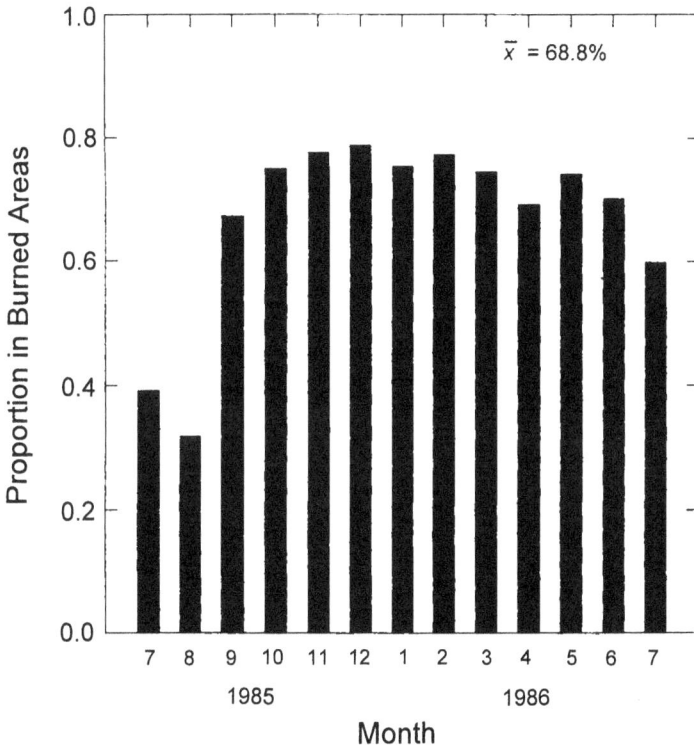

Figure 9.7. Proportion of observations of western grey kangaroos in systematic surveys that occurred in burned areas (December 1984 burn) as opposed to unburned areas that contained a tall, dry grass stand.

Feeding Concentrations

The earliest feeding concentrations of red and western grey kangaroos occurred on former holding paddocks for sheep, where the effects of manure and the churning of hooves had resulted in greater soil formation. In such old paddock areas, and other isolated sites where hay or other material had been stacked, barley grass (*Hordeum,* an annual) germinated and grew most rapidly (fig. 2.12). An "A horizon" of organic matter was present in the soils of these sites. Despite removal of fencing and other cultural effects, these sites were apparent, based on their soil and vegetation characteristics (fig. 2.12). Although obvious, these areas were small, covering only a few hectares total on Yathong. The organic layer is absent over most of Yathong—soil color and texture being unchanged over several meters' depth—presumably because dry conditions disfavor decomposition and frequent fires, droughts, and termites and ants eliminate litter and mulch.

At greenup, bright green patches on paddock areas contrasted strongly with the pale green over most of the area, and red and western grey kangaroos gathered there and fed heavily. As measured by step-point transects, these areas were significantly different from other grassland areas in herbaceous cover (table 9.1; chi square = 118.31, df = 5, $P < 0.001$). Green pick grasses were twice the cover of either burned or unburned areas. Litter was statistically higher, and bare ground lower. Another notable characteristic of the *Hordeum* sites was the relatively high shrub cover of black bluebush (*Maireana pyramidata*). The step-point transects in the *Hordeum* concentration area sampled showed a shrub layer canopy cover of 4.6% (23 hits out of 500). Apparently, this shrub occupies disturbed sites on Yathong, for other small stands occurred around Shearers' Quarters where sheep had been concentrated in paddocks. We never observed kangaroos feeding on black bluebush, although Barker (1987) reported this for red kangaroos at Kinchega, and we believe it is consumed at Yathong during severe droughts. Feeding concentrations of kangaroos in *Hordeum* areas persisted for several weeks until green growth became widespread, and kangaroos dispersed broadly over Yathong. Once barley grass began to head out and dry, these areas were no longer used for feeding.

The main plant species consumed by kangaroos at Yathong follow a seasonal succession based upon phenology. For example, barley grass was first to green up following rains. With general greenup and scattering of kangaroos, dominant grasses such as *Stipa* then became the predominant forage species. Although a number of forbs were palatable and consumed regularly in the green stage, it seems grasses were preferred, always showing the greatest removal where both were in approximately equal abundance.

As vegetation dried across Yathong, feeding concentration areas again appeared in mid-November 1985. The first concentration areas were shallow swales, where water drained to the bottom of shallow basins, and green feed persisted after most of the countryside had dried considerably. These sites were unique for a number of reasons other than concentrations of feeding kangaroos. In night spotlight surveys, we found them to be 4 to 6 °C warmer than adjacent upland areas, even though the elevational difference was less than a meter. Warm air settled there much like water in shallow pools. White moths and other insects formed clouds that obscured vision. When we stopped the vehicle there during night surveys, we would place the spotlight facing upward on the roof to direct moths away from the note-recorder. The masses of moths in the beam rose in the sky like an alabaster column. Another feature of these sites is that crickets chirped in a continuous chorus. The concentration of insects attests to the residual moisture in the soil, and this moisture supported the residual green feed that attracted kangaroos.

Although both red and western grey kangaroos used the same concentration areas, they sorted themselves by microhabitat. Red kangaroos fed near the center (or bottom) of these swales, primarily on late-germinating grasses such as panic grass. Western grey kangaroos fed on the periphery of the swales on *Stipa* grass bunches that still contained green growth. These sites showed pronounced grazing removal due to concentrated feeding by kangaroos. Whereas most of Yathong was grazed so lightly that it was hard to find visual evidence of removal, here plants were eaten down to ground level. Panic grass adopted a prostrate life form in response to red kangaroo grazing while *Stipa* appeared clipped as western grey kangaroos took the mixed green and dry tops. Many *Stipa* bunches were reduced to stubble. The complete lack of observed agonistic encounters between the two species, despite their close proximity in concentration areas, suggests that feeding niche separation rather than competitive exclusion was the basis for the different microsites used for feeding.

By mid-December 1985, swales had completely dried and red and western grey kangaroos gathered in the last remaining green feed patches, shallow drainages. These areas contained tall stands of dried thistles (predominantly saffron thistle) that ranged from one-half to one meter in height. Kangaroos often were completely concealed while feeding, and were observed only when their heads popped out above the dried thistle thickets to watch the observer. Feeding site examination, a most prickly and unpleasant task in the thistles, showed that both red and western grey kangaroos were feeding on green growth at ground level—predominantly panic grass and sow thistle—but also some Russian thistle, another late-season plant. The latter remained green in other areas as well, but it seemed to be not very palatable, showing minor grazing effects at the same sites where most of the panic grass and sow thistle had been consumed. Most of these feeding concentration areas had been burned, and one of the most heavily used by kangaroos was assessed by step-points (table 9.1). It contained substantial green pick of grasses and forbs, despite herbaceous material on most of Yathong being dry. These sites had almost no bunch grass growth because of shading by the "overstory" of thistles, and much of the ground was bare. Most of the litter encountered was recently dropped from the dry thistle layer (table 9.1).

Total production of green plants under the thistle stand was small and sparsely distributed. Given that kangaroos tend to avoid thistle patches under other circumstances, feeding in such situations must cause discomfort. When these last green herbaceous plants dried out in late February, kangaroos scattered and fed in the uplands, primarily on the abundant and easily obtained dried *Stipa* seed stalks. Moreover, occasional small rains produced sparse flushes of green growth that defied our detection, but were apparently found and eaten by red kangaroos, as evident from changes in the form of their droppings. Generally, kangaroo droppings become drier as green feed disappears and water is conserved (Dawson 1995). Red kangaroo droppings, however, became more moist in response to small rains, demonstrating their adeptness at locating and consuming ephemeral emergence of green growth.

Form of Feces

In this study, 10 samples of dropping of each species of kangaroo were collected monthly for nitrogen content determination, and the sample collection process proved instructive about green feed availability. It soon became obvious that kangaroos defecate primarily where they feed. Searches for fresh dropping around resting sites or travel paths proved largely fruitless. However, when kangaroos were observed actively feeding, the site invariably yielded fresh droppings.

During periods of abundant green feed, the droppings of the three species could be easily distinguished by physical form. We did not encounter the problem reported by Barker (1987) of red and western grey kangaroo droppings being so amorphous as to be indistinguishable. Although more amorphous than under dry feed conditions, droppings during green feed periods still could be identified to species. Red kangaroo droppings were lighter green in color, and tended to be fusiform, long, and tapered at either end. Western and eastern grey kangaroo droppings were darker and browner in color. Western grey droppings had approximately equal dimensions in all axes, and tended towards cubical form. Eastern grey droppings, by contrast, tended to be pressed together and flattened in the manner of a stack of thick poker chips. As dry feed conditions persisted, however, these distinctions between species declined, as form and color converged. Even red kangaroo droppings became indistinguishable from the two species of grey kangaroos, and correct identification required direct observation of defecation, or exclusive occupation of a feeding site by one species, and extremely fresh droppings in either case.

During the long dry period (March through mid-July 1986), changes in the form of red kangaroo droppings occurred about a week after light rains fell. These droppings assumed the more typical form and color of the red kangaroo that clearly showed they were finding and consuming green feed that we could not locate. No feeding concentrations of red kangaroos were identified at these times, nor were there recognizable changes in the droppings of the grey kangaroos. Obviously, red kangaroos are extremely adept at finding and gathering the sparsest scattering of green growth.

Fecal Nitrogen

Dropping samples, 10 for each species of kangaroo, were collected monthly (except for August 1985 when only eight were obtained for eastern greys) from August 1985 through September 1986 for analysis of percent fecal nitrogen to assess food quality. However, fecal nitrogen has had a checkered history as an index to diet quality. Some ungulate researchers found a good correlation between diet quality and fecal nitrogen (Raymond 1948; Lancaster 1949; Hebert 1973; Stallcup et al. 1975; Holloway et al. 1981; Leslie and Starkey 1985; Mubanga et al. 1985). Similarly, Brown and Main (1967) reported a close correlation between feed and fecal nitrogen in the euro. Arnold et al. (1991) reported fecal nitrogen results for western grey kangaroos that tracked feed conditions in a general way. Brown (1968) reported that kangaroo fecal nitrogen levels were lower than those of ungulates, correlating with their lower nitrogen requirement than ungulates.

Others criticized the approach because fecal nitrogen is derived from sources other than diet (Mould and Robbins 1981), because secondary plant metabolites in some forages artificially raise fecal nitrogen (Robbins et al. 1987; Robbins 1983), or because of some statistical difficulties (Hobbs 1987).

Despite these limitations, many researchers have found fecal nitrogen to be a useful method to track broad seasonal changes in diet quality (Wehausen 1980; Kie and Burton 1984; Renecker and Hudson 1985; Hodgeman and Bowyer 1986; Beier 1987; Goldsmith 1988; Massey et al. 1994), establish differences between sites (Seip and Bunnell 1985; Gogan and Barrett 1994), or differences between species (Leslie and Starkey 1985, 1987; Goldsmith 1988; Berbach 1991). This seems particularly true of comparisons based upon herbaceous diets in which the tannins found in woody species are not a confounding factor, as was the case with the kangaroo species compared here.

Over all samples, red kangaroos had significantly higher percent fecal nitrogen ($n = 140$, $\bar{x} = 1.239$) than either eastern grey ($n = 138$, $\bar{x} = 1.043$; $t = 5.41$, $P < 0.001$) or western grey kangaroos ($n = 140$, $\bar{x} = 1.102$; $t = 3.69$, $P < 0.001$). There was no significant difference between eastern and western grey kangaroo fecal nitrogen ($t = 1.74$, $P = 0.08$), although the low P-value suggests a tendency for the western grey to be higher; the difference may not be significant simply due to an insufficient sample size.

Fecal nitrogen of all three species showed seasonal patterns that were similar (fig. 9.8) and all showed significant differences by month

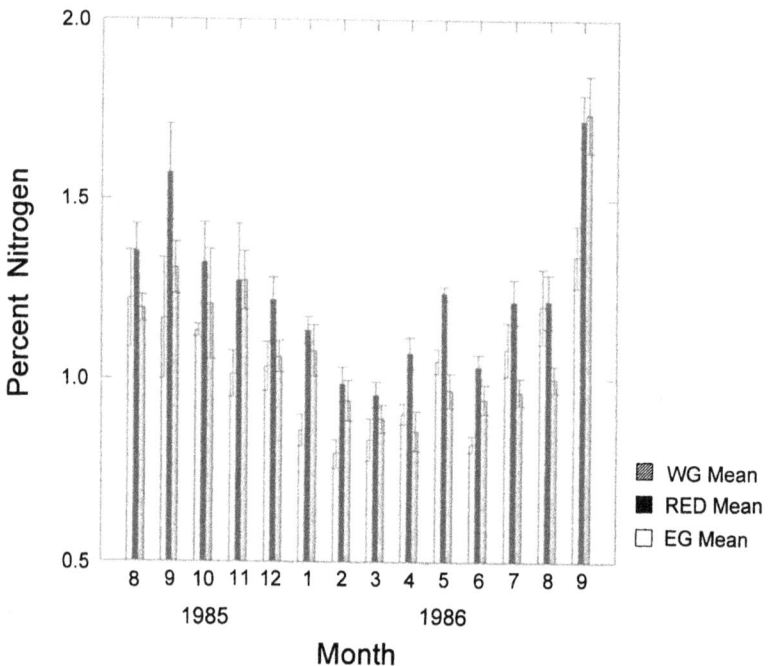

Figure 9.8.   Percent fecal nitrogen in fecal samples by month for the three species of kangaroos.

(eastern grey $F = 13.33$, df $= 13$, $P < 0.001$; red $F = 15.578$, df $= 13$, $P < 0.001$; western grey $F = 15.075$, df $= 13$, $P < 0.001$). The seasonal trends in fecal nitrogen matched the pattern of green feed availability (fig. 2.9) with high values during green feed availability and low values during dry feed periods. Thus, fecal nitrogen seemed to track food quality over time reasonably well. This agreed with the fecal nitrogen results on western grey kangaroos in southwestern Australia reported by Arnold et al. (1991). However, our fecal nitrogen results for western greys, on average, were only about two-thirds of what they reported. We don't know if this difference is real, or perhaps due to differences in analytical technique.

As expected, red kangaroos, the species best adapted to locating and consuming green feed, had the highest fecal nitrogen under all conditions (fig. 9.8), except for September 1986 when western grey values were virtually equal. The relationship between eastern grey and western grey kangaroos, however, was confused. Whereas some results suggested that eastern grey kangaroos varied more over time—being higher during green feed periods and lower during dry feed periods—this trend broke down with the September 1986 values.

The extremely high fecal nitrogen results for western greys for September 1986 (fig. 9.8) raised questions as to whether these samples might have been mishandled or mislabeled. However, with the assistance of David Spratt of CSIRO, Division of Wildlife and Rangelands Research in Canberra and Ian Beveridge of the Institute of Medical and Veterinary Science in Adelaide, subsamples of these droppings were tested for *Eimeria purchasi*, a coccidial parasite (Mykytowycz 1964) that occurs only in western grey kangaroos (David Spratt, personal communication). Their results showed that every sample containing *Eimeria* was from a sample labeled as western grey kangaroo, thus confirming that no errors had been made, and that our results were correct. Overall, given these ambiguous results, it is probably best to consider the trends between the two species of grey kangaroos as too similar to differentiate without a much larger sample.

Differences by sex could not be explored in this study because too few of the dropping samples could be assigned to sex. However, in nearly every case where the sex producing the sample was identified (a total of 30, 10 of which were from males), the fecal nitrogen of males was lower than females of the same species. Given the large sexual dimorphism of kangaroos, this result was not unexpected, and sex almost certainly was one variable contributing to variance in fecal nitrogen.

Discussion of Feeding Habits

Previous reports of the red kangaroo being a green feed specialist (Bailey et al. 1971; Newsome 1977a; Low et al. 1973) were borne out by this study. Their propensity to gather in isolated pockets of green feed, their greener and more moist droppings, and their higher fecal nitrogen all point to their ability to seek out and consume green growth. Furthermore, studies by Short (1985, 1986, 1987) showed that red kangaroos were more efficient than western greys in gathering any kind of feed.

The two species of grey kangaroos, by contrast, consumed more coarse and dry material. Western greys consumed dry *Stipa* seedheads in considerable quantity, even during periods when green feed was widespread and abundant. Although the differences were small, and at times the data were contradictory, it was our impression that, overall, western grey kangaroos eat a slightly higher quality of diet than do eastern grey kangaroos. They also sought out green feed patches when vegetation dried up, but tended to feed on the margins of such areas, whereas the red kangaroos occupied the centers. They used burned areas more than eastern grey kangaroos, and they had higher, but narrowly not significant, levels of fecal nitrogen.

Anatomically, both red and western grey fed easily on green pick at the ground surface. The eastern grey, by contrast, appeared awkward in the posture required to gather green pick. This agrees with their reputation as a coarse grass feeder. Differences in body form among the species result in somewhat different postures during feeding. When feeding on low grasses or forbs, red kangaroos appear more "comfortable," because their hindlegs are placed farther to the rear. Their knees do not reach their back line in profile, and their tails assume a relaxed S-shape (fig. 9.9). The western grey kangaroo more nearly approximates the posture of the red than does the eastern grey. The western grey appears to crouch more easily than the eastern grey, and appears to have relatively shorter hindlegs, whereas in the eastern grey, the hindlegs appear too long for comfortable feeding at ground level. The knees are raised above the profile of the back, and the tail assumes a serpent-like, flexed S-shape (fig. 9.9).

Although Prince (1976) suggested that digestive efficiencies of eastern and western grey kangaroos were similar, we believe that they are functionally different in their feeding ecology. Eastern greys seem to be much more strongly confined to grasses. This species evolved in the eastern mountains where grasses compose 99 percent of the diet (Jarman and Phillips 1989). Although eastern greys select green feed if available

Figure 9.9. Comparison of feeding posture of red kangaroos (above) with eastern grey kangaroos (below) when feeding on new growth. The red kangaroo seems better proportioned to feed at the ground surface. The pictures were taken within a few minutes of each other about 100 m apart in burned pine-box vegetation type.

(Jarman and Phillips 1989), as they did on Yathong in this study, such feed is not available very frequently. Eastern greys usually are forced to feed on the coarse tall grasses that constitute the great bulk of their forage on an annual basis (Griffiths et al. 1974).

Western grey kangaroos are much more flexible in their use of forages. Although not nearly as efficient in finding and gathering green feed as the red kangaroo, they seek out green grasses and forbs when available. However, our studies showed them to be better at finding and gathering green feed than eastern greys. Western greys are best at utilization of shrubs (Barker 1987). Their tendency to occupy low-quality habitat such as mallee, and detoxify secondary compounds as toxic as sodium fluoroacetate demonstrates their ability to survive in low-quality habitats (i.e., those dominated by trees and shrubs characteristic of much of their distributional range in southern and western Australia). On Yathong, their capacity to use low-quality forages was expressed not so much by consumption of shrubs as by consumption of large amounts of dried seedheads of *Stipa,* even when green feed was abundant.

10

HABITAT AND NICHE RELATIONSHIPS

Animals do not use landscape in a random fashion. Rather, they differentially choose to live their lives in those parts of the landscape that are more likely to satisfy their life prerequisites. We say that this is the habitat of the species, which implies that natural selection has resulted in adaptation of the species to the characteristics of those parts of the landscape occupied. Habitat refers primarily to vegetation, but of course, vegetation is the product of climate, topography, soils, and other factors; so, by extension, habitat is a broad representation of the characteristics of the landscape in the occupied area.

Niche refers to the way in which a species utilizes its habitat. Thus, by occupying different niches, an array of species can coexist in the same habitat as a complex community. Habitat has been likened to an animal's address whereas niche is its occupation.

Macropods are known to sort out by habitat characteristics (Pople 1989) and this is true of kangaroo species as well (Taylor 1985; Southwell 1987; Coulson 1990). Yathong Nature Reserve contained a mosaic of habitat types and gradients, which allowed kangaroos to select preferred conditions.

Habitat Use

Habitat use was calculated from radio telemetry locations for each species by sex. Each location was assigned to the major habitat types as derived from the type map produced by CSIRO (chapter 2, fig. 2.10).

Because type blocks were large in relation to probable error of telemetry location, few misplacements were likely, and those that may have occurred should be equal across species and sex comparisons. The radio telemetry data have the advantage of not being subject to observer or visibility bias; and they were taken at all times of day and night. Furthermore, sampling effort across all individuals was equal, so the differences in sample size between species and sexes are solely due to length of time individuals were collared because of date of capture, death, and so forth.

The strong bias inherent in visual counts may be seen by comparing data on habitat use by species and sexes of kangaroos as derived from radio telemetry (table 10.1) with that derived from systematic surveys (table 10.2). Radio telemetry indicated that habitat use was distributed rather broadly over the habitats available, with the exception of mulga, a very low-quality habitat for kangaroos (table 10.1). Systematic surveys, by contrast, showed the predominant use in the most open habitats, especially grassland (table 10.2). Moreover, habitats with high amounts of cover, such as mallee, bimble-grey box (these two types were combined because of their similar physical structure and because grey box was so limited in extent) and wilga-belah, showed little use by this method.

The influence of these biases in observability has the serious consequence of making use of habitats by various species and sex categories appear to be much more uniform than radio telemetry shows them to be (compare tables 10.1 and 10.2). In the absence of independent data from radio telemetry, it is easy to accept the reasonableness of observational

Table 10.1. Percent use of different habitat types by species and sex categories of kangaroos as based upon radio locations. Differences between species-sex categories were highly significant ($P < 0.001$).

Habitat Type	Eastern Grey Female $N = 414$	Eastern Grey Male $N = 436$	Red Female $N = 561$	Red Male $N = 382$	Western Grey Female $N = 480$	Western Grey Male $N = 451$
Grassland	0.2	17.2	31.0	25.1	14.8	17.1
Wilga-belah	0.0	14.9	17.8	0.5	20.2	1.8
Bimble-grey box	0.0	17.9	12.1	8.9	1.1	6.2
Pine-box	51.2	36.9	14.5	33.0	25.0	35.0
Pine	43.5	9.6	18.7	24.9	18.1	12.9
Mulga	1.7	0.0	0.0	0.0	0.0	0.0
Mallee	3.4	3.5	5.9	7.6	20.8	27.0

Table 10.2. Percent use of different habitat types by species and sex categories of kangaroos as based upon systematic surveys.

Habitat Type	Eastern Grey Female N = 194	Eastern Grey Male N = 86	Red Female N = 194	Red Male N = 86	Western Grey Female N = 1160	Western Grey Male N = 498
Grassland	40.2	31.4	67.0	56.3	61.0	65.9
Wilga-belah	2.1	4.7	1.9	1.9	3.1	2.2
Bimble-grey box	11.8	11.6	6.1	7.8	6.0	5.4
Pine-box	34.5	36.1	13.7	20.1	16.1	10.8
Pine	9.3	11.6	10.5	13.2	12.1	12.7
Mulga	0.0	0.0	0.0	0.0	0.0	0.0
Mallee	2.1	4.6	0.8	0.7	1.7	3.0

data. It was our impression that the relatively open nature of habitat on Yathong Nature Reserve would result in relatively minor biases due to observation method. The data in tables 10.1 and 10.2 should carry a sobering message to field ecologists. We usually cherish our results, derived from much hard work, so it is easier to imagine they are reliable than to imagine they are not.

All species and sex categories of kangaroos used different combinations of habitat types (table 10.1). In all cases, chi-square tests of independence showed that differences in distribution of habitats selected (exclusive of mulga, which had low cell frequencies) were highly significant ($P < 0.001$ in all cases). Red kangaroos of both sexes used open grasslands more than the other two species (table 10.1). However, red females used wilga-belah heavily whereas red males used this type hardly at all. Sinclair (1980) reported that a mosaic of types was important for red kangaroos.

Eastern grey kangaroos in our study preferred woodland habitat, as has been reported elsewhere (Taylor 1985; Southwell 1987; Coulson 1990, 1999). Eastern grey females avoided open grassland and were found predominantly in pine-dominated habitat types. They were also the only kangaroos recorded in mulga, albeit infrequently (1.7%, table 10.1). This type, which occurred on rocky, poor slopes associated with ridges, contained few kangaroos. Eastern grey males, by contrast, showed much greater use of open cover types such as grassland, wilga-belah, and bimble-grey box. Even among types dominated by pine, they used the more open type, pine-box, and used pine much less than females for which it was a favored type (table 10.1). Western grey kangaroos of both sexes made the heaviest use of mallee, which was only moderately used by other species-sex categories. Short and Grigg (1982) described mallee as poor habitat for kangaroos.

The differences between species and sex categories remained significant when broken down by seasons or environmental states, with dry periods being most different ($P < 0.001$ in all cases) and wet periods least different ($P < 0.05$ in all cases). There were no differences within species-sex category between seasons or environmental states ($P > 0.05$), demonstrating that habitat use was fairly consistent across the seasons and independent of green feed availability, although differences were less when green feed was abundant. During those times, kangaroos were released from food constraints because green feed was plentiful everywhere. Still, the species and sexes remained separated, illustrating the influence of shade, shelter, concealment cover, and variables other than food in kangaroo habitat use.

When the data were examined by time of day (morning, afternoon, and night), western grey kangaroos showed no difference ($P > 0.05$), occupying the same habitat types at all times of day and night. Eastern grey and red kangaroos, however, showed significant differences ($P < 0.05$) in habitat use between afternoon and nighttime locations, but not between morning and either afternoon or nighttime locations ($P > 0.05$). The difference between afternoon and nighttime habitat use is attributable to differences in habitats with good feeding conditions, and those with good cover for resting during the heat of day. For both sexes of eastern grey kangaroos, the shift was toward greater use of grassland at night, and pine and pine-box during the afternoon. Both sexes of red kangaroos increased use of grassland at night, but whereas females used bimble-grey box and wilga-belah in the afternoon for resting areas, males used pine-box.

Mapping habitat types on the scale of Yathong Nature Reserve inevitably results in a coarse-scale effort (fig. 2.10), and considerable variation occurs within a mapped type. For example, grasslands have scattered trees and clumps of trees too small to map. Mallee has patches of open grassland, and some mallee species of eucalyptus grow as clumps in other habitat types. Given this variation, each mapped type contains a fine-scale mosaic within which kangaroos may differentially select favored patches.

Consequently, we wanted to examine kangaroo habitat use on a finer scale. The categories of tree and shrub density recorded at the specific spots at which kangaroos were observed during systematic surveys (chapter 3) were used for this finer discrimination. Herbaceous cover at the specific site has already been addressed (chapters 3, 9). Unlike the radio telemetry locations, the direct observations of the systematic surveys are biased by concealment cover. Nevertheless, they can be used to examine the relative preferences of the three species of kangaroos. There is no reason to suspect that relative observability varied by species or sex of kangaroo. By comparison to the more cryptic New World deer, kangaroos are easily observed. Therefore, the relative use of fine-scale, within habitat variation can be compared across the species and sexes from the systematic survey data.

Tree densities at kangaroo observation sites are given in table 10.3. These results illustrate the mixed nature of the habitats as mapped. Mallee showed instances of open areas (trees more than 100 m apart), although most kangaroo observations in that habitat (mostly western greys) were in typical mallee. Grassland showed observations of kangaroos in dense trees and mallee, although most of the observations occurred at low tree densities. Bimble-grey box had the highest tree densities, although wilga-belah was nearly as high because of the relatively common inclusion of patches of mallee in this type (table 10.3). Overall, the order of the habitat types in tree density, from high to low, was mallee, wilga-belah, bimble-grey box, pine, pine-box, and grassland. Too few observations were obtained in mulga to include it in table 10.3, but its tree density resembles wilga-belah and bimble-grey box.

Tree and shrub density estimates were categorical, so tests for differences were made with chi-square statistics with cells of less than five comprising more than one-fifth of the sample being removed (Wilkinson et al. 1992). Two cases, in which marginally significant differences ($P < 0.05$) occurred for eastern grey kangaroos, were arbitrarily discarded due to the small sample size, and the generally recognized questionable reliability of chi-square tests on small samples. No tests could be done for the mulga habitat because of the low overall use of this type.

The tree densities used by the sexes of the same species were not significantly different ($P > 0.05$) for most habitat types (table 10.4). The one exception was in open grassland, where female western greys were significantly more associated with trees (chi-square $= 10.44$, 2 df, $P = 0.005$) than were males.

In cross species tests there were three significant differences. In bimble-grey box, eastern grey kangaroos were found in higher tree densities significantly more than were western greys (chi-square $= 8.60$, 2 df, $P = 0.014$). In wilga-belah, western grey kangaroos were more associated with trees than were red kangaroos (chi-square $= 7.69$, 2 df,

Table 10.3. Tree density categories (percentages) by habitat type for all kangaroo observational sites in systematic surveys.

				Habitat type				
Tree Density Category[1]	Mallee N = 86	Bimble-grey Box N = 353	Wilga-belah N = 131	Pine-box N = 855	Pine N = 645	Grassland N = 3444	Total N = 5514	N
0	1.16	8.50	11.45	8.19	10.08	70.33	47.21	2603.00
1	9.30	53.26	56.49	80.23	73.18	27.90	43.33	2389.00
2	3.49	32.01	19.85	10.76	16.12	1.36	6.98	385.00
3	0.00	5.67	1.53	0.82	0.47	0.15	0.67	37.00
4	86.05	0.57	10.69	0.00	0.16	0.26	1.81	100.00

[1]Tree density categories: 0 = >100 m apart; 1 = 25–50 m apart; 2 = 10–24 m apart; 3 = 0–9 m apart; and 4 = mallee.

Table 10.4.　Percentage of kangaroo observations by species and sex in different tree density categories, as derived from systematic surveys. Lower sample sizes than in Table 10.3 reflect the exclusion of mixed sex groups within species and mixed species groups.

Tree Density Category[1]	Eastern Grey Female N = 194	Eastern Grey Male N = 86	Red Female N = 2038	Red Male N = 433	Western Grey Female N = 1160	Western Grey Male N = 498
0	30.93	25.58	52.06	44.11	46.72	51.61
1	54.64	50.00	40.43	44.34	45.34	38.55
2	10.82	18.60	5.99	9.93	5.34	6.22
3	1.55	1.16	0.54	0.92	0.26	0.40
4	2.06	4.65	0.98	0.69	2.33	3.21

[1] Tree density categories: $0 = >100$ m apart; $1 = 25-50$ m apart; $2 = 10-24$ m apart; $3 = 0-9$ m apart; and $4 =$ mallee.

$P = 0.021$). And in grasslands, female western greys (but not males) were more likely to be in denser trees than were red kangaroos (chi-square $= 7.42$, 2 df, $P = 0.025$).

Shrub density categories from the systematic surveys were analyzed in the same way as tree density. Shrub densities at kangaroo sites for different habitat types are given in table 10.5. The total from all habitats (see table 10.5) shows that kangaroos were observed mainly in areas with no shrubs (78.4%) or scattered shrubs (16.0%). Only about six percent of the kangaroo observations were made in moderate or dense shrub areas. Of the habitat types, mallee especially, and to a lesser extent wilga-belah, had common shrub components.

Table 10.6 gives the shrub densities at the observation sites of kangaroos in the systematic surveys. Most kangaroos were distributed in areas with either no or sparse occurrence of shrubs. As noted in the discussion of behavior (chapter 8), kangaroos are susceptible to tripping and tend to avoid shrubby areas. There were two significant differences between the sexes for shrub density. In pine habitat, female western grey kangaroos were significantly (chi-square $= 15.42$, 2 df, $P < 0.001$) more likely to be observed in greater shrub density than males, and in grasslands, male red kangaroos were observed in higher shrub density (chi-square $= 14.45$, 2 df, $P = 0.001$) more than females.

There were four significant differences between species, two in bimble-grey box and two in grassland. In bimble-grey box, red kangaroos were in areas with fewer shrubs than both eastern (chi-square $= 12.15$, 2 df, $P = 0.002$) and western grey (chi-square $= 7.89$, 1 df, $P =$

Table 10.5. Shrub density categories (percentages) by habitat type for all kangaroo observational sites in systematic surveys.

| Shrub Density Category[1] | Habitat type | | | | | | Total | |
	Mallee N = 86	Bimble-grey Box N = 353	Wilga-belah N = 131	Pine-box N = 855	Pine N = 645	Grassland N = 3444	N = 5514	N
0	9.30	77.05	30.53	76.02	64.50	85.31	78.42	4324.00
1	27.91	18.13	52.67	18.01	27.91	11.41	16.03	884.00
2	50.00	3.68	16.03	4.80	5.89	2.76	4.55	251.00
3	12.79	1.13	0.76	1.17	1.71	0.52	1.00	55.00

[1] Shrub density categories: 0 = no shrubs; 1 = a few scattered shrubs; 2 = moderate shrub density; and 3 = dense shrubs obscuring visibility.

Table 10.6. Percentage of kangaroo observations by species and sex in different shrub density categories, as derived from systematic surveys. Lower sample sizes than in Table 10.5 reflect an exclusion of mixed sex groups within species and mixed species groups.

Shrub Density Category[1]	Eastern Grey Female N = 194	Eastern Grey Male N = 86	Red Female N = 2038	Red Male N = 433	Western Grey Female N = 1160	Western Grey Male N = 498
0	78.87	74.42	82.09	74.64	76.64	72.69
1	13.40	19.77	13.79	17.78	17.76	19.88
2	6.19	4.65	3.34	6.24	5.00	6.22
3	1.55	1.16	0.79	1.39	0.60	1.20

[1] Shrub density categories: 0 = no shrubs; 1 = a few scattered shrubs; 2 = moderate shrub density; and 3 = dense shrubs obscuring visibility.

0.005) kangaroos. In grassland, once again, red kangaroos were in more shrub-free areas than eastern grey (chi-square = 6.21, 1 df, P = 0.013) or western grey (chi-square = 13.52, 2 df, P = 0.001) kangaroos.

Comparison of shrub densities across species and sex categories gave six significant differences, two for pine and four for grassland habitats. In the pine habitat, red kangaroos (sexes combined because of no significant difference) were found in areas with fewer shrubs significantly more often than were female (chi-square = 6.00, 2 df, P = 0.05) or male (chi-square = 7.22, 2 df, P = 0.27) western grey kangaroos. In grassland, red kangaroos of both sexes combined were significantly different from western grey female kangaroos (chi-square = 7.42, 2 df, P = 0.025), again the latter being found in greater shrub density. Red females were significantly different from western grey (combined sexes, chi-square = 20.74, 2 df, P < 0.001), and red males were different from eastern greys (combined sexes, chi-square = 8.11, 2 df, P = 0.017). As in all cases, red kangaroos used areas with fewer shrubs. Finally, eastern grey females were different from western grey kangaroos (sexes combined, chi-square = 7.42, 2 df, P = 0.024) with eastern greys located in more shrubs.

Niche Relationships and Resource Partitioning

These results can be expressed in terms of niche relationships, better termed resource partitioning, given that sex within species as well as interspecific relationships are addressed. Niche breadth, the degree to which resources are used widely in response to their availability, can be expressed by Smith's (1982) measure (chapter 3). According to Krebs's

(1989) review, the choice between various expressions is determined by the character of the comparison being made. Smith's measure seems most suitably weighted for the present comparison, and it has the advantage of having a known variance.

What we refer to as dispersion (table 10.7) measures the spread of kangaroos by species and sex over Yathong by applying Smith's measure to the numbers of kangaroos observed in the systematic surveys in given km² grids. In the systematic surveys, kangaroos were observed in a total of 124 km² grids. The most widely dispersed species and sex was red kangaroo females, which occurred in 113 of the 124 grids, and the least dispersed was eastern grey males, which occurred in 53 grids (table 10.7). Dispersions of red and western grey kangaroos were similar and the 0.95 confidence limits overlapped. However, dispersion of eastern grey kangaroos of both sexes was substantially less, and the 0.95 confidence limits did not overlap with those of red or eastern grey kangaroos of either sex, indicating a significant difference (table 10.7).

Niche breadth was also calculated by Smith's (1982) measure, but was applied to habitat use derived from radio telemetry locations (table 10.7). For habitats, eastern grey male kangaroos were much more like the other species and sexes and nonoverlap of 0.95 confidence limits indicate no significant difference (table 10.7). Thus, whereas eastern grey males were least widely dispersed, within their confined space they selected habitats similar to those of other species and sexes (table 10.7). Eastern grey females, conversely, did not overlap in 0.95 confidence

Table 10.7.   Dispersion based upon km² grids and niche breadth for the species and sex categories of kangaroos calculated by Smith's (1982) measure of niche breadth.

	Dispersion [1] Value (0.95 CL) [2]	Niche Breadth [3] Value (0.95 CL)
Eastern grey females	0.69 (+ 0.04)	0.66 (+ 0.04)
Eastern grey males	0.61 (+ 0.06)	0.88 (+ 0.02)
Red females	0.85 (+ 0.01)	0.90 (+ 0.02)
Red males	0.82 (+ 0.02)	0.84 (+ 0.03)
Western grey females	0.84 (+ 0.01)	0.88 (+ 0.02)
Western grey males	0.83 (+ 0.02)	0.86 (+ 0.02)

[1] Based upon numbers of a given species and sex of kangaroo occurring over the 124 grids in which any kangaroos were observed.

[2] CL = Confidence Limit.

[3] Based upon numbers of a given species and sex of kangaroo occupying each of seven major habitat types derived from radio telemetry.

Table 10.8. Niche overlap in habitat use between species and sex categories of kangaroos as calculated by Morisita's (1959) index of similarity from numbers by habitat type derived from radio telemetry.

	EG Female	EG Male	Red Female	Red Male	WG Female
Eastern grey male	0.84				
Red female	0.49	0.82			
Red male	0.82	0.88	0.69		
Western grey female	0.83	0.82	0.66	0.82	
Western grey male	0.70	0.89	0.71	0.82	0.89

limits with any other species or sex, including eastern grey males (table 10.7). Thus, not only did they have narrow distribution in space, but they selected habitats that were significantly different from other kangaroos.

Niche overlap using data from habitat use from radio telemetry locations were calculated by Morisita's (1959) index of similarity, as recommended by Krebs (1989). The highest overlap value (0.89) was obtained between the sexes of western grey kangaroos and between eastern and western grey males (table 10.8). The lowest overlap (0.49) was between red and eastern grey females, and that between red and western grey females was also low (0.66). The overlap between females across species (0.49 to 0.83) was generally lower than that of males (0.82 to 0.88). Although no expression of variance is available for Morisita's measure, these differences are probably real. Comparisons between sexes within species produced mixed results. Although the sexes showed reasonable overlap for eastern and western grey kangaroos, reds were relatively low in overlap of sexes (0.69, table 10.8).

Discussion of Habitat and Niche

The results from tree density. and shrub density observations from systematic surveys yielded no surprises. They paralleled exactly the results on habitat use of the various species and sexes of kangaroos derived from radio telemetry. Red kangaroos used the most open habitats, and within different habitats, selected areas with the fewest trees and shrubs. These results reinforce the prevailing view that red kangaroos are a species of the openlands. Nevertheless, this characterization is relative to other species of kangaroos, because radio telemetry shows that red kangaroos make substantial use of woodland types (table 10.1) in accordance with Sinclair's (1980) conclusion that they prefer a mosaic of types. As shown by table 10.2, part of their reputation as an openland animal is a bias due to their observability in open areas, but not in more closed habitats.

Both eastern and western grey kangaroos used relatively more closed habitats, and selected greater tree and shrub densities within major habitat types. Similar use of woodland types was reported for western grey kangaroos by Arnold et al. (1994). Eastern grey kangaroos were most extreme in this regard, being a woodland species at Yathong much as elsewhere. However, it is noteworthy that western grey kangaroos make heaviest use of mallee (table 10.1), a type with the greatest concealment cover, based on both tree stem density and shrub density. Eastern grey kangaroos use this type even less than red kangaroos (table 10.1), pointing to their preference for forested areas with a generally open understory. The long, low trajectory of their hopping behavior bespeaks their preference for a ground surface with few obstructions.

The overlap indices we calculated were based on year-long data. If they were calculated on a seasonal basis, overlap would be substantially lower, but smaller sample sizes would result in less confidence in the results. In the absence of variance estimates, judgment must be used in assessing differences. We believe that differences of 0.05 in table 10.8 are probably real. Therefore, we believe that no species or sex showed complete overlap in niche, and females were much more separated by niche than males between species. In the case of red kangaroos, the difference between sexes was as great as many of the cross comparisons between species and sex. This suggests that differences between sexes within a species of kangaroo approach those between species. Consequently, in an ecological context, this assemblage of kangaroos resembled a six-species system, whereas taxonomically, only three were present.

11

PREDATORS AND COMPETITORS

Kangaroos on Yathong live in a faunal assemblage that is greatly altered from that of pre-European times. Most larger native mammals other than kangaroos have been extirpated, and exotic species are common. It is not possible, therefore, to put interspecific relationships in their original perspective. Today, it appears that, by and large, interspecific interactions play a small role in the lives of kangaroos at Yathong. They experience many other kinds of interaction, however. For example, insect harassment is a nuisance, if not a minor draw on the physiology of the animals, as constantly tossing the head and flopping the ears requires some energy. Kangaroos also clearly recognize and are affected by alarm calls of the abundant and varied avifauna. During the nesting season, it is difficult for researchers to move through areas of woody vegetation without kangaroos being alerted by such bird calls. The most important interspecific interactions, however, relate to predation and competition.

Current Predation

During this study, predation on kangaroos at Yathong was minor. Current predators are small and are effective mainly against newly emerged pouch young. The wedge-tailed eagle (fig. 2.15) is large enough to kill small young-at-foot when they first briefly emerge from the pouch (Leopold and Wolfe 1970; Brooker and Ridpath 1980). Once they reach the size of permanent emergence, young-at-foot are probably too large for the eagle to kill easily, plus they are too heavy to carry off. Over the course of this study, we did not verify a single case of predation. However, we did observe two cases of wedge-tailed eagles swooping down on small young-at-foot. In neither case did the eagle make contact with the young kangaroo, although in one case it came within a meter. Kangaroo mothers showed a surprising lack of alertness to these attacks. Their behavior could better be described as curiosity rather than alarm or protectiveness. If eagle predation were common, one would expect a pronounced response—either the young animal quickly retreating to the mother or the mother coming forward to protect the young. This behavior suggests that either eagle attacks are so rare as to have not selected for a defense response, or that defense is not possible. The latter explanation seems unlikely given the kangaroos' large size and capability of fighting with the forelimbs. Apparently the eagles are not much of a threat because kangaroos, by the time they emerge from the pouch, are simply too large for eagles to kill.

Exotic European foxes (fig. 2.15) and feral cats are also potential predators on young, but again, no evidence of predation was obtained. However, foxes definitely were important scavengers of kangaroos dying of other causes. Foxes were extremely abundant on Yathong; we observed them daily. Kangaroos largely ignored them, even when they passed nearby. Apparently, both foxes and cats concentrate their predation on the superabundant supply of European rabbits.

Kangaroo alarm behavior is a reflection of both aboriginal and recent human predation, and the ghosts of thylacines and dingoes past. Similarly, aborigines hunted on Yathong, at least seasonally, and kangaroos were undoubtedly major prey. Shooting by Europeans has occurred since occupation of the land, and continued until the establishment of Yathong Nature Reserve as a protected area in 1978. Even yet some poaching occurs along Belford Road, a north-south thoroughfare. These activities have made kangaroos wary of humans. Kangaroos on Yathong seemed every bit as wild as those in the surrounding areas. Overall though, during our study, predation of kangaroos on Yathong Nature Reserve was negligible and could not have exerted more than a trivial effect on population growth (see chapter 4).

Historic Predation

The high-speed gait of kangaroos bespeaks selection for escape from coursing predators. Over much of their evolutionary history, their primary predator was probably the thylacine, a large marsupial coursing predator. The thylacine became extinct on the mainland about 1000 years ago, surviving into historic times only in Tasmania. The first humans arrived on the continent about 40,000 years ago, but they didn't introduce the dingo until about 4000 years ago (Archer 1981; Gollan 1984). It is thought that the dingo competitively displaced the thylacine and occupied its niche as a predator on kangaroos. Prior to introduction of domestic sheep by European settlers, kangaroos would have been the major biomass supporting dingoes, although undoubtedly, medium- to small-sized wallabies also contributed (Robertshaw and Harden 1989). Introduction of exotic species has led to the widespread elimination of these intermediate-sized macropods by competition and predation (Calaby and Grigg 1989).

There is considerable evidence of predation by dingoes on kangaroos. Studies of dingo food habits show that kangaroos make an important contribution to their diet (Whitehouse 1977; Shepherd 1981; Robertshaw and Harden 1985, 1986; Marsack and Campbell 1990; Thomson 1992). Wright (1993) and Jarman and Wright (1993) report on observed attacks of dingoes on eastern grey kangaroos. Thomson (1992) reported that capture of kangaroos was more successful when dingoes hunted in groups. In addition, there is indirect evidence from kangaroo behavior. Formation of groups by kangaroos in openlands was reported by Low (1979), Oliver (1986), and Heathcote (1987), and Coulson (1999) noted that mixed-species groups may form in response to predators because the additional numbers may be of value, independent of the species. An inverse relationship between dingo and kangaroo population density was reported by Jarman and Denny (1976) and Caughley et al. (1980). Not only are dingoes quite capable of killing kangaroos, they do so commonly enough to depress kangaroo populations.

Because of their depredation on sheep, dingoes were heavily persecuted by Europeans, and have been eliminated from the area around Yathong. The most recent report we heard was from Mr. Rod Forsyth, manager of Stanniford Station adjacent to Yathong, who told us that two dingoes were seen on his property in 1957. So, dingoes have been absent from this area for nearly 30 years prior to this study, and their reestablishment is prevented by the dingo fence. Nevertheless, one would expect kangaroos' retention of escape response to predators to be conservative, and they exhibit the same basic high-speed flight response to both four-legged and human threats. Continued encounters with humans very likely perpetuate behavior selected in response to coursing predators.

Analysis of habitat use (chapter 10) showed that kangaroos prefer relatively open environments, and high-speed flight for escape is an obvious adaptation in animals occupying open areas where crypsis and stealth are not feasible. Nevertheless, we documented one case of the use of stealth by a kangaroo to avoid a human intruder. On November 23, 1985, at 1000 hours, the senior author located the radio-collared red female kangaroo known as Scarlett O'Hara. Upon locating her, I was upwind, and a moderate breeze was carrying my scent in her direction. She was with four other kangaroos, three females and a young male. As they ran away, Scarlett separated from this group. After a while, curious as to whether she and the other kangaroos reformed in a social group, I moved toward her radio signal. She had first been located in pine habitat, with a visibility of about 100 to 150 m, but after alarm, had moved into heavier cover patches within the pine. I walked directly toward her in the expectation that she would remain where she was until I came into view, at which time I could verify a social group and retreat with minimal disturbance to the animals. It soon became apparent from the radio signals that she was moving away—keeping out of sight, but not simply taking alarm. Although it is not possible to determine precise distance with only one azimuth, based on experience with approaching kangaroos, I judged her to be less than 500 m distance, and probably around 200 to 300 m.

I then skirted to the right, intending to approach her perpendicular to the direction of the wind. To my surprise, she too moved in that direction, remaining downwind, and out of sight several hundred meters ahead of me. Thinking that this may have been a chance move on her part, I then continued to try to flank her to the right. Again, she moved to maintain her position relative to mine. By this time, it became apparent that she was using my scent and her concealed position to foil my attempt to approach her. I then tried to flank her in the opposite direction, and to this maneuver, she reversed course and maintained her position relative to mine. The lateral movements included 1.5 km actual ground distance, and left no doubt in my mind that she was using scent to monitor my position, and moving in order to maintain her advantage. This behavior is common for ungulates living in heavy concealment cover. But such stealth shown by a kangaroo—especially a species that ordinarily occupies open areas and typically escapes by putting distance between itself and an intruder—expanded our appreciation of the behavioral flexibility of kangaroos. Other than humans, there does not seem to have been an ambush predator that hunted by stealth in the pre-European fauna since the Pleistocene. Scarlett's ancestors probably foiled aboriginal hunters over the millennia in the same manner she did me.

Competition with Emus

Discounting termites and ants, both major consumers of plant biomass on Yathong, most of the competition with kangaroos comes from exotic rabbits, goats, and pigs. The only native vertebrate competitors of any consequence on Yathong are emus (*Dromaius novaehollandiae*). Emus were part of the original Australian fauna, with distribution over most of the continent; thus, they coevolved with kangaroos. These large, ostrichlike birds are herbivores, and prefer new green growth (fig. 11.1). During greenup they can be seen grazing extensively on the low growth of grasses and forbs. As herbaceous plants dried, emus began coming to the water tanks. This coincided with kangaroos watering at the tanks, but whereas emus came to tanks at midday, kangaroos came in the morning, evening, or at night.

The form of emu droppings indicates green food, but not to the same extent as those of kangaroos. Emu droppings resemble those of domestic cows, only much smaller, and with a tendency to be drier, more elongated, and with a somewhat rippled surface. It is of interest that the relatively common short, stumpy lizard of this region, known as the "shingle-back" (*Trachydosaurus rugosus*) has the dubious distinction

Figure 11.1. Two emus feeding on new green grass in a site near where the kangaroos shown in Figure 9.9 were observed on the same day.

of having evolved as an emu dropping mimic, a connection we made when we found ourselves slowing down along roads to make the call: shingle-back or emu dropping? Aesthetics are not a necessary criterion for evolutionary design, so long as it works.

Early in this study, when green feed was available (fig. 2.9), emus were often seen in large aggregations of 20 or more birds, grazing over the open landscape. Only casual field notes were kept on emus during this period. Some young birds, produced in 1985, were present in these flocks. When we began our study at the end of May, these young birds had a juvenile plumage similar to that of adults, having already lost the watermelon-striped pattern (reminiscent of newborn wild boar, *Sus scrofa*) of early life.

Beginning in January 1986, data on emus paralleling those of kangaroos were kept on all daylight systematic surveys. Emus were rarely seen in the night spotlight surveys, suggesting that they are predominantly diurnal in activity. Taking systematic records was prompted by belated recognition of possible competition between kangaroos and emus, and that emu behavior was changing. Whereas, earlier in the study, kangaroos and emus were regularly seen in proximity to each other, now they no longer were. Also, the large emu aggregations, formerly so common, were no longer observed. And finally, there was no evidence of breeding activity among the emus, normally seasonal (winter) breeders. As it turned out, the 1986 young-of-the-year class was totally absent at Yathong—not a single nest or young bird was observed. Although it is possible to overlook nests, the fact that we did not record a single young among 461 emus observed in systematic surveys, plus nearly the same number noted during other activities, is overwhelming evidence of reproductive failure. Also, there is no evidence for opportunistic nesting among emus, so drought conditions can totally prevent breeding activities (O'Brien 1990).

Emu group sizes in 1986 were relatively small (fig. 11.2) with an overall mean of 2.49 for 185 groups. The largest group observed was nine birds, much smaller than the large aggregations observed the previous year.

Emus selected the most open habitats (table 11.1). About 70 percent of the emu observations were made in the grassland habitat, and another 16 percent in pine, one of the next most open habitats. These values exceeded the use of open habitat types by female red kangaroos—the species and sex that selected openlands the most (table 10.2). Similarly, specific sites where emus were observed (table 11.2) were characterized by lower tree density (table 10.3) and lower shrub density (table 10.5)

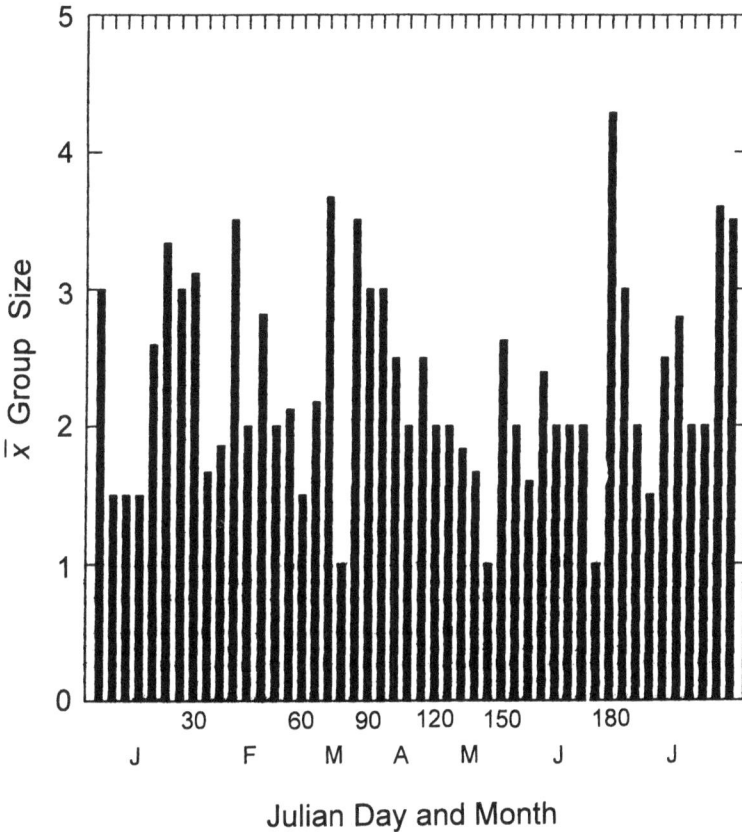

Figure 11.2.   Relationship of mean group size of emus and Julian day in 1986 derived from daylight systematic surveys.

than those of kangaroos. These results show that emus were selecting for open characteristics of habitat even more so than kangaroos.

Emu population size was estimated from perpendicular distances of emu groups from the road, the center of the 87.7-km transect represented by the four systematic routes sampled each fortnight. Each fortnight was considered a replication. Examination of the frequency distribution of perpendicular distances showed an approximately equal probability of sighting up to 150 m from the transect (fig. 11.3). Breakdown of sightings within 150 m into smaller intervals showed that the relative consistency of observation held for finer units as well. Consequently, as recommended by Caughley (1977), the sample was treated as a strip of 300 m (150 m on either side of the road) by 87.7 km (the length of the

Table 11.1.   Use of different habitat types by emus as based upon systematic surveys in 1986.

Habitat Type	No. Emus Observed	% of All Emus Observed	% of Emus at High Numbers[1]	% of Emus at Low Numbers[2]
Grassland	324	70.3	69.6	72.4
Wilga-belah	7	1.5	1.7	0.9
Bimble-grey box	4	0.9	1.2	0.0
Pine-box	48	10.4	6.6	21.5
Pine	74	16.0	19.7	5.2
Mulga	0	0.0	0.0	0.0
Mallee	4	0.9	1.2	0.0
Totals	461	100.0	100.0	100.0

[1] Data from Jan 1 to Mar 15 and Jun 16 to Jul 31 ($N = 345$; see Table 11.3).
[2] Data from Mar 16 to Jun 15 ($N = 116$; see Table 11.3).

Table 11.2.   Tree and shrub density at emu observational sites in all habitats combined for the systematic counts in 1986.

Category[1]	Tree Density		Shrub Density	
	N	%	N	%
0	293	63.6	385	83.5
1	142	30.8	49	10.6
2	20	4.3	14	3.1
3	1	0.2	13	2.8
4	5	1.1	—	—
Total	461	100.0	461	100.0

[1] Categories for tree density: $0 = >100$ m apart; $1 = 25–50$ m apart; $2 = 10–24$ m apart; $3 = 0.9$ m apart; and $4 =$ mallee. Categories for shrub density: $0 =$ no shrubs; $1 =$ a few scattered shrubs; $2 =$ moderate shrub density; and $3 =$ dense shrubs obscuring visibility.

combined transects) for a total area of 26.31 km². Strip counts are preferable to line transect methods, which depend on a number of assumptions that may not be met. Thus, the data included in the population estimates were the number of individuals within 150 m of the transect, instead of the number of groups. Of the 461 total individual emus observed in the systematic routes, 287 (62 percent) were within 150 m of the road.

Mean density was about 0.8 emus/km², and density ranged from about 1.5 to 0.1 over the sample period (table 11.3). Mean number in the intensive study area was 225. Population size of emus was similar to that

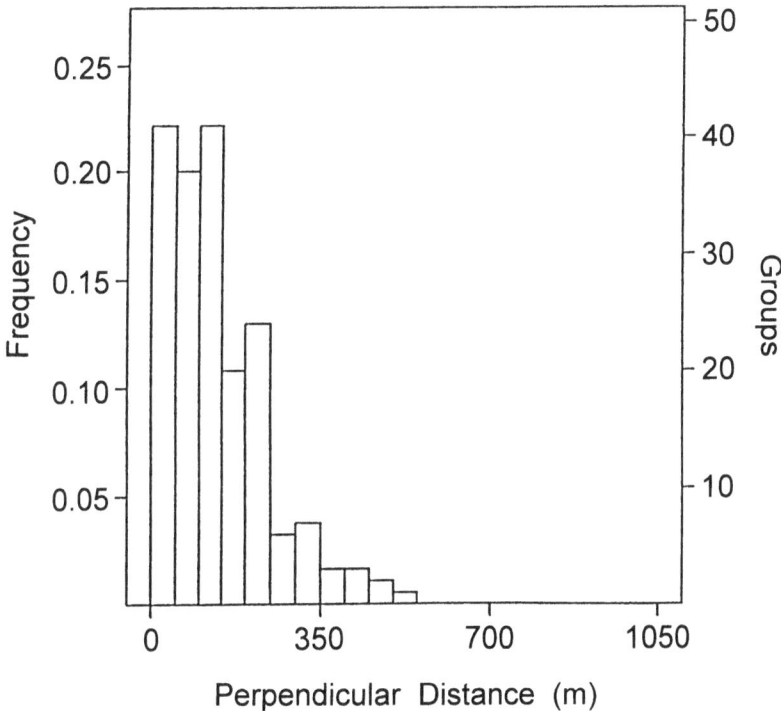

Figure 11.3. Frequency distribution of emu perpendicular sighting distances in daylight systematic surveys in 1986.

of eastern grey kangaroos, but substantially lower than that of red and western grey kangaroos (chapter 4). Nevertheless, emus were a significant grazer in the Yathong herbivore complex.

The numbers of emus observed varied from highs early and late in the sample, and lows in the middle, from the second half of March to the first half of June (table 11.3). We do not know where the missing emus were during this time. In view of their general avoidance of heavy cover (tables 11.1 and 11.2), and the fact that use of habitat types with heavier cover did not increase during the low period, use of concealment cover did not seem to account for the decline in numbers observed. From the comparison of habitat type use at high versus low population estimate periods (table 11.1), the major shift was from pine to pine-box, two types with roughly equal visibility. These differences are insufficient to account for the large decline in population estimates. Therefore, we believe that emus moved out of the intensive study area during this time.

Table 11.3.   Estimated emu density and numbers by fortnight in the intensive study area.

Fortnight	Total N	N <150 m	N/km²	N in Study Area [1]
Jan. 1–15	24	18	0.68	196
Jan. 16–31	68	38	1.44	415
Feb. 1–15	42	34	1.29	372
Feb. 16–28	62	40	1.52	438
Mar. 1–15	41	27	1.03	297
Mar. 16–31	14	2	0.08	23
Apr. 1–15	10	5	0.19	55
Apr. 16–30	20	10	0.38	109
May 1–15	28	17	0.65	187
May 16–31	22	15	0.57	164
June 1–15	23	5	0.19	55
June 16–30	52	32	1.22	350
July 1–15	24	20	0.76	219
July 16–31	31	24	0.91	263
	$N = 461$	$N = 287$	$\bar{x} = 0.78$	225
			$SD = \pm 0.48$	± 137

[1] Number of emus estimated to occur in the 288 km² that includes the systematic survey routes in the intensive study area.

If serious competition occurred between kangaroos and emus on Ya-thong, it probably occurred during dry periods, because green feed is widespread and abundant during green growth periods. For competition to occur, by definition a resource must be in limited supply, i.e., a shortage that results in detriment to one or both species. Given the abundance of green feed early in this study (fig. 2.9), and the fact that emus and kangaroos were observed in proximity, competition did not seem to occur. From all appearances, emus were taking the same foods as kangaroos (compare figs. 9.9 and 11.1). No aggressive interactions were observed between kangaroos and emus, however. For the most part, they seemed to ignore each other except when escape behavior of one alarmed the other. Emus seem to be similar to red kangaroos in using space rather than cover to escape from disturbance.

Under the assumption that competition, if present, would occur during dry feed conditions—which occurred over most of the study in 1986—we calculated overlap in space (by km² grid) of emus and the combinations of species and sexes of kangaroos observed in systematic surveys using Morisita's (1959) method (table 11.4, fig. 11.4). Recall that these data are biased because of variable observability by habitat

Table 11.4. Niche overlap in space use between kangaroos by sex and species and between kangaroos and emus as calculated by Morisita's (1959) index of similarity from number of kangaroos and emus observed on 124 km² grids during daylight systematic surveys.

	EG Female	EG Male	Red Female	Red Male	WG Female	WG Male
Eastern grey male	0.82					
Red female	0.66	0.47				
Red male	0.28	0.53	0.94			
Western grey female	0.27	0.59	0.87	0.83		
Western grey male	0.54	0.58	0.82	0.80	0.96	
Emu	0.20	0.33	0.64	0.62	0.64	0.66

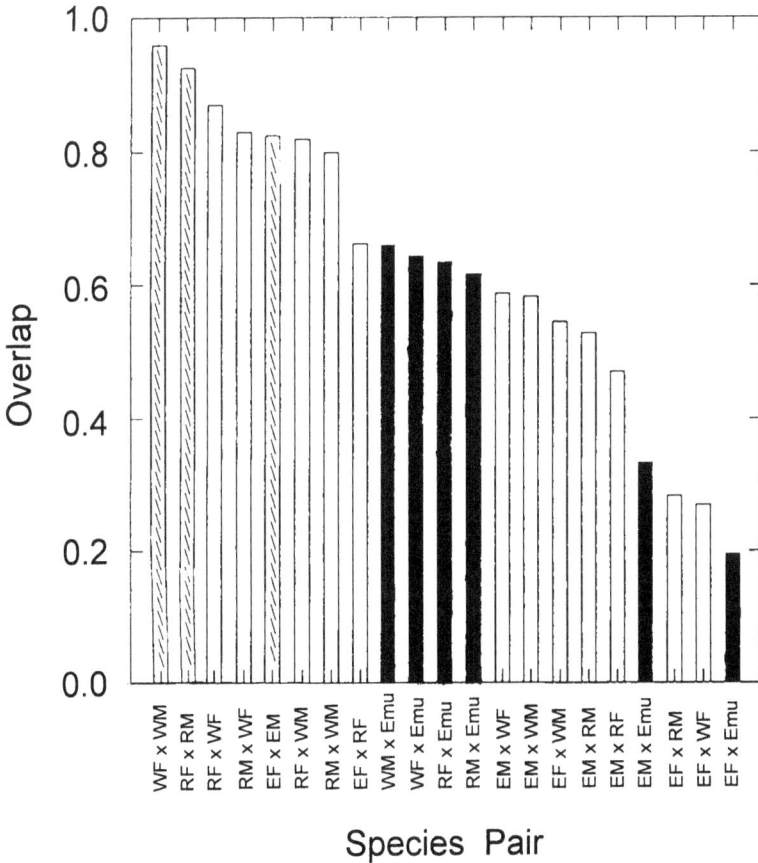

Figure 11.4. Ranked overlap values between kangaroo species by sex, and kangaroos and emus calculated by the Morisita (1959) method. Cross-hatched bars are overlap values for kangaroo sexes within species and solid bars are overlap values for emu by kangaroo species by sex combinations.

type (compare table 11.4 to table 10.8). Therefore, overlap values are relative, not absolute. By the systematic survey data, the sexes within species of kangaroos had greater similarity than by the unbiased radio telemetry data (fig. 11.4). Western grey kangaroos had the highest overlap (0.96), but the other two species were only slightly lower (reds 0.94, eastern greys 0.82, table 11.4). Similarly, overlap values between both red and western grey kangaroos, which make heavy use of openland areas, and eastern grey kangaroos, were relatively low (concentrated on the right side of fig. 11.4).

As expected, overlap between emus and eastern grey kangaroos was low (table 11.4, fig. 11.4) because the emu is an openland species whereas the eastern grey prefers woodlands. Overlap values between emus and red and western grey kangaroos ranged from 0.62 to 0.66 whereas overlap between red and western greys ranged from 0.80 to 0.87 (table 11.4). Thus, there was greater overlap between these two species of kangaroos than between either of these species and emus. Emus were the most strongly associated with openlands. The order of this set of species from open grasslands to woodlands was emus, red kangaroos, western grey kangaroos, and eastern grey kangaroos (compare table 11.4 with tables 10.4 and 10.6). However, emus were less different in mean overlap value from reds (15.5) and western greys (18.5) than reds and western greys were from eastern greys (red = 34.5, western grey = 35). This suggests greater separation of eastern greys than emus from reds and western greys, although in opposite directions in terms of habitat use.

Exotic Competitors

Exotic species were recorded on data sheets and in field notes as encountered. The exotic competitor with the largest impact by far was the European rabbit. Rabbit warrens were ubiquitous on Yathong—one seldom was more than a few hundred meters from an active warren. Despite predation by wedge-tailed eagles, foxes, and cats, and the impact of myxomatosis (which resulted in stuporous, lumpy-headed rabbits along the roads in the summer), European rabbits were extremely abundant. Extermination is not possible. Attempts at control have proven futile. Old "rabbit-proof" fencing on Yathong and rabbit resistance to myxomatosis attest to failed attempts. Poisoning campaigns using Compound 1080 have failed to have more than a local impact and may cause undesirable secondary poisoning of nontarget species. During our study, New South Wales National Parks and Wildlife Service made some attempts at control on Yathong by deep plowing with a tractor to collapse warren burrow systems. This method was expensive, and only a small portion of the area was treated. Even if rabbit populations on Yathong were reduced, the surrounding properties would serve as a source of continuous recolonization, which is true of the other problem exotic species as well.

The impact of rabbits was confined to the immediate vicinity of their warrens where feeding eliminated perennial grasses and palatable forbs. During greenup periods, annual grasses and forbs grow in the warren areas, and kangaroos often feed there. We think that soil disturbance by rabbits creates a favorable seedbed for annuals, and production was probably higher there than on surrounding undisturbed sites. However, feeding by rabbits kept these areas fairly well mowed, and accumulation of biomass was small. After greenup, warrens are dominated by unpalatable thistles (fig. 2.13).

Feral goats also were common on Yathong (fig. 2.13). Periodically, they were mustered (rounded up) and sold, and this kept their numbers in reasonable check. Goats were found mainly in a few regions of the reserve with heavier cover. After roundup, goats quickly infiltrated from surrounding properties and from areas of Yathong not mustered. During this study, more than 100 goats in one group were observed in an area of Yathong where a complete roundup had been carried out less than a month previously.

Feral pigs are common on Yathong but occur in relatively low numbers due to the scarcity of wet sites and lack of heavy shrub areas for escape cover, as well as to ongoing control programs (trapping and shooting) by the New South Wales National Parks and Wildlife Service. Pigs had a trivial impact on kangaroos during the study period. Similar to goats, they were found mainly in heavy shrub cover areas where kangaroos were seldom found. Pig populations would have to be substantially higher than were noted during this study to have any measurable impacts on kangaroos. It is unclear, however, what those impacts would be.

Discussion of Predation and Competition

Kangaroos of Yathong were subjected to relatively little predation pressure—both during this study and in the recent past—since creation of the Nature Reserve. Other than the wedge-tailed eagle, no native predators were present, and neither the eagle nor the exotic red fox, from all indications, was a very successful predator of kangaroos. Both are too small to be a threat to kangaroos other than newly emerged pouch young. The lackadaisical response of mother kangaroos to the presence of these predators suggests that predation was not a serious threat. Occasional poaching by humans was present at Yathong, but it seemed to be isolated and sporadic. Overall, predation probably had a trivial impact on kangaroo population dynamics on Yathong in recent times; population dynamics were driven primarily by rainfall and periodic drought.

Competition also has probably had relatively little influence on population dynamics of kangaroos at Yathong in recent years. Prior to acquisition of Yathong as a nature reserve, kangaroos probably competed for food with sheep, as reported by Dawson (1989), Edwards (1990), Dawson and Ellis (1994), and Edwards et al. (1996). However, sheep were completely removed from Yathong in 1978.

Given the abundance of rabbit warrens, it was apparent that the presence of rabbits altered kangaroo habitat on Yathong. Given that the diets of kangaroos and rabbits overlap appreciably (Dawson and Ellis 1994), it would be easy to assume that rabbits have reduced the carrying capacity for kangaroos. Such an inference would be logical for most temperate zone ecosystems where large grazing herbivore populations tend to be at or near the carrying capacity of the environment. Seasonal predictability in temperate zones has resulted in evolution of means in ungulate herbivores for coping with periods of shortfall by migration, fattening, and timing demands for growth and reproduction to favorable periods. Introduction of a competitor into such a system invariably results in lowered carrying capacity for the native species. However, the applicability of this logic to kangaroos at Yathong is beset with problems. In fact, kangaroo populations are seldom in precise relationship with carrying capacity (Caughley 1987b). Their populations often are far below carrying capacity during favorable periods, and far above carrying capacity during drought. Thus, kangaroo populations in this area are constantly shifting to adjust to a wildly fluctuating carrying capacity.

The impact of rabbits—profound in most ecological aspects—probably has had minimal influence on kangaroo populations at Yathong. In contrast to Fowler's Gap, where Dawson and Ellis (1994) found that rabbits in the dry season consumed twigs, roots, and bark, rabbits at Yathong have relatively little access to woody material. Still, we seriously doubt whether total removal of rabbits would result in an appreciable change in kangaroo populations. When green feed is available, it is abundant everywhere, and the entire complex of grazers, both invertebrate and vertebrate, cannot consume it all. When drying occurs, forage quality drops; biomass is widespread at first, but is lost over time to consumption and weathering. If rabbits have a negative impact on kangaroos, it likely would be during this period—by reducing availability of low-quality dry standing crop of perennial grasses, particularly *Stipa*. But this effect probably is small, because dry grass consumption by mammalian herbivores is only one of the variables in standing crop depletion; others include consumption by insects and loss by wind, weathering, and fire.

Rabbits may at times actually benefit kangaroos to some extent in that, similar to old sheep paddocks, rabbit warrens produce soil disturbance where annual grasses and forbs germinate earliest, following rains, and also following rainfalls too light to produce a general greenup. At these times, kangaroos can be found on rabbit warrens, feeding on green growth. Conversely, rabbits have no impact on kangaroo feeding congregations at the end of green periods, because these feeding concentrations are in low-lying areas where rabbit warrens would be drowned out. Warrens are found only on higher sites where flooding is uncommon.

As with rabbits, goats have little overall impact on kangaroo populations because of the boom-and-bust nature of the herbaceous vegetation. However, they may have a larger overall impact than rabbits because of their heavy consumption of woody vegetation. Goats were the primary consumers of palatable shrubs and trees on Yathong. If shrubs showed evidence of browsing, goat droppings abounded. Direct observation of goat feeding also showed considerable consumption of woody species. Indeed, goats concentrated in those areas with a woody plant composition within reach of the ground, but sometimes even climbed trees to feed. Regeneration by some palatable species of woody plants (e.g., kurrajong, fig. 2.14) has been eliminated on Yathong by voracious goats.

The probable impact of goats on kangaroos occurs as a consequence of their reduction of woody forage, which kangaroos usually do not use. However, during extended drought, kangaroos consume woody plants as an emergency food. Browse is a source of green feed to supplement dry grass. Indeed, there was relatively little in the way of shrubs on Yathong (table 10.5), and feeding by goats reduced shrub production (particularly palatable species) even more. Thus, by the time kangaroos become dependent on browse, it has already been browsed by goats, thereby reducing an important hedge against the extremes of the outback environment.

Although goats can be rounded up effectively on Yathong, extirpation would require similar elimination on surrounding sheep properties, or construction of a goat-proof boundary fence. This is not likely to happen. Station owners view goats as a useful part of a station operation because they require no labor or capital investment to raise, and are available to be rounded up and sold periodically when the need for cash arises. They are used much like a biological savings account.

The competitive relationship of emus to kangaroos is unclear. Emus search for green feed as do red kangaroos, and they overlap in space with red and western grey kangaroos by 64 percent (table 11.4). There is little overlap between emus and eastern grey kangaroos because the former is selective for open areas and the latter for wooded areas. It is notable that red and western grey kangaroos show the greatest overlap (83 percent) whereas emus are separated by greater use of open habitats, and eastern greys by greater use of wooded habitats. Thus, these four species are partially separated on habitat gradients with the red and western grey being most similar in use of space. Nevertheless, as pointed out in chapter 9, these two species tend to sort on a feeding microhabitat scale.

It appears that the native emu has evolved along with kangaroos into a structured large herbivore community in Australia, with at least partial separation by habitat. However, it is notable that the niche separation of emus from red and western grey kangaroos is not as great as that between red and western grey and eastern grey kangaroos. Thus, some interclass differences were exceeded by differences among species within a genus. This points out how phylogeny intertwines with ecology over time to produce feeding guilds within communities of plants and animals.

We never observed antagonistic behavior within kangaroos or between kangaroos and emus that could be attributed to food resources per se. Individual space or mates were the only things we observed kangaroos to contest. We believe that competition during the time we studied at Yathong was not important either among kangaroos, or between kangaroos and emus.

Perhaps competition becomes more severe during extended droughts. But it also may be that the high uncertainty of resource abundance over time and space, along with high amplitude in population sizes of kangaroos and emus, favored scrambling ability and disfavored competitive interrelations, thereby lessening the role of competition as an evolutionary force in structuring this community.

12

KANGAROOS, UNGULATES, AND ECOLOGICAL
AND EVOLUTIONARY MODELS

In this study, we have described a number of differences between the species and sexes of kangaroos—differences that beg for an explanation. Knowing what kangaroos do is interesting in itself; but the ultimate question is always, "Why do they do what they do?" In this chapter, we explore the possible evolutionary pathways that led to their current behavior. We also compare and contrast kangaroo behavior and ecology with that of ungulates to examine the robustness of paradigms developed from the latter. The real test, however, is in the accuracy of the predictions of evolutionary and behavioral models. Have comparable environmental conditions produced similarities in behavior and ecology of large herbivores across distantly related taxa?

Differences between the Sexes

The most obvious difference between kangaroo sexes is the extreme disparity in body size (fig. 1.5). Males are much larger and more robustly constructed (figs. 1.2 to 1.4). They have skeletal structures and musculature exceeding that due to body scaling. Males are not simply females multiplied by two or three. Sexual dimorphism is most extreme in red kangaroos, not only in size, but in morphology and coat color as well. Dimorphism is still impressive in the two species of grey kangaroos, but compared to red kangaroos, the difference is more a matter of scale and less of separate blueprints.

These differences can be accounted for by the model put forward for ungulates that competition between males is for mates, which favors large size and fighting ability, whereas females compete mainly for resources in order to produce the greatest number of surviving young (McCullough 1979; Clutton-Brock et al. 1982). Males compete for resources only to the degree that this contributes to the ultimate goal of securing mates. Ultimately, these arguments trace back to sexual selection as first described by Darwin (1871).

We agree with the basic premises of the above model, and believe it to be the most parsimonious accounting for evolution of dimorphism in kangaroos. Yet, we think that the specific framing of the hypothesis somewhat misplaces the emphasis. The role of competition is stated in ways that suggest that competition for resources is less for males than females, because the former are actually competing for females. We suggest that both sexes compete aggressively for resources, as well as strive to be reproductively successful. The key variable that is different between the sexes is parental care. Males contribute none; for them the fitness struggle in each reproductive bout ends with insemination. Females contribute all; the fitness struggle in the current reproductive effort begins at insemination and the major costs lie in the future. Consequently, the time of insemination is a watershed in the life history of polygynous species, with the interests of males originating with events earlier in time whereas the interests of females flow mainly to events later in time.

However, it is not as if each sex has no fitness considerations on the opposite side of the insemination time divide. The need to survive is a constant over time. Both sexes must survive to the time of conception and both must survive to engage in subsequent reproductive efforts. The female must pay the costs of physiological changes and production of eggs prior to insemination, and a male may forego some competitive advantage to avoid competing with females bearing his young, a contribution made after insemination.

This decoupling of the interests of the sexes over time has profound consequences in a seasonal world. In temperate zones, the breeding season of ungulates typically occurs in the fall, so that the long gestation required by these large-bodied species results in birth of offspring in the following spring at the beginning of favorable food conditions. Consequently, late gestation and lactation, the most costly parts of reproduction for the female, are borne at a time when resources are most abundant. The transition of the young from nursing to independent feeding usually occurs before the favorable season ends, and this adds to the probability of survival of the young and success of the reproductive effort. Similar circumstances seem to pertain to kangaroos. It is widely believed that young are born at a time when weaning will occur at the most favorable season (Short et al. 1983; Norbury et al. 1988; Quin 1989). Russell (1989) and Dawson (1995) reported that early pouch young growth was slow, with most rapid weight gain occurring in the last trimester of pouch life.

This schedule, driven by the needs of the females, works, probably inadvertently, to the advantage of the males. Favorable food conditions occur over a long period during which the male can grow and fatten prior to the breeding season. This fortuitous circumstance allows males to take advantage of nature's largesse, and fosters the selection for even greater sexual dimorphism in the species.

We believe that the emphasis on different kinds of competition between the sexes has been viewed in a simplistic light. We agree with the general proposition that the competitive agendas of the sexes have diverged, and that direct competition between the sexes has been substantially reduced. However, natural selection theory implies that individuals should compete if a fitness advantage is thereby conferred to one of the parties. We suggest that the widely held belief that males do not compete for resources to the degree that females do is misplaced. A more inclusive consideration of competitive theory is in order.

Competition, by definition, occurs only when some resource is in limited supply, so that one (or both to different degrees) of the competing parties suffers some detriment. "Not enough to go around" is a key point in competition theory. Ungulates and kangaroos compete for food quality as much or more than for food quantity. For herbivores, food quality oscillates seasonally, being highest during the green growth stage and lowest during the dry, dormant stage. During much of the green period, there is an abundance of high-quality forage, more than can be consumed; consequently, no competition occurs. Similarly, during the dry period, there is typically an abundance of uniformly low-quality food, and competition again does not occur. Competition tends to be most severe at those transition periods in the phenological sequence when quality food is still present but is limited in distribution. In a system such as Yathong, this is a fairly narrow window in time during which competition is high.

We suggest that the appearance of less resource competition among males in large herbivorous mammals is an artifact that originated because most of the important activity of growth, fighting, and struggle for dominance by males occurs during a favorable period of plant growth.

We submit that the same favorable circumstances apply to the females' perinatal activities as well and that competition during this period is low in both sexes.

Competition is greater for both sexes in the postinsemination period that coincides with the seasonal low in food resources, but the evidence is rather ambiguous that it is greater in females than in males. Males can subsist on poorer quality food because of the scaling effects of a large body (Kleiber 1961). Because the males and females of polygynous ungulate species usually occupy separate areas (Main et al. 1996), a direct comparison is difficult. Usually males are in equal or lower quality habitats (Wrangham and Rubenstein 1986), but they also usually occur at much lower density. How these various factors play out is unclear, and it seems to us that no generalization can be made about relative competition between the sexes in ungulates during the dormant plant period.

Certainly, it can be said that the "strategies" of the sexes differ in exploiting the seasonal quality of resources. Females follow a more stable, sustained approach across seasons, whereas males (older ones particularly) show great flux in body mass. Males accumulate mass during favorable conditions prior to the breeding season, expend mass during the breeding season, and attempt to recover during the poor conditions following the breeding season (McCullough 1969). Young males, whose mating prospects are poor because of the prevalence of more dominant large males, show much less flux in mass. Their best prospect is to minimize current risk and thereby increase their survivorship to reach older age, larger size, and dominant status. According to Dawson (1995), red kangaroo males at Fowler's Gap do not contribute to breeding much before five years of age. In contrast, dominant males nearing the end of their physiological lifespan are selected to go "all out" to secure copulations, for their probability of further life is low in any event; and extreme exhaustion of body resources makes the strategy a self-fulfilling prophecy. For example, Stuart-Dick and Higginbottom (1989) reported that in kangaroos, the largest males obtain 75 percent of the matings, but their tenure lasts only one year.

Collapsing the argument to "males compete for mates and females compete for resources" obscures the fact that both sexes need food *and* mates. Whereas obtaining sufficient food can be a burdensome task for both sexes, securing mates by females usually is not difficult by virtue of male-male competition in the polygynous system. Because males can inseminate many females, and some males are able to exclude other males (resulting in great variance in fitness between males), male-male competition is great, and males are not in limited supply for females.

Furthermore, females need not exert much selectivity for mate quality, because males that succeed in the face of this competition are already the most fit, and likely to produce offspring equally capable of future success.

The time of insemination also divides the expenditure of energy and nutrients between the sexes in the reproductive effort. For both sexes, the direct costs of sperm and egg production are small. The large cost to the male is the energy expended and the hazard of all events leading to insemination. In males, major energy expenditure goes toward physical growth and hierarchical and reproductive behaviors necessary for successful inseminations. Body growth in males does not reach an adult size and stop (i.e. it is not determinant). Fastest growth occurs during young adulthood but continues thereafter until senescence. A male juvenile red kangaroo weighs around 27 kg and by ten years of age may weigh around 85 kg, for an average annual increment of about 6 kg. Males expend much time and energy moving about a large home range, maintaining hierarchical positions, and pursuing and sequestering females in tending bonds.

The cost to the female through the point of fertilization is small compared to the costs of gestation, birth, and lactation. A female kangaroo's annual production of an offspring — grown from gestation through weaning — is roughly comparable to the average annual male increase in body size. For both ungulates and kangaroos, the female invests about 35 to 37 percent of her body mass in production of offspring to weaning (Russell 1982). Maynes (1976) suggested that macropodids are more advanced at pouch emergence than ungulates at birth. Still, they are roughly comparable in this regard. In contrast to males, however, females (particularly among docile red kangaroos) expend little time and energy on intrasexual competitive behavior directed toward reproductive success.

Although adult males may derive some benefits from economy of scale due to their larger body sizes, the combined costs of body growth and behavioral energy expenditure do not seem less than those of adult females for reproduction. Again, the assumed lesser need of males than females to compete for resources is debatable. We believe that the sexes compete similarly for resources. Whereas their acquisition of energy and nutrients is comparable, the energy expenditure of each sex is different. Males avoid parental care, but instead spend energy on body growth (and its absolute higher maintenance cost) and on competing for breeding opportunities. Because male-male competition determines male fitness, females need not make expenditures on that activity. Instead, females expend their energy on producing and parenting offspring. With these modest amendments, we believe that the ungulate model for evolution of different roles by the sexes is applicable to kangaroos at Yathong.

Three Species, Six Niches?

Kangaroo species (and sexes) show differences on Yathong that appear to be organized into separate niche spaces in a coevolved community. In addition to sexual dimorphism, it is notable that in virtually every respect, the biology of the sexes of kangaroos at Yathong was substantially different. Female kangaroos have smaller home ranges, move shorter distances, select habitat types in different proportions, and form independent social groups except when joined by males checking their reproductive condition. Often, these differences equaled or exceeded differences between species. In other words, in many ways the same sex across species was more similar than the two sexes within species. Even for the eastern and western grey sibling species, the sexes within species were more alike in appearance than in biology.

This complex of three species and two sexes behaves in an ecological context like a set of six "species," each with its own niche space that separates it from the other five. The original native, large herbivore community on Yathong included the emu and the rock-dwelling euro as well, for an eight "species" system. This number is similar to that found in large herbivore communities in temperate regions of the world.

Conventional wisdom holds that such niche spacing represents community structuring—the result of selection favoring reduction of interspecific competition. Rather than current competition, the niche separation is a reflection of the "ghosts of competition past." This paradigm is appealing because it ties together so much natural history and biology across levels from the species to the community, and also melds ecological and evolutionary theory.

Approaching this study with the competition paradigm in mind, we were not surprised when the data showed that the three species of kangaroos were separating out on various niche dimensions. The congruence of empirical results and niche theory at first seemed a good fit. Yet, the longer we contemplated the question—were kangaroos shaped by interspecific competition?—and the more we explored the larger context of the evolution of kangaroos in Australia, the more we began to doubt what seemed to be a pat answer. Could it be that competition theory, and separation of niches, although perfectly "explaining" the results, was wrong? Was it possible that this complex fit was not the outcome of cause and effect but instead the outcome of biogeography?

This question has been debated in the past. Endler (1982) pointed out the problem of distinguishing historical from ecological factors in species differentiation. Endler was responding to and disagreed with Cracraft's (1982) use of vicariance to account for evolution of the Australian avifauna. Caughley et al. (1987b) argued that kangaroos were a product of climate, and that biological interactions were not important, but Fox (1989), in his review, concluded that competition was a major source of separation of habitat between kangaroo species. If Caughley (1987b) and McLeod (1997) are correct (as we believe) that a stable-equilibrium carrying capacity is not very meaningful for widely fluctuating kangaroo populations, then competition as an organizing force would be expected to be reduced.

The role of biogeography, as opposed to competition among co-evolved species, must be evaluated in the light of the climatic history of the continent. Australia appears ancient and timeless. In its vastness and flatness, it seems worn and eroded, a remnant of the ravages of ages past. Its harsh dry climate, bleak aspect, and fugitive plant and animal communities suggest a forlorn landscape where time has stood still. We associate it with the ancient continent, Gondwanaland, and we talk of its primitive mammalian fauna, composed of egg-laying monotremes and the only slightly more advanced marsupials. In its isolation and unchanged nature, it is as if the Australian continent were a relic from the past.

From a broad perspective, this impression is correct. Australia's ancient mountains have largely been worn away. But its flatness is due not only to time but also to its location, far removed from the junctures of tectonic plates where mountains are built. And the isolation of the continent by vast oceans spared it from invasions by plants and animals produced by evolutionary experiments in other lands. But unchanging it is not. As previously noted (see chapter 1), prior to the Pleistocene, the continent was home to many other species of large mammals, including other species of kangaroos (Flannery 1984). Only 10,000 years ago, there were vast lakes, a different fauna, and moist forests where almost lifeless deserts now exist (Frakes et al. 1987). The continent has endured repeated cycles of alternating cold and warm periods, and wet and dry periods (Bowler 1978, 1983; Galloway and Kemp 1981). The impression of timelessness and stability of the continent is an illusion.

In continents dominated by topography, shifts in climate are spatially constrained by patterns of valleys and mountains. On a nearly featureless expanse such as outback Australia, by contrast, a comparable climate change would extend its effects over broad areas, just as water poured into a shallow pan has a much greater surface area than the same amount poured into a bowl. Consequently, biogeographic shifts due to changing climate would have had a much greater spatial scale in the Australian environment than would be the case in most continents.

If the past could be replayed in fast-forward, we would see plants and animals shifting over the continent in flows and ebbs on a major scale until 10,000 years ago, and on a smaller scale up to 2000 to 3000 years ago. Frakes et al. (1987) reported a wet period between 8000 and 6000 years ago, and Wasson (1976) reported a dry period and wind-blown dune building in the Yathong area 3000 to 2000 years ago. Recall that Yathong is the Aboriginal word for sandhills, and that the western side of Yathong is characterized by such a landscape.

The "ghosts of competition past" model for the evolution of niche relationships assumes a long coevolutionary history over which selection for traits that reduced interspecific competition occurred. A coevolved community implies that the species evolved together over long periods of time. However, there is serious doubt that the current assemblage of kangaroos consistently coexisted over the various expansions and contractions of species distributions that accompanied periods of cold and warm, wet and dry. Indeed, it seems reasonable that the sibling species, eastern and western grey kangaroos, diverged during a period of very different environmental circumstances—when there was a large separation between their respective ranges, with western greys confined to the southwestern area of the continent, and the eastern greys confined to the eastern mountains (Main 1987; Heatwole 1987; Clemens et al. 1989; Maynes 1989). Although directional selection for adaption to different environments occurred, ultimately, it was genetic differentiation that accounted for this speciation event. Templeton (1981) has termed this mechanism transilience, or a genetic separation; the alternative, divergence based on separation, is inadequate since eastern and western greys would have hybridized upon secondary contact, if genetically compatible.

The observation of people in the Yathong area that western grey kangaroos used to be dominant, and that red kangaroos had increased recently, as well as Dawson's (1995) observation that both eastern and western grey kangaroos first appeared at Fowler's Gap as late as the 1970s indicate the labile distribution of the three species in the overlap zone in western New South Wales. Periodic drought patterns in the historical records (Robertson et al. 1987) undoubtedly influenced the three species differently. Shifts in ranges even into recent times raise questions about the stability of coexistence. Thus, a coevolutionary model to account for niche relationships among the three kangaroo species is problematic.

An alternative model that involves geographically separate strongholds and periodic overlap seems more parsimonious to explain the apparent niche relationships of kangaroos. Rather than "ghosts of competition past" it invokes "ghosts of adaptation elsewhere" to explain relationships between species. Instead of attributing the observed differences entirely to ecology, this view considers separate histories and short times of coexistence to explain the differences between species, and why the assemblage is not an ecological optimum in the current environment. Consider the situation where three species each have their respective strongholds in different regions, as shown schematically in figure 12.1A. Each species would be selected under different circumstances and would adapt to the characteristic habitats of the stronghold environment.

Now let us assume that the continental climate changed in ways that resulted in overlap of the habitats of the three species (fig. 12.1B). Some elements of the habitats amalgamate into new species compositions, but others retain much of their original structure in the microenvironments

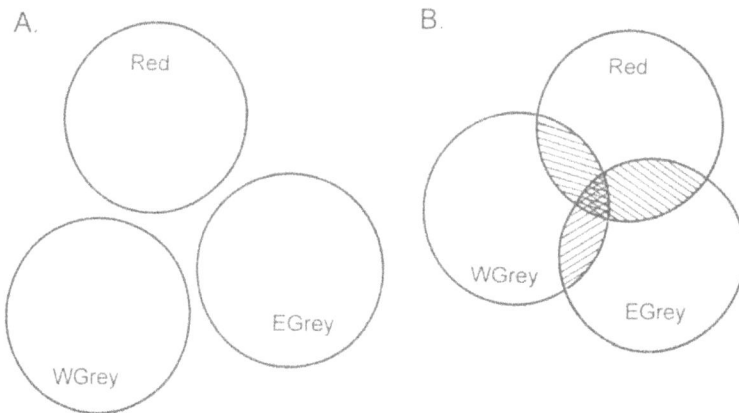

Figure 12.1. Schematic diagram of stronghold and periodic overlap model. Panel A shows the three species, each in its own stronghold during climatic conditions when the distributions are separate. Panel B shows periods of climatic conditions when the three species overlap.

that resemble the stronghold environment. Each species would expand into the overlap area along with the habitats of its choice to create an overlap of species, but with each preferring the habitat characteristic of its own stronghold in the nonoverlap part of its range. Because each species remains best adapted to its own kind of habitat, an apparent "community structuring" and "niche separation" would result.

In fact, with climate changes, kangaroo distributions surely shift more rapidly than does vegetation. However, the two are not tightly coupled, and therefore distributions of specific habitats do not necessarily limit distributions of kangaroos. If preferred habitat, similar to that in the stronghold, is present in the expansion area, kangaroos would occupy that habitat. If not, they would occupy the most similar habitats that would support the species. For example, although the western limit of eastern grey kangaroos is set more by climate than habitat, their habitat becomes sparser and more confined to stream courses to the west. The southern distribution of red kangaroos is set by winter cold, not suitable habitat. The variables controlling the eastern and northern limits of the western grey kangaroo are not so obvious, and may be either habitat or climate, or a combination of both.

Cairns et al. (1988) noted considerable shift in the numbers of red and western grey kangaroos in South Australia due to drought. Similarly, Fox (1965) presented isoclines on the relative abundance of red and western grey kangaroos, and Caughley et al. (1984b) of western and eastern grey kangaroos, both of which suggest substantial shifts over drought cycles. In 1965, red kangaroo numbers were 25 percent those of western greys in the Yathong area as compared to 155 percent obtained in this study (table 4.3). The two species of grey kangaroos first appeared at Fowler's Gap in the 1970s (Dawson 1995). These changes in the relative abundance of red kangaroos to western grey kangaroos agree with local sheep station owners' impressions in the Yathong area, and they illustrate how labile the overlap zone is. Indeed, from Fox's (1965) isoclines, one need move only 100 km east of Yathong to find no red kangaroos. The transitions in rainfall amounts are steep in the Yathong area, and it is likely that trends in rainfall on the scale of decades could, and probably have, shifted or eliminated the overlap zone. Eastern grey kangaroo distributions also retreat eastward considerable distances during drought (Denny 1975).

Competition among kangaroo species in the overlap zone might occur, but if the overlap zone were small in extent or brief in time, selection would be swamped by gene flow from the much larger stronghold populations. Also, selection due to competition would be interrupted both by large regional trends of climate and short-term drought cycles. The former would shift the balance among the species' habitats, and the latter would reduce population numbers (thereby reducing intensity of competition during the subsequent increase phase) and differentially affect the three species. Divergence by competitive displacement in the overlap zone would be less likely in a fluctuating environment. Kangaroo populations seldom appear to be at carrying capacity in outback environments. Furthermore, as noted earlier, the seasonal window for competition is narrow. Competition is minimal when green feed is abundant and widespread, or when the vegetation is dry and of uniformly low quality. Competition would be expected to be a strong process mainly during the transition between seasons, when quality green feed is concentrated in microsites that retain moisture for the longest time.

Keddy (1989) noted that disturbance alters competitive relationships and may allow the coexistence of more species than undisturbed communities, which are dominated by fewer species with the greatest competitive ability. In aquatic systems, this phenomenon has been termed the "intermediate disturbance hypothesis" (Ward and Stanford 1983). In temperate zones, disturbance is usually thought of as occurring as discrete events (e.g., storm, fire, and so forth) on a limited scale. In outback Australia, however, shifting climates might be regarded as continuous disturbance over a broad geographic scale. This raises the possibility that the species richness of areas like Yathong may be greater than it would be if climatic conditions were stable and competitive interactions were allowed to proceed over longer periods of time. Competitive exclusion might then occur with the elimination of one or more species from the overlap zone.

Still, some level of productivity may be necessary before competitive structuring of communities can occur. The outback is a severe environment, and low population abundance and a narrow seasonal competitive window may not be sufficient for competition to act as an important evolutionary force. Keddy (1989) noted a paradox that competition must have a limited resource to occur, whereas in resource-poor environments, competition does not occur because populations do not reach sufficient levels to interact competitively. We think that Yathong, and perhaps most of the Australian outback, is below a threshold of productivity for competition to be a major animal community-shaping process. If density of kangaroos is too low, the impact of scramble competition is spread too thinly over a uniform resource base to be important, and the gain in additional resources obtained by interference competition does not offset the energy and risk of aggression.

All things considered, we propose that biogeography accounts for most of the differences between the kangaroo species we studied at Yathong. The red kangaroo's stronghold is in the large, central area of Australia where summer rains predominate. The eastern grey's stronghold is in the mesic eastern Great Dividing Range with cold winters and mild summers. The western grey's stronghold is in the southwestern part of the continent with its cool, wet winters and hot, dry summers. These species come together in an overlap zone in western New South Wales (Caughley et al. 1984b, 1987a), with Yathong near the center—that is, the three-species overlap area shown schematically in figure 12.1B. There was little competition between kangaroos at Yathong because of their distinct separation by habitat. Even in areas where they apparently competed, such as at concentrated green feed sites, there was microscale separation: the red kangaroos were mainly in the center and western greys around the edges. Significantly, no interference competition (aggressive displacement) was observed in these situations.

As stated earlier, we observed that Yathong is a transition area where elements of the large eastern mountain and southern winter rainfall zones of Australia come together. So, too, do the characteristic denizens of those environments—the three species of kangaroos. Although some elements of the flora mix, to a considerable degree the dominant elements are interdigitated in patchlike fashion according to topography, soils, moisture regime, and other factors of the microenvironment. Even further patchiness, induced by fire and floods, influences plant establishment and mortality. This diversity is what drew us to Yathong. As an intergrade zone, Yathong is not stable in the long term. Species composition can easily be shifted in favor of any of the three kangaroo species, and against the others, by periodic shifts in climate. Longer term trends, global warming for instance, would favor the red kangaroo and disfavor the two species of greys, particularly the eastern grey. But the overlap zone itself would be expected to shift to the south and east as well.

We do not wish to dispute the general importance of competition and coevolution as forces shaping community structure. Still, it is best to remember that just because results fit a model does not prove the model correct. Complex interrelations may have historical and phylogenetic roots that override current ecological conditions. Before one accepts ghosts of competition past, it is advisable to establish that the plant and animal members of the community have indeed had a long history of coevolution in the same community. Biogeography therefore needs to be eliminated as an alternative explanation. Finally, it is not necessarily all one cause or the other. The ghosts of competition past and the ghosts of distributions past may have danced together.

Predation and Group Formation

Formation of groups is a major means for openland dwelling herbivores to avoid predation (McCullough 1969; Hamilton 1971; Alexander 1974; Coulson 1999). Geist (1974) and Jarman (1974) have pointed out that group size in ungulates is related to their use of escape and concealment cover. Species that occupy closed habitats, such as mule deer and white-tailed deer, are either solitary or form small groups and escape predation by stealth and crypsis. Open habitat species, such as elk, bison, and caribou, form larger groups and escape predation by flight or mutual defense. Even white-tailed deer (Hirth 1977; McCullough 1982) and mule deer (Bowyer 1984) form larger groups when they venture into open areas to feed.

There is some ambiguity about the impact of predation on kangaroo social organization. Low (1979) and Heathcote (1987) reported larger group sizes of kangaroos in open areas, and suggested predation as a cause. Oliver (1986) reported that female red kangaroos with young-at-foot remained closer to cover than other classes. Clark et al. (1995) found smaller group sizes at night than in daytime for eastern greys; however, because surveillance time did not increase at night, they suggested that smaller groups are more cryptic to predators. Coulson (1999) suggested that formation of mixed-species groups was in response to predation. Jarman and Wright (1993) found that eastern grey kangaroos in large groups benefited from higher total vigilance, and were seldom surprised by dingoes, as were some small groups. Most kangaroos fled from dingoes, and large groups fled at greater distances than small groups. Coulson (1999) suggested that mixed-species groups gained advantages in surveillance for predators. In their review, Jarman and Coulson (1989) concluded that kangaroos had been subject to predation to a degree that well could have shaped their behavior. On the other hand, Russell (1984) and Lee and Cockburn (1985) questioned the role of predation in shaping social behavior in kangaroos.

Kangaroos at Yathong formed very small groups, whether in woodland habitats or in the open grasslands (chapter 7). Group sizes for all three species were similar and low ($\bar{x} = 2.08$ for eastern greys, 2.06 for reds, and 1.91 for western greys). Mean group sizes at Yathong were even smaller than for most other studies of these species. We noted seasonal changes in mean group size, but these appeared to be related to breeding and aggregation at concentrated green feed patches rather than predator avoidance.

Kangaroos differ from large herd-forming ungulates in the use of groups for predator avoidance. Herd-forming ungulates use the social group to organize use of space, and groups move over a common home range shared by the group (e.g., McCullough 1969). Kangaroos, by contrast, occupy individual home ranges and lack such a fluid social structure. Nevertheless, they use open areas to a substantial degree, especially the reds and western greys, and this behavior must have been allowed by past predation costs. Kangaroos are not particularly cryptically colored; in fact, they are readily observable in all habitats, especially in openlands where they are often observed at distances greater than a kilometer.

Predators can readily see kangaroos in the habitats they favor on Yathong. Therefore, it must be the case that kangaroos can successfully avoid predator attacks not to depend on either crypsis or group formation. Although Yathong's predator fauna is not so varied as in other parts of the world, the dingo probably posed a significant threat, and kangaroos must have learned to successfully outrun them or fend them off.

Both high speed escape and defense were probably responsible for kangaroos being able to avoid predation without utilizing the benefits of a group in environments such as Yathong. Their bounding speed is quite rapid and exceeds the upper ranges of elk, caribou, or bison, but does not match that of pronghorn antelope, perhaps the most fleet ungulate. The immediate response of kangaroos to disturbance is to take flight and put distance between themselves and the source of disturbance. The favorable energetics of hopping means that sustained flight can be maintained at less cost by kangaroos than comparable-sized ungulates. As noted in chapter 8, the two species of greys tend to head for wooded cover, whereas the red kangaroo usually remains in open areas during flight. Furthermore, larger groups and mixed-species groups tend to break up upon disturbance. Even if they flee in the same direction, they seldom close ranks in the manner of ungulates. Thus, for example, the open country–adapted tule elk, in order to maintain group cohesion, runs in groups at speeds slower than each individual comprising the group can sustain (McCullough 1969).

Among kangaroos at Yathong, alarm behavior essentially appeared to be selfish. Particularly interesting is the local folklore: when severely pressed by a pursuer, kangaroo females sacrifice larger pouch young by dumping them out to be taken by the predator while the mother escapes. Evidence in support of this belief derives from field researchers who say it is necessary to tape around the pouch to retain larger young before releasing a female kangaroo captured by stunning (Bevan Brown and Jeff Short, personal communication). Apparently, their pouch muscles are relaxed when female kangaroos are about to be captured; thus larger pouch young are sacrificed in favor of survival of the female. Robert-shaw and Harden (1986) reported similar ejection of pouch young by female swamp wallabies (*Wallabia bicolor*) pursued by humans.

In addition to escape by running, kangaroos are much better adapted to ward off attacks by predators than comparably sized ungulates. Both have dangerous hind limbs that can deliver severe kicks. Ungulates can strike with the forelegs, and some have horns or antlers; however, these weapons require a head-on posture that brings vital areas close to the predator. Headgear requires close approach, and the range of forelimb use in ungulates is restricted mainly to forward and backward motion. Lack of clavicular articulation, and the rotation that this allows, limits sideways motion—one cost of high-speed running on four legs.

Kangaroos, in contrast, have highly dexterous forelimbs because the bipedal gait has freed them from specialization for running (Jarman 1983b). Forepaws are excellent weapons for fighting, particularly if armed with sharp claws. Large body size confers protection against predation for eastern grey and red kangaroos. On the other hand, western grey kangaroos, the smallest kangaroos at Yathong, possess razor sharp claws, which are formidable weapons against each other, and probably against predators. Although dingoes can probably simply overpower western grey juveniles if they catch them, adults of both sexes would seem to be a fair match, face-to-face with a lone dingo. These dangerous weapons, along with their belligerent nature, make western greys, despite their small size, serious adversaries for would-be predators, and probably accounts for their ability to occupy open areas, frequently as lone individuals.

In any event, kangaroos at Yathong do not use group formation as a means of thwarting predators, and yet they can still effectively use openland environments. Again, kangaroos do not conform to models of evolution based on ungulates. Ironically, the situation may even be the reverse of that of ungulates. The kangaroos showing largest social groups are eastern greys in the eastern mountains where escape cover is high, and according to ungulate models, one would expect crypsis and small group sizes.

Polygyny and Breeding Group Size

Alexander et al. (1979) and Clutton-Brock et al. (1982) both reported a strong positive correlation between degree of sexual dimorphism and the size of breeding groups (or harems). Kangaroos in this study displayed a notable lack of conformation to the ungulate model of male reproductive success being correlated with breeding group size. Kangaroos did not form larger breeding groups despite sexual dimorphism greatly exceeding that of most ungulates (Weckerly 1998). The most dimorphic species, the red kangaroo, did not have larger group sizes than the less dimorphic two species of grey kangaroo (chapter 7). All three species formed consort pairs, or tending bonds, and multifemale groups did not play a role in the breeding system. Other females, rival males, and young-at-foot were sometimes present, but did not figure into the male kangaroo strategy to obtain copulations.

Several variables seem to contribute to this difference. First, in ungulates, female-young groups form in open habitats, apparently in response to predation, and males in those species are able to capitalize on this behavior by defending such groups as harems. Males are not successful in retaining harems if the disposition of females is to disperse, which is typical of ungulates in habitats with cover. The poor visibility in such habitats also makes it physically difficult for males to control harems. Thus, the size of the breeding group is largely determined by the behavior of females, and males are able to contain and defend such groups only if allowed to do so by females. Kangaroo females do not form large groups except when concentrations occur in response to concentrated green food sources—generally when green feed is drying out. However, no harem behavior was observed at these times either.

Second, the breeding season in kangaroos, particularly red kangaroos, which are virtually aseasonal, is much longer than is typical of ungulates. Ungulates in temperate zones usually have relatively short breeding seasons that are constrained by selection so that young will be born at the most favorable period for survival. This concentrates female receptivity to a narrow time period, and increases the number of females likely to be receptive simultaneously. These factors favor harem systems as the most successful way for dominant males to monopolize females, and this places an even greater premium on dimorphism. Ungulates in which the females are dispersed usually have breeding systems based on consort pairs or territoriality; in these systems, a few males are less able to so completely dominate reproduction with consequent lower variance in reproductive success between males and weaker selection for dimorphism.

In kangaroos, by contrast, more dispersed females can be bred by dominant males in consort pairs because breeding is spread over a longer time interval and more females can be reached in serial consort pairs than is typical of temperate zone ungulates. Furthermore, by indicating their estrus state, female behavior may further contribute to the success of dominant males (Ganslosser 1995). This success is greatest in the nearly aseasonal breeding of the red kangaroo, which probably yields reproductive success equivalent to or exceeding harem-breeding ungulates. In fact, sexual dimorphism in kangaroos exceeds that of ungulate species with the largest harem groups, and rivals that of pinnipeds, which gather in dense breeding colonies on beaches (Weckerly 1998). Hume et al. (1989) claim that kangaroos are more dimorphic than pinnipeds and therefore are the most dimorphic of mammals, but data presented by Weckerly (1998) suggest the opposite. For ungulates, the maximum dimorphism (for red deer and elk) for males is about 1.7 times the mass of females (McCullough 1969; Alexander et al. 1979; Clutton-Brock et al. 1982). In kangaroos, males have two to three times the body mass of females (Frith and Calaby 1969; Newsome 1977b; Poole et al. 1982a, 1982b; Taylor 1983). We think that the degree of sexual dimorphism in kangaroos likely reflects the relative success of a few males in dominating breeding opportunities. Therefore, the ungulate correlate of sexual dimorphism with breeding group size certainly does not apply to the consort pair breeding system of kangaroos. This discrepancy raises questions about the robustness of the correlation. It seems to apply primarily to temperate zone species; inclusion of tropical and aseasonal breeding species reduces its strength.

Body Size, Habitat Disturbance, and *r*- and K-Selection

Among ungulates, and large mammals in general, there is a broad correlation between increasing body size and degree of K-selectedness (McCullough 1979, 1992). Nevertheless, within large mammals, species vary in their relative position on the *r*- to K-selected continuum. Multiple births and early sexual maturity are characteristics of the smaller species; they have relatively short life spans and are the most *r*-selected species. The great whales and elephants have but a single offspring per birth, do not reach sexual maturity for many years, and have very long lifespans.

Still, among all species of mammals, there are many exceptions to the correlation of body size and K-selection. Humans are highly K-selected, but have relatively small body sizes. Sexual maturity is greatly delayed, as is rate of growth and maturation of offspring. These life history traits are apparently linked with the long learning period required to utilize the highly developed brain. Intelligence and the behavioral adaptability it confers probably increase survivorship and lifespan without requiring a large body, and thereby allow slow maturation, and consequently, lower reproductive rate. Bats provide another exception to the body size and K-selection correlation (Myers 1978). The constraints of flight limit body and litter size; thus, many species of bats have long lifespans and low reproductive rates, and hence are K-selected despite small body size.

Among cervids, moose (*Alces americana*) depart somewhat from the body size correlation also. They are the largest member of the Cervidae, but show multiple births, with twinning common. This deviation seems to relate to their preference for frequently disturbed habitats (Peterson 1955). Indeed, disturbance species—those that favor secondary successional stages of vegetation—are often more *r*-selected than comparably sized species occupying climax habitats. Selection favors rapid population growth in species adapted to habitats periodically created by unpredictable events (fires, windstorms, floods, logging, and so on). Thus, successional species would be expected to vary from a simple correlation of body size with K-selected attributes. More generally, *r*-selected attributes would be more prevalent in those species living in boom-and-bust environments where good and bad conditions are due to habitat stages in a successional series or to drought-rainfall oscillations as with kangaroos.

The closest North American ungulate counterparts to kangaroos, the odocoileid deer, are adapted to secondary successional habitats. The white-tailed deer is more *r*-selected than the mule deer, and has an early age of sexual maturity, higher mean litter sizes in young females, and a lower average life expectancy (McCullough 1987). Both species do best under disturbed conditions, and can be considered primarily adapted to subclimax habitat, although both can sustain viable populations in some climax habitats. The degree to which these species are *r*-selected is dependent upon the frequency of habitat creation (i.e., how frequently the climax or advanced successional stages are set back to early successional stages), and the amplitude of change in carrying capacity from pre- to postdisturbance. These variables control the circumstances under which and over how long a period rapid population growth can occur, and thus, the frequency with which genes promoting rapid production of offspring are favored over those promoting slower production of young and longer life span in a competitive environment.

Both the frequency and particularly the amplitude of variation in carrying capacity is very great in the range of eastern white-tailed deer. Under poor habitat conditions, or in dense populations in good habitat, few yearling females reach sexual maturity, and many fawns of adult females fail to survive. Under good conditions, female fawns reach sexual maturity at seven months and produce an offspring at thirteen months (Haugen 1975; McCullough 1979). Twins and even triplets are commonly recorded in fawn females (Haugen 1975; McCullough 1979). Under poor conditions, female fawns weigh approximately 18 kg, whereas they often exceed 36 kg under good conditions. By lowering density of the George Reserve population in Michigan, McCullough (1982) produced male fawns that exceeded 45 kg in the fall; also, yearling males, which under poorer conditions had simple spike antlers, had eight points.

Such pronounced phenological variation has not been recorded for mule deer, and only a few cases of female fawns reaching sexual maturity have been reported (Connolly 1981; McCullough 1997). The Pacific northwestern mule deer (the black-tailed races) live in forested areas, and show strong growth responses to creation of successional stages (Einarsen 1946). However, the frequency of such disturbance, primarily by fire, must have been low to account for the large expanses of uninterrupted old growth Douglas-fir (*Pseudotsuga menziesii*), and coast redwood (*Sequoia sempervirens*) found by the first European explorers to reach California, Oregon, and Washington. Indeed, the sitka deer (*O. h. sitkensis*) of southeastern Alaska appears to be more of a climax than subclimax-adapted animal (Kirchoff and Schoen 1987). The mule deer is less *r*-selected than the white-tailed deer.

The situation in Australian kangaroos presents an intriguing contrast to that of ungulates. The counterpart of the white-tailed deer, the eastern grey kangaroo, is the most K-selected, whereas the red kangaroo is the most *r*-selected (Richardson 1975; Poole 1983a). The mule deer counterpart, the western grey kangaroo, is intermediate, judging from life history traits (Lee and Ward 1989) (table 8.5). Reproductive rate is not a simple function of body size; at Yathong, the red kangaroo is the largest of the three species, with the eastern grey nearly as large and the western grey kangaroo the smallest. Consequently, two of the three species studied deviate from the body size and K-selection correlation. Only the eastern grey kangaroo is large and K-selected, whereas the small western grey is also relatively K-selected, and the largest species, the red kangaroo, is by far the most *r*-selected.

Part of the difference is probably due to the fact that the western grey kangaroo is a seasonal breeder, whereas the eastern grey kangaroo is weakly aseasonal, and the red kangaroo is aseasonal. The combination of a single offspring per litter and a favorable season for birth limits the western grey kangaroo to one offspring produced per female per year (Lee and Ward 1989). Although eastern grey females are to some extent aseasonal breeders in environments such as Yathong, in their prime range in the eastern mountains, they are seasonal because of cold winters, and are also limited in reproduction to one offspring per female per year (Pearse 1981; Lee and Ward 1989).

The overall conservative reproductive strategy of the eastern grey kangaroo may be a function of infrequent creation of disturbed areas, difference between good and poor habitats, or a combination of both. The heavy consumption of coarse, low-quality grasses by eastern greys in situations where sheep or other species of kangaroo are more selective (Griffiths et al. 1974) suggests that low habitat amplitude is the more important variable.

The relatively *r*-selected red kangaroo, by contrast, lives in a zone of frequent and severe drought. Populations frequently suffer severe declines (Frith and Calaby 1969; Newsome 1971; Caughley 1987b). Its reproductive biology of simultaneously having an embryo in utero, a young in pouch, and a suckling young-at-foot shortens the interval between successive offspring as well as the start-up time after drought conditions break (Newsome 1975; Low 1978; Poole 1983a). These variables maximize rate of population increase under favorable conditions, which occur when rains follow drought periods. The potential annual production of young of the red kangaroo is 1.5 as compared to 1.0 for the two species of grey kangaroos (Lee and Ward 1989).

The situation with western grey kangaroos appears to be intermediate between the eastern grey and red kangaroo, although more similar to the eastern grey than red. Its early maturation (table 8.5; Lee and Ward 1989) is offset by its seasonal breeding habits that limit offspring production to one per year. It seems to be less associated with disturbed vegetation, instead living a conservative (i.e., relatively K-selected) life in relatively low-quality habitats. It can successfully occupy mallee habitat, which the other two species use only sporadically. It has broad food habits and can detoxify secondary compounds that are common in plants of low-quality habitat. Its habitat requirements are more variable than those of the other two species.

The heart of its range is more frequently subjected to drought than that of the eastern grey. It is a tough survivor. Its reproductive rate is lower than that of the red kangaroo, and it responds less rapidly following the return of favorable conditions. However, the amplitude of fluctuation between good and bad is less extreme in the habitat of the western grey than in that of the red kangaroo.

These kangaroos also deviate from the usual pattern of small species being the most selective of diet quality (Hofmann 1973; Jarman 1974; Jarman and Phillips 1989). The red kangaroo, the largest species at Yathong, was clearly the most selective for high-quality diet, and the work of Short (1985, 1987) showed its superiority to the western grey in gathering a meal when forage is limited.

We conclude that kangaroos do not fit the r- and K-selection model based upon ungulates. The largest species, the red kangaroo, is the most r-selected, and the smallest species, the western grey, is the most K-selected. The eastern grey, also a large species, is intermediate but similar to the western grey kangaroo.

Discussion of Kangaroos and Ungulate Comparisons

It is no longer tenable to consider kangaroos as primitive animals. The growing body of research on kangaroos indicates that they are highly evolved in adaptation to a dry, harsh, and unpredictable environment. Both physiologically and anatomically, they are more efficient than ungulates in such environments. Our results show kangaroos to be well adapted behaviorally and ecologically to environments like Yathong. Their abundance today, after more than a century of competition with massive numbers of domestic stock (despite hindrance to kangaroos and assistance to livestock) and exotic rabbits, is pragmatic evidence for rejecting the assumption that they are competitively inferior to grazing placental mammals from other parts of the world.

Although species richness of large, grazing mammals in the Australian fauna is low, and perhaps still evolving from a major extinction period in the Pleistocene, the current four species at Yathong largely fill the grazing niche. The three kangaroo species studied, eastern and western grey and red, show considerable differentiation in size, behavior, and ecology; this, in addition to differences between the sexes, results in the large grazer fauna at Yathong functioning much as six species, ecologically. The rock-dwelling euro, too uncommon on Yathong to be included in the study, extends the complex to seven species (or eight if differences by sex in euros are similar to those of the species studied), and the emu makes eight. This list is comparable in number to that of ungulates in major grassland areas at similar latitudes in the Northern Hemisphere.

Kangaroo species in this study showed niche differentiation that is usually attributed to past competitive interactions in comparable studies of ungulates. It is easy to view kangaroo interspecific relationships as community structuring, if this paradigm is applied. However, the broad, flat landscape of Australia, climatic shifts (i.e., short-term drought cycles and long-term global climatic oscillations) present compelling evidence that climatic adaptation was the major evolutionary force shaping kangaroo species. Relatively small changes in rainfall result in large shifts in spatial distribution of kangaroos, a consequence of a nearly level landscape. Overlap of the kangaroo species occurred at some times in some places, as currently at Yathong, and undoubtedly, competition was a weak selective force during these intervals. Yet the overlap zone is relatively small, and from all appearances, shifts frequently. If competition were a strong force historically, one would expect stronger evidence of competitive interactions currently. But despite the three species of kangaroos sorting significantly by habitat characteristics, there is considerable overlap, and a notable lack of interspecific strife beyond defense of individual space. Mixed-species groups, which rarely occur in the comparable grazing assemblages of large mammals in the Northern Hemisphere, are common. Our results suggest that evolution of each species of kangaroo in the stronghold of its range, followed by secondary contact, is a more parsimonious explanation of apparent structuring of the three species in their common overlap zone.

Overall, the fit of kangaroos to the evolutionary, ecological, and behavioral models of ungulates is poor. Based on these models, most of our expectations of what we would see in kangaroos on Yathong were not met. Although arguments justifying the deviation of kangaroos from the expected may be made, they are weak overall and easily subject to counterargument. It may be best to confront the fact that kangaroos do not fit ungulate models very well. Although our comparison of kangaroos was made primarily with white-tailed and mule deer, the fact that kangaroos did not conform to the evolutionary models for deer, and deer conform to other ungulates, makes a broader interspecific comparison superfluous.

We conclude that, if applied to grazing systems generally, these evolutionary models are not robust. Their failure to predict kangaroo ecology and behavior is partly due to the unique physiology and anatomy of kangaroos. Kangaroos are not simply oddly constructed ungulates, and this allows them to behave in ways different from ungulates. But in addition, these models are products of the environments in which they were derived. They have traveled well to comparable environments in other places. However, Australia really is a different sort of place. Kangaroos and the Australian outback, both being extremes, are the acid test for these models. If the models had predicted kangaroos correctly, we could have confidence in their robustness and generality. The fact that they did not is not reason enough to totally reject them, for they do fit ungulates in an impressive array of environments. Furthermore, over broader scales of body size from smallest to largest mammals, they may have validity for macropods (Kaufman 1974b, Jarman 1983b, Jarman and Coulson 1989). Consequently, kangaroos may represent local "noise" in a global "signal." Still, these models have their limits, and kangaroos should be a cautionary tale; correlations are problematical, and it behooves us to look more carefully at extreme cases to challenge their power to explain. Contrary to the old saying, exceptions do not prove the rule. Instead, they allow refinement or limitation of the rule in ways that perhaps make the rule less general, but ultimately more useful.

LITERATURE CITED

Alexander, R. D. 1974. Evolution of social behavior. *Annual Review of Ecology and Systematics* 5:325–383.

Alexander, R. D., J. L. Hoogland, R. D. Howard, M. Noonan, and P. W. Sherman. 1979. Sexual dimorphism and breeding systems in pinnipeds, ungulates, primates, and humans. In N. A. Chagnon and W. Irons, eds., *Evolutionary Biology and Human Social Behavior: An Anthropological Perspective*, pp. 402–435. North Scituate, Mass.: Duxbury Press.

Anderson, A. E., D. E. Medin, and D. C. Bowden. 1974. Growth and morphometry of the carcass, selected bones, organs, and glands of mule deer. *Wildlife Monographs* 39:1–122.

Anderson, D. R. 1980. *Information on Management Programs for Red, Eastern Grey, and Western Grey Kangaroos in Relation to the U.S. Endangered Species Act.* Logan, Utah: Utah Cooperative Wildlife Research Unit.

Anderson, D. R. and C. Southwell. 1995. Estimates of macropod density from line transect surveys relative to analyst expertise. *Journal of Wildlife Management* 59:852–857.

Archer, M. 1981. A review of the origins and radiations of Australian mammals. In A. Keast, ed., *Ecological Biogeography of Australia*, pp. 1435–1488. The Hague, Netherlands: W. Junk.

———. 1984a. The Australian marsupial radiation. In M. Archer and G. Clayton, eds., *Vertebrate Zoogeography and Evolution in Australasia (Animals in Space and Time)*, pp. 633–808. Carlisle, Australia: Hesperian Press.

———. 1984b. Evolution of arid Australia and its consequences for vertebrates. In M. Archer and G. Clayton, eds., *Vertebrate Zoogeography and Evolution in Australasia (Animals in Space and Time)*, pp. 97–108. Carlisle, Australia: Hesperian Press.

Archer, M. and G. Clayton, eds. 1984. *Vertebrate Zoogeography and Evolution in Australasia (Animals in Space and Time)*. Carlisle, Australia: Hesperian Press.

Arnold, G. W., A. Grassia, D. E. Steven, and J. R. Weeldenberg. 1991. Population ecology of western grey kangaroos in a remnant of wandoo woodland at Baker's Hill, southern Western Australia. *Wildlife Research* 18: 561–575.

Arnold, G. W., W. Lecrivain, K. G. Johnson, and A. Grassia. 1988. Effects of weather conditions in summer on the maintenance behavior of western grey kangaroos, *Macropus fuliginosus*. *Australian Wildlife Research* 15: 129–138.

Arnold, G. W., D. E. Steven, and J. R. Weeldenberg. 1994. Comparative ecology of western grey kangaroos (*Macropus fuliginosus*) and euros (*M. robustus erubescens*) in Durokoppin Nature Reserve, isolated in the central wheatbelt of Western Australia. *Wildlife Research* 21: 307–322.

Bailey, N. T. J. 1951. On estimating the size of mobile populations from recapture data. *Biometrika* 38: 293–306.

Bailey, P. 1992. A red kangaroo, *Macropus rufus*, recovered 25 years after marking in north-western New South Wales. *Australian Mammalogy* 15: 141.

Bailey, P. T. 1971. The red kangaroo, *Megaleia rufa* (Desmarest), in north-western New South Wales. I. Movements. *CSIRO Wildlife Research* 16: 11–28.

Bailey, P. T., P. N. Martensz, and R. Barker. 1971. The red kangaroo, *Megaleia rufa* (Desmarest), in north-western New South Wales. II. Food. *CSIRO Wildlife Research* 16: 29–39

Baker, M. W. de C. and D. B. Croft. 1993. Vocal communication between the mother and young of the eastern grey kangaroo, *Macropus giganteus*, and the red kangaroo, *M. rufus* (Marsupialia: Macropodidae). *Australian Journal of Zoology* 41: 257–272.

Baker, R. H. 1984. Origin, classification and distribution. In L. K. Halls, ed., *White-tailed Deer: Ecology and Management*, pp. 1–17. Harrisburg, Penn.: Stackpole Books.

Banks, J. 1962. *The Endeavour Journal of Joseph Banks 1768–1771*. J. C. Beaglehole, ed. Sydney: Trustees Public Library and Angus and Robertson Publishers.

Barker, R. D. 1987. The diet of herbivores in the sheep rangelands. In G. Caughley, N. Shepherd, and J. Short, eds., *Kangaroos: Their Ecology and Management in the Sheep Rangelands of Australia*, pp. 69–83. Cambridge: Cambridge University Press.

Barker, S., G. D. Brown, and J. H. Calaby. 1963. Food regurgitation in the Macropodidae. *Australian Journal of Science* 25:430–432.

Barnard, C. J. and T. Burk. 1979. Dominance hierarchies and the evolution of "individual recognition." *Journal of Theoretical Biology* 81:65–72.

Bartholomai, A. 1972. Aspect of the evolution of the Australian marsupials. *Proceedings of the Royal Society of Queensland* 82:V–XVIII.

———. 1975. The genus *Macropus* Shaw (Marsupialia: Macropodidae) in the upper Cenozoic deposits of Queensland. *Memoirs of the Queensland Museum* 17:195–235.

Baudinette, R. V. 1994. Locomotion in macropodoid marsupials: Gaits, energetics and heat balance. *Australian Journal of Zoology* 42:103–123.

Bayliss, P. 1987. Kangaroo dynamics. In G. Caughley, N. Shepherd, and J. Short, eds., *Kangaroos: Their Ecology and Management in the Sheep Rangelands of Australia*, pp. 119–134. Cambridge: Cambridge University Press.

Beadle, N. C. W. 1981. The vegetation of the arid zone. In A. Keast, ed., *Ecological Biogeography of Australia*, pp. 695–731. The Hague, Netherlands: W. Junk.

Beal, A. M. 1987. Effect of diet and mineralocorticoid administration on the concentration of anions in parotid saliva from the red kangaroo, *Macropus rufus*. *Australian Journal of Zoology* 35:133–145.

———. 1989. Differences in salivary flow and composition among kangaroo species: implications for digestive efficiency. In G. Grigg, J. Jarman, and I. Hume, eds., *Kangaroos, Wallabies, and Rat-kangaroos*, pp. 189–195. Chipping Norton, Australia: Surrey Beatty.

Beier, P. 1987. Sex differences in quality of white-tailed deer diets. *Journal of Mammalogy* 68:323–329.

Beier, P. and D. R. McCullough. 1988. Motion-sensitive radio collars for estimating white-tailed deer activity. *Journal of Wildlife Management* 52:11–13.

———. 1990. Factors influencing white-tailed deer activity patterns and habitat use. *Wildlife Monographs* 109:1–51.

Bell, H. M. 1973. The ecology of three macropod marsupial species in an area of open forest and savannah oak woodland in north Queensland, Australia. *Mammalia* 37:527–544.

Bentley, P. J. 1960. Evaporative water loss and temperature regulation in the marsupial *Setonyx brachyurus*. *Australian Journal of Experimental Biology* 38:301–306.

Berbach, M. W. 1991. Activity patterns and range relationships of tule elk and mule deer in Owens Valley. Ph.D. diss., University of California, Berkeley.

Bowler, J. M. 1978. Glacial age aeolian events at high and low latitudes: a Southern Hemisphere perspective. In E. M. Van Zinderen Bakker, ed.,

Antarctic Glacial History and World Paleoenvironments, pp. 149–172. Rotterdam: A. A. Balkema.

———. 1980. Quaternary chronology and paleohydrology in the evolution of mallee landscapes. In R. R. Starrier and M. E. Stannard, eds., *Aeolian Landscapes in the Semi-arid Zone of Southeastern Australia,* pp. 17–36. Wagga Wagga, Australia: Riverine Society of Soil Science.

———. 1983. Lunettes as indices of hydrologic change: a review of Australian evidence. *Proceedings of the Royal Society of Victoria* 95: 147–168.

Bowyer, R. T. 1984. The socioecology of southern mule deer: correlates of group living in a large herbivore. Ph.D. diss., University of Michigan, Ann Arbor.

Bridgewater, P. B. 1987. The present Australian environment—terrestrial and freshwater. In G. R. Dyne and D. W. Walton, eds., *Fauna of Australia: Volume 1A, General Articles,* pp. 69–100. Canberra: Australian Government Publishing Service.

Brooker, M. G. and M. G. Ridpath. 1980. The diet of the wedge-tailed eagle, *Aquila audax,* in Western Australia. *Australian Wildlife Research* 7: 433–452.

Brown, B. A. and D. H. Hirth. 1979. Breeding behavior in white-tailed deer. *Proceedings of the Welder Wildlife Foundation* 1: 83–95.

Brown, G. D. 1968. The nitrogen and energy requirements of the euro (*Macropus robustus*) and other species of macropod marsupials. *Proceedings of the Ecological Society of Australia* 3: 106–112.

Brown, G. D. and T. J. Dawson. 1977. Seasonal variations in the body temperatures of unrestrained kangaroos (Macropodidae: Marsupialia). *Comparative Biochemistry and Physiology* 56A: 59–67.

Brown, G. D. and A. R. Main. 1967. Studies on marsupial nutrition. V. The nitrogen requirements of the euro, *Macropus robustus. Australian Journal of Zoology* 15: 7–27.

Burney, D. A. 1993. Recent animal extinctions: recipes for disaster. *American Scientist* 81: 530–541.

Burnham, K. P., D. R. Anderson, and J. L. Laake. 1980. Estimation of density from line transect sampling of biological populations. *Wildlife Monographs* 72: 1–202.

Burt, W. H. 1943. Territoriality and home range concepts as applied to mammals. *Journal of Mammalogy* 24: 346–352.

Cairns, S. C. and G. C. Grigg. 1993. Population dynamics of red kangaroos (*Macropus rufus*) in relation to rainfall in the South Australian pastoral zone. *Journal of Applied Ecology* 30: 444–458.

Cairns, S. C., A. R. Pople, and G. C. Grigg. 1988. Habitat associations of kangaroos in the South Australian pastoral zone. Australian National Parks and Wildlife Service unpublished report.

Calaby, J. H. 1971. The current status of Australian Macropodidae. *Australian Zoology* 16:17–29.

———. 1983a. Antilopine wallaroo *Macropus antilopinus*. In R. Strahan, ed., *Complete Book of Australian Mammals*, p. 252. Sydney: Angus and Robertson.

———. 1983b. Black wallaroo *Macropus bernardus*. In R. Strahan, ed., *Complete Book of Australian Mammals*, p. 254. Sydney: Angus and Robertson.

Calaby, J. H., and G. C. Grigg. 1989. Changes in macropodoid communities and populations in the past 200 years, and the future. In G. Grigg, J. Jarman, and I. Hume, eds., *Kangaroos, Wallabies, and Rat-kangaroos*, pp. 813–820. Chipping Norton, New South Wales: Surrey Beatty.

Caughley, G. 1964. Social organization and daily activity of the red kangaroo and the grey kangaroo. *Journal of Mammalogy* 45:429–436.

———. 1977. *Analysis of Vertebrate Populations*. Chichester, U.K.: John Wiley and Sons.

———. 1987a. Introduction to the sheep rangelands. In G. Caughley, N. Shepherd, and J. Short, eds., *Kangaroos: Their Ecology and Management in the Sheep Rangelands of Australia*, pp. 1–13. Cambridge: Cambridge University Press.

———. 1987b. Ecological relationships. In G. Caughley, N. Shepherd, and J. Short, eds., *Kangaroos: Their Ecology and Management in the Sheep Rangelands of Australia*, pp. 159–187. Cambridge: Cambridge University Press.

Caughley, G., P. Bayliss, and J. Giles. 1984a. Trends in kangaroo numbers in western New South Wales and their relation to rainfall. *Australian Wildlife Research* 11:415–422.

Caughley, G., B. Brown, P. Dostine, and D. Grice. 1984b. The grey kangaroo overlap zone. *Australian Wildlife Research* 11:1–10.

Caughley, G., B. Brown, and J. Noble. 1985a. Movement of kangaroos after a fire in mallee woodland. *Australian Wildlife Research* 12:349–353.

Caughley, G. and G. C. Grigg. 1982. Numbers and distribution of kangaroos in the Queensland pastoral zone. *Australian Wildlife Research* 9:365–371.

Caughley, G., G. C. Grigg, J. Caughley, and G. J. E. Hill. 1980. Does dingo predation control the densities of kangaroos and emus? *Australian Wildlife Research* 7:1–12. ·

Caughley, G., G. C. Grigg, and J. Short. 1983. How many kangaroos? *Search* 14:99–108.

Caughley, G., G. C. Grigg, and L. Smith. 1985b. The effect of drought on kangaroo populations. *Journal of Wildlife Management* 49:679–685.

Caughley, G., N. Shepherd, and J. Short, eds. 1987a. *Kangaroos: Their*

Ecology and Management in the Sheep Rangelands of Australia. Cambridge: Cambridge University Press.

Caughley, G., J. Short, G. C. Grigg, and H. Nix. 1987b. Kangaroos and climate: an analysis of distribution. *Journal of Animal Ecology* 56: 751–761.

Caughley, G., R. G. Sinclair, and G. R. Wilson. 1977. Numbers, distribution and harvesting rate of kangaroos on the inland plains of New South Wales. *Australian Wildlife Research* 4:99–108.

Clancy, T. F. and D. B. Croft. 1989. Space-use pattern of the common wallaroo *Macropus robustus erubescens* in the arid zone. In G. Grigg, J. Jarman, and I. Hume, eds., *Kangaroos, Wallabies, and Rat-kangaroos,* pp. 603–609. Chipping Norton, Australia: Surrey Beatty.

Clancy, T. F., A. R. Poole, and L. A. Gibson. 1997. Comparison of helicopter line transects with walked line transects for estimating densities of kangaroos. *Wildlife Research* 24:397–409.

Clark, M. J. and W. E. Poole. 1967. The reproductive system and embryonic diapause in the female grey kangaroo, *Macropus giganteus. Australian Journal of Zoology* 15:441–459.

Clarke, J. L., M. E. Jones, and P. J. Jarman. 1989. A day in the life of a kangaroo: activities and movements of eastern grey kangaroos *Macropus giganteus* at Wallaby Creek. In G. Grigg, J. Jarman, and I. Hume, eds., *Kangaroos, Wallabies, and Rat-kangaroos,* pp. 611–618. Chipping Norton, New South Wales: Surrey Beatty.

———. 1995. Diurnal and nocturnal grouping and foraging behaviours of free-ranging eastern grey kangaroos. *Australian Journal of Zoology* 43: 519–529.

Clemens, W. A., B. J. Richardson, and P. R. Baverstock. 1989. Biogeography and phylogeny of the metatheria. In D. W. Walton and B. J. Richardson, eds., *Fauna of Australia: Volume 1B, Mammalia,* pp. 527–548. Canberra: Australian Government Publishing Service.

Clegg, S. M., P. Hale, and C. Moritz. 1998. Molecular population genetics of the red kangaroo (*Macropus rufus*): mtDNA variation. *Molecular Ecology* 7:679–686.

Clutton-Brock, T. H., F. E. Guinness, and S. D. Albon. 1982. *Red Deer: Behavior and Ecology of Two Sexes.* Chicago: University of Chicago Press.

Colagross, A. M. L. and A. Cockburn. 1993. Vigilance and grouping in the eastern grey kangaroo, *Macropus giganteus. Australian Journal of Zoology* 41:325–334.

Coman, B. J. 1983. Fox *Vulpes vulpes.* In R. Strahan, ed., *Complete Book of Australian Mammals,* p. 486. Sydney: Angus and Robertson.

Connolly, G. E. 1981. Assessing populations. In O. C. Wallmo, ed., *Mule and Black-tailed Deer of North America,* pp. 287–345. Lincoln: University of Nebraska Press.

Corbett, L. K. and A. E. Newsome. 1987. The feeding ecology of the dingo. III. Dietary relationships with widely fluctuating prey populations in arid Australia: an hypothesis of alternation of prey. *Oecologia* 74:215–227.

Costermans, L. 1983. *Native Trees and Shrubs of South-eastern Australia.* Adelaide: Rigby.

Cottam, G. and J. T. Curtis. 1956. The use of distance measures in phytosociological sampling. *Ecology* 37:451–460.

Coulson, G. M. 1983. Comparative behavioural ecology of the eastern grey kangaroo and western grey kangaroo. Abstract No. 61. *Proceedings 18th International Ethological Conference, Brisbane, Australia.*

———. 1989. Repertoires of social behaviour in the Macropodoidea. In G. Grigg, J. Jarman, and I. Hume, eds., *Kangaroos, Wallabies, and Rat-kangaroos,* pp. 457–473. Chipping Norton, Australia: Surrey Beatty.

———. 1990. Habitat separation in the grey kangaroos, *Macropus giganteus* Shaw and *M. fuliginosus* (Desmarest) (Marsupialia: Macropodidae), in the Grampians National Park, Western Australia. *Australian Mammalogy* 13:33–40.

———. 1993a. Use of heterogeneous habitat by the western grey kangaroo, *Macropus fuliginosus. Wildlife Research* 20:137–149.

———. 1993b. The influence of population density and habitat on grouping in the western grey kangaroo, *Macropus fuliginosus. Wildlife Research* 20:151–162.

———. 1997a. Male bias in road-kills of macropods. *Wildlife Research* 24:24–25.

———. 1997b. Repertoires of social behaviour in captive and free-ranging grey kangaroos, *Macropus giganteus* and *Macropus fuliginosus* (Marsupialia: Macropodidae). *Journal of Zoology,* London 242:119–130.

———. Management of overabundant macropods—are there conservation benefits? In *Managing Marsupial Abundance for Conservation Benefits. Occasional Papers of the Marsupial CRC* 1:37–48.

———. 1999. Monospecific and heterospecific grouping and feeding behavior in grey kangaroos and red-necked wallabies. *Journal of Mammalogy* 80:270–282.

Coulson, G., P. Alviano, D. Ramp, S. Way, N. McLean, and V. Yazgin. 1999. The kangaroos of Yan Yean: issues for a forested water catchment in a semi-arid matrix. In J. L. Craig, N. Mitchell, and D. A. Saunders, eds., *Nature Conservation 5—Managing the Matrix,* In press. Chipping Norton, Australia: Surrey Beatty.

Coulson, G., G. Norbury, and B. Walters. 1990. Forage biomass and kangaroo populations (Marsupialia: Macropodidae) in summer and winter in Hattah-Kulkyne National Park, Victoria. *Australian Mammalogy* 13:219–221.

Coulson, G. M. and J. A. Raines. 1985. Methods for small-scale surveys of grey kangaroo populations. *Australian Wildlife Research* 12:119–125.

Cracraft, J. 1982. Geographic differentiation, cladistics, and vicariance biogeography: reconstructing the tempo and mode of evolution. *American Zoologist* 22:411–424.

Croft, D. B. 1981. Behaviour of red kangaroos *Macropus rufus* (Desmarest, 1822), in north-western New South Wales, Australia. *Australian Mammalogy* 4:5–58.

————. 1982. Radio-telemetry studies on arid zone kangaroos. CSIRO Arid Zone Newsletter (as cited by Clancy and Croft 1989).

————. 1983. Fighting behaviour in the red kangaroo, *Macropus rufus*. Abstract No. 70. *Proceedings 18th International Ethological Conference, Brisbane, Australia.*

————. 1989. Social organization of the Macropodoidea. In G. Grigg, J. Jarman, and I. Hume, eds., *Kangaroos, Wallabies, and Rat-kangaroos*, pp. 505–525. Chipping Norton, Australia: Surrey Beatty.

————. 1991. Home range of the red kangaroo, *Macropus rufus*. *Journal of Arid Environments* 20:83–98.

Croft, D. B. and F. Snaith. 1991. Boxing in red kangaroos, *Macropus rufus*: aggression or play? *International Journal of Comparative Psychology* 4: 221–236.

Damuth, J. 1981. Home range, home range overlap, and species energy use among herbivorous mammals. *Biological Journal of the Linnean Society* 15:185–193.

Darwin, C. 1871. *The Descent of Man, and Selection in Relation to Sex.* London: J. Murray.

Dawson, T. J. 1977. Kangaroos. *Scientific American* 237:78–89.

————. 1983. *Monotremes and Marsupials: The Other Mammals.* The Institute of Biology Studies in Biology, No. 150., London: Edward Arnold.

————. 1989. Diets of macropodoid marsupials: general patterns and environmental influences. In G. Grigg, J. Jarman, and I. Hume, eds., *Kangaroos, Wallabies, and Rat-kangaroos*, pp. 129–142. Chipping Norton, Australia: Surrey Beatty.

————. 1995. *Kangaroos: Biology of the Largest Marsupials.* Sydney: University of New South Wales Press.

Dawson, T. J. and G. D. Brown. 1970. A comparison of the insulative and reflective properties of the fur of desert kangaroos. *Comparative Biochemistry and Physiology* 37:23–38.

Dawson, T. J. and M. J. S. Denny. 1969. A bioclimatological comparison of the summer day microenvironments of two species of arid zone kangaroo. *Ecology* 50:328–332.

Dawson, T. J. and B. A. Ellis. 1994. Diets of mammalian herbivores in Australian arid shrublands: seasonal effects on overlap between red kangaroos, sheep and rabbits and on dietary niche breadths and electivities. *Journal of Arid Environments* 26:257–271.

Dawson, T. J. and A. J. Hulbert. 1970. Standard metabolism, body temperature, and surface areas of Australian marsupials. *American Journal of Physiology* 218:1233–1238.

Dawson, T. J., D. Robertshaw, and C. R. Taylor. 1974. Sweating in the kangaroo: a cooling mechanism during exercise, but not in the heat. *American Journal of Physiology* 227:494–498.

Dawson, T. J. and C. R. Taylor. 1973. Energetic cost of locomotion in kangaroos. *Nature* 240:313–314.

Dellow, D. W. 1982. Studies on the nutrition of macropodine marsupials. III. The flow of digesta through the stomach and intestines of macropodines and sheep. *Australian Journal of Zoology* 30:751–765.

Dellow, D. W., and J. V. Nolan, and I. D. Hume. 1983. Studies on the nutrition of macropodine marsupials. V. Microbial fermentation in the forestomach of *Thylogale thetis* and *Macropus eugenii*. *Australian Journal of Zoology* 31:433–443.

Demment, M. W., and P. J. Van Soest. 1985. A nutritional explanation for body size patterns of ruminant and non-ruminant herbivores. *American Naturalist* 125:641–672.

Denny, M. J. S. 1975. The occurrence of the eastern grey kangaroo (*Macropus giganteus*) west of the Darling River. *Search* 6:89–90.

———. 1982. Adaptations of the red kangaroo and euro (Macropodidae) to aridity. In W. R. Barker and P. J. M. Greenslade, eds., *Evolution of Flora and Fauna of Arid Australia,* pp. 179–183. Frewville, South Australia: Peacock Publications.

Denny, M. J. S. and T. J. Dawson. 1975. Comparative metabolism of tritiated water by macropodid marsupials. *American Journal of Physiology* 228:1794–1799.

Dobson, F. S. 1982. Competition for mates and predominant juvenile male dispersal in mammals. *Animal Behaviour* 30:1138–1192.

Dudzinski, M. L., A. E. Newsome, J. C. Merchant, and B. L. Bolton. 1977. Comparing the two usual methods for aging Macropodidae on toothclasses in the agile wallaby. *Australian Wildlife Research* 4:219–221.

Dyne, G. R. and D. W. Walton, eds. 1987. *Fauna of Australia: Volume 1A, General Articles.* Canberra: Australian Government Publishing Service.

Ealey, E. H. M. 1967. Ecology of the euro, *Macropus robustus* (Gould), in north-western Australia. II. Behavior, movements, and drinking patterns. *CSIRO Wildlife Research* 12:27–51.

Edwards, G. P. 1989. The interaction between macropodids and sheep: a review. In G. Grigg, J. Jarman, and I. Hume, eds., *Kangaroos, Wallabies, and Rat-kangaroos,* pp. 795–804. Chipping Norton, Australia: Surrey Beatty.

———. Competition between red kangaroos and sheep in arid New South Wales. Ph.D. thesis, University of New South Wales, Sydney.

Edwards, G. P., D. B. Croft, and T. J. Dawson. 1994. Observations of differential sex /age class mobility in red kangaroos (*Macropus rufus*). *Journal of Arid Environments* 27: 169–177.

———. 1996. Competition between red kangaroos (*Macropus rufus*) and sheep (*Ovis aries*) in the arid rangelands of Australia. *Australian Journal of Ecology* 21: 165–172.

Einarsen, A. S. 1946. Management of black-tailed deer. *Journal of Wildlife Management* 10: 54–59.

Eisenberg, J. F. and I. Golani. 1977. Communication in metatheria. In T. A. Seebeok, ed., *How Animals Communicate,* pp. 575–599. Bloomington: Indiana University Press.

Ellis, B. A., E. M. Russell, T. J. Dawson, and C. J. F. Harrop. 1977. Seasonal changes in diet preferences of free-ranging red kangaroos, euros and sheep in western New South Wales. *Australian Wildlife Research* 4: 127–144.

Endler, J. A. 1982. Problems in distinguishing historical from ecological factors in biogeography. *American Zoologist* 22: 441–452.

Erskine, D. J. and R. C. Smith. 1983. Griffith weather data summarized by percentiles: 1962 to 1981. *CSIRO Center for Irrigation Research Technical Report Number 1*. New South Wales, Australia: Griffith.

Evans, R. A. and R. M. Love. 1957. The step point method of sampling—a practical tool in range research. *Journal of Range Management* 10: 208–212.

Fitzgibbon, C. D. 1990. Mixed-species grouping in Thomson's and Grant's gazelles: the antipredator benefits. *Animal Behaviour* 39: 1116–1126.

Flannery, T. F. 1984. Kangaroos: 15 million years of Australian bounders. In M. Archer and G. Clayton, eds., *Vertebrate Zoogeography and Evolution in Australasia (Animals in Space and Time),* pp. 817–835. Carlisle, Australia: Hesperian Press.

———. 1989. Phylogeny of the Macropodoidea; a study of convergence. In G. Grigg, J. Jarman, and I. Hume, eds., *Kangaroos, Wallabies, and Rat-kangaroos,* pp. 1–46. Chipping Norton, Australia: Surrey Beatty.

Forbes, D. K. and D. E. Tribe. 1970. The utilization of roughages by sheep and kangaroos. *Australian Journal of Zoology* 18: 247–256.

Ford, J. F., ed. 1985. *Fire Ecology and Management in Western Australia Ecosystems*. WAIT Environmental Studies Group Bulletin 14. Perth: Western Australian Institute of Technology.

Fowler, G. S. 1985. An evaluation of line transect methods for estimation of large mammal populations in heterogeneous habitats. M. Sc. thesis, University of California, Berkeley.

Fox, A. M. 1965. The red kangaroo (*Megaleia rufa*) in New South Wales. *Wildlife Service* 3(4):1–8.

Fox, B. J. 1989. Community ecology of macropodoids. In G. Grigg, J. Jarman, and I. Hume, eds., *Kangaroos, Wallabies, and Rat-kangaroos*, pp. 89–104. Chipping Norton, Australia: Surrey Beatty.

Frakes, L. A., B. McGowran, and S. M. Bowler. 1987. Evolution of Australian environments. In G. R. Dyne and D. W. Walton, eds., *Fauna of Australia: Vol. 1A, General Articles*, pp. 1–16: Canberra: Australian Government Publishing Service.

Freudenberger, D. O., I. R. Wallis, and I. D. Hume. 1989. Digestive adaptations of kangaroos, wallabies and rat-kangaroos. In G. Grigg, J. Jarman, and I. Hume, eds., *Kangaroos, Wallabies, and Rat-kangaroos*, pp. 179–187. Chipping Norton, Australia: Surrey Beatty.

Frith, H. J. 1964. Mobility of the red kangaroo, *Megaleia rufa*. *CSIRO Wildlife Research* 9: 1–19.

Frith, H. J. and J. H. Calaby. 1969. *Kangaroos*. Melbourne: R. W. Cheshire.

Galloway, R. W. and E. M. Kemp. 1981. Late Cainozoic environments in Australia. In A. Keast, ed., *Ecological Biogeography of Australia*, pp. 53–80. The Hague, Netherlands: W. Junk.

Ganslosser, U. 1980. An annotated bibliography of social behaviour in kangaroos (Macropodidae). *Saugetierkundliche Mitteilungen* 38: 138–148.

———. 1989. Agonistic behavior in Macropodoids—a review. In G. Grigg, J. Jarman, and I. Hume, eds., *Kangaroos, Wallabies, and Rat-kangaroos*, pp. 475–503. Chipping Norton, Australia: Surrey Beatty.

———. 1995. Courtship behaviour in Macropoidea (kangaroos, wallabies and rat-kangaroos)—phylogenetic and ecological inferences on ritualization. *Mammal Review* 25: 131–157.

Gavin, T. 1978. Status of Columbian white-tailed deer (*Odocoileus virginianus leucurus*): some quantitative uses of biogeographic data. In *Threatened Deer*, pp. 185–202. Morges, Switzerland: International Union for Conservation of Nature and Natural Resources.

Gavin, T. A., L. H. Suring, P. A. Vohs, Jr., and E. C. Meslow. 1984. Population characteristics, spatial organization, and natural mortality in the Columbian white-tailed deer. *Wildlife Monographs* 91: 1–41.

Geist, V. 1974. On the relationship of social evolution and ecology in ungulates. *American Zoologist* 14: 205–220.

Gill, A. M., R. H. Groves, and I. R. Noble, eds. 1981. *Fire and the Australian Biota*. Canberra: Australian Academy of Sciences.

Gogan, P. J. P. and R. H. Barrett. 1994. Roosevelt elk dietary quality in northern coastal California. *California Fish and Game* 80: 80–83.

Goldsmith, A. E. 1988. Behavior and ecology of pronghorn after reintroduction to Adobe Valley, California. Ph.D. diss., University of California, Berkeley.

Gollan, K. 1984. The Australian dingo: in the shadow of man. In M. Archer and G. Clayton, eds., *Vertebrate Zoogeography and Evolution in Australasia (Animals in Space and Time)*, pp. 921–927. Carlisle, Australia: Hesperian Press.

Grant, T. R. 1973. Dominance and association among members of a captive and free-ranging group of grey kangaroos (*Macropus giganteus*). *Animal Behaviour* 21:449–456.

Gray, A. P. 1972. *Mammalian Hybrids.* Slough, U.K.: Commonwealth Agricultural Bureaux, Farnham Royal.

Green, B. 1984. Composition of milk and energetics of growth in marsupials. *Symposium of the Zoological Society of London* 51:369–387.

———. 1989. Water and energy turnover in free-living macropodids. In G. Grigg, J. Jarman, and I. Hume, eds., *Kangaroos, Wallabies, and Rat-kangaroos,* pp. 223–229. Chipping Norton, Australia: Surrey Beatty.

Green, B., K. Newgrain, and J. Merchant. 1980. Changes in milk composition during lactation in the tammar wallaby (*Macropus eugenii*). *Australian Journal of Biological Science* 33:35–42.

Greenwood, P. J. 1980. Mating systems, philopatry, and dispersal in birds and mammals. *Animal Behaviour* 28:1140–1162.

Griffiths, M. and R. Barker. 1966. The plants eaten by sheep and by kangaroos grazing together in a paddock in south-western Queensland. *CSIRO Wildlife Research* 11:145–167.

Griffiths, M., R. Barker, and L. MacLean. 1974. Further observations on the plants eaten by kangaroos and sheep grazing together in a paddock in south-western Queensland. *Australian Wildlife Research* 1:27–43.

Grigg, G., P. Jarman and I. Hume, eds. 1989. *Kangaroos, wallabies, and Rat-kangaroos.* Chipping Norton, Australia: Surrey Beatty.

Hallam, S. J. 1985. The history of aboriginal firing. In J. R. Ford, ed., *Fire Ecology and Management of Western Australian Ecosystems,* pp. 7–20. WAIT Environmental Studies Group Bulletin 14. Perth: Western Australia Institute of Technology.

Hamilton, W. D. 1971. Geometry for the selfish herd. *Journal of Theoretical Biology* 31:295–311.

Harden, G. J., ed. 1990, 1991, 1992. *Flora of New South Wales, Vols. 1–3.* Kensington, Australia: New South Wales University Press.

Harden, R. H. 1985. The ecology of the dingo in north-eastern New South Wales. I. Movements and home ranges. *Australian Wildlife Research* 12:25–37.

Harestad, A. S. and F. L. Bunnell. 1979. Home range and body weight—a reevaluation. *Ecology* 60:389–402.

Haugen, A. O. 1975. Reproductive performance of white-tailed deer in Iowa. *Journal of Mammalogy* 56:151–159.

Hawkins, R. E. and W. D. Klimstra. 1970. A preliminary study of the social organization of white-tailed deer. *Journal of Wildlife Management* 34: 407–419.

Heady, H. F. 1956. Changes in a California annual plant community induced by manipulation of natural mulch. *Ecology* 37:798–812.

Heathcote, C. F. 1987. Grouping of eastern grey kangaroos in open habitat. *Australian Wildlife Research* 14:343–348.

Heatwole, H. 1987. Major components and distributions of the terrestrial fauna. In G. R. Dyne and D. W. Walton, eds., *Fauna of Australia, Vol. 1A, General Articles,* pp. 101–135. Canberra: Australian Government Publishing Service.

Hebert, D. M. 1973. Altitudinal migration as a factor in the nutrition of bighorn sheep. Ph.D. diss., University of British Columbia, Vancouver.

Hill, G. J. E. 1981. A study of habitat preferences in the grey kangaroo. *Australian Wildlife Research* 8:245–254.

———. 1982. Seasonal movement patterns of the eastern grey kangaroo in southern Queensland. *Australian Wildlife Research* 9:373–387.

Hirth, D. H. 1977. Social behavior of white-tailed deer in relation to habitat. *Wildlife Monographs* 53:1–55.

Hobbs, N. T. 1987. Fecal indices to dietary quality: a critique. *Journal of Wildlife Management* 51:317–320.

Hodgeman, T. P. and R. T. Bowyer. 1986. Fecal crude protein relative to browsing intensity by white-tailed deer on wintering areas in Maine. *Acta Theriologica* 31:347–353.

Hofmann, R. R. 1973. *The Ruminant Stomach: Stomach Structure and Feeding Habits of East African Game Ruminants.* Nairobi: East African Literature Bureau.

Holloway, J. W., R. E. Estell II, and W. T. Butts, Jr. 1981. Relationships between fecal components and forage consumption and digestibility. *Journal of Animal Science* 52:836–848.

Horton, D. R. 1984. Red kangaroos: last of the megafauna. In P. S. Martin and R. G. Klein, eds., *Quaternary Extinctions: A Prehistoric Revolution,* pp. 639–679. Tucson: University of Arizona Press.

Hume, I. D. 1978. Evolution of the Macropodidae digestive system. *Australian Mammalogy* 2:37–42.

———. 1982. *Digestive Physiology and Nutrition of Marsupials.* Cambridge: Cambridge University Press.

———. 1984. Principal features of digestion in kangaroos. *Proceedings of the Nutrition Society of Australia* 9:76–81.

Hume, I. D., P. J. Jarman, M. B. Renfree, and P. D. Temple-Smith. 1989. Macropodidae. In D. W. Walton and B. J. Richardson, eds., *Fauna of Australia: Volume 1B, Mammalia,* pp. 679–715. Canberra: Australian Government Publishing Service.

Hutchinson, G. E. 1957. Concluding remarks. *Cold Spring Harbor Symposium on Quantitative Biology* 22:415–427.

Jaremovic, R. V. 1983. Living in macropolis: grouping behaviour in a dense population of eastern grey kangaroos, *Macropus giganteus*. Abstract No. 149. *Proceedings 18th International Ethological Conference, Brisbane, Australia.*

———. 1984. Space and time related behaviour in eastern grey kangaroos (*Macropus giganteus* Shaw). Ph.D. thesis, University of New South Wales, Sydney.

Jaremovic, R. V. and D. B. Croft. 1987. Comparison of techniques to determine eastern grey kangaroo home range. *Journal of Wildlife Management* 51:921–930.

———. 1991. Social organization of eastern grey kangaroos in southeastern New South Wales. II. Associations within mixed groups. *Mammalia* 55:543–554.

Jarman, P. J. 1974. The social organization of antelope in relation to their ecology. *Behaviour* 48:215–267.

———. 1983a. Behavioural strategies of acquiring mates in macropodid marsupials. Abstract No. 150. *Proceedings 18th International Ethological Conference, Brisbane, Australia.*

———. 1983b. Mating systems and sexual dimorphism in large, terrestrial mammalian herbivores. *Biological Reviews* 58:485–520.

———. 1987. Group size and activity in eastern grey kangaroos. *Animal Behaviour* 35:1044–1050.

———. 1989. Sexual dimorphism in Macropodoidea. In G. Grigg, J. Jarman, and I. Hume, eds., *Kangaroos, Wallabies, and Rat-kangaroos,* pp. 433–447. Chipping Norton, Australia: Surrey Beatty.

———. 1994. The eating of seedheads by species of Macropodidae. *Australian Mammalogy* 17:51–63.

Jarman, P. J. and G. Coulson. 1989. Dynamics and adaptiveness of groupings in macropods. In G. Grigg, J. Jarman, and I. Hume, eds., *Kangaroos, wallabies, and Rat-kangaroos,* pp. 525–547. Chipping Norton, Australia: Surrey Beatty.

Jarman, P. J. and M. J. S. Denny. 1976. Red kangaroos and land use along the New South Wales, Queensland, and South Australian borders. In P. Jarman, ed., *Agriculture, Forestry or Wildlife: Conflict or Co-existence?,* pp. 55–67. Armidale, New South Wales, Australia: University of New England.

Jarman, P. J. and C. M. Phillips. 1989. Diets in a community of macropod species. In G. Grigg, J. Jarman, and I. Hume, eds., *Kangaroos, wallabies, and Rat-kangaroos,* pp. 143–149. Chipping Norton, Australia: Surrey Beatty.

Jarman, P. and C. J. Southwell. 1986. Grouping, associations, and reproductive strategies in eastern grey kangaroos. In D. I. Rubenstein and R. W. Wrangham, eds., *Ecological Aspects of Social Evolution,* pp. 399–428. Princeton, N.J.: Princeton University Press.

Jarman, P. J. and R. J. Taylor. 1983. Ranging of eastern grey kangaroos and wallaroos on a New England pastoral property. *Australian Wildlife Research* 10:33–38.

Jarman, P. J. and S. M. Wright. 1993. Macropod studies at Wallaby Creek. IX. Exposure and response of eastern grey kangaroos to dingoes. *Wildlife Research* 20:833–843.

Johnson, C. N. 1983a. Variations in group size and composition in red and western grey kangaroos, *Macropus rufus* (Desmarest) and *M. fuliginosus* (Desmarest). *Australian Wildlife Research* 10:25–31.

———. 1983b. Ranging behaviour and social organization of the red-necked wallaby, *Macropus rufogriseus.* Abstract No. 151. *Proceedings 18th International Ethological Conference, Brisbane, Australia.*

———. 1986. Philopatry, reproductive success of females and maternal investment in the red-necked wallaby. *Behavioral Ecology and Sociobiology* 19:143–150.

———. 1989. Dispersal and philopatry in the Macropodoids. In G. Grigg, J. Jarman, and I. Hume, eds., *Kangaroos, Wallabies, and Rat-kangaroos,* pp. 593–601. Chipping Norton, Australia: Surrey Beatty.

Johnson, C. N. and P. G. Bayliss. 1981. Habitat selection by sex, age and reproductive class in the red kangaroo, *Macropus rufus,* in western New South Wales. *Australian Wildlife Research* 8:465–474.

Johnson, K. A., A. A. Burbridge, and N. L. McKenzie. 1989. Australian Macropodoidea: status, causes of decline and future research and management. In G. Grigg, J. Jarman, and I. Hume, eds., *Kangaroos, Wallabies, and Rat-kangaroos,* pp. 641–657. Chipping Norton, Australia: Surrey Beatty.

Johnson, C. N. and P. J. Jarman. 1983. Geographic variation in offspring sex ratios in kangaroos. *Search* 14:152–154.

Kaufmann, J. H. 1974a. Habitat use and social organization of nine sympatric species of macropodid marsupials. *Journal of Mammalogy* 55: 66–80.

———. 1974b. The ecology and evolution of social organization in the kangaroo family (Macropodidae). *American Zoology* 14:51–62.

———. 1974c. Social ethology of the whiptail wallaby, *Macropus parryi,* in northeastern New South Wales. *Animal Behaviour* 22:281–309.

———. 1975. Field observations of the social behaviour of the eastern grey kangaroo, *Macropus giganteus. Animal Behaviour* 23:214–221.

Keast, A., ed. 1981. *Ecological Biogeography of Australia.* The Hague, Netherlands: W. Junk.

Keddy, P. A. 1989. *Competition.* New York: Chapman and Hall.

Kie, J. G., J. A. Baldwin, and C. J. Evans. 1994. *CALHOME: Home Range Analysis Program, a User's Manual.* General Technical Report PSW-000. Albany, Cal.: U.S. Forest Service, Pacific Southwest Research Station.

———. 1996. CALHOME: a program for estimating animal home ranges. *Wildlife Society Bulletin* 24:342–344.

Kie, J. G. and T. S. Burton. 1984. Dietary quality, fecal nitrogen, and 2,6 diaminopimelic acid in black-tailed deer in northern California. Berkeley, Cal.: U.S. Forest Service, Pacific Southwest Forest and Range Experiment Station.

King, D. R., A. J. Oliver, and R. J. Mead. 1978. The adaptation of some Western Australian mammals to food plants containing fluoroacetate. *Australian Journal of Zoology* 26:699–712.

Kirchoff, M. D. and J. W. Schoen. 1987. Forest cover and snow: implications for deer habitat in southeast Alaska. *Journal of Wildlife Management* 51:28–33.

Kirkpatrick, T. H. 1965a. Molar progression and macropod age. *Queensland Journal of Agriculture and Animal Science* 21:163–165.

———. 1965b. Studies of Macropodidae in Queensland. 3. Reproduction in the grey kangaroo (*Macropus major*) in southern Queensland. *Queensland Journal of Agriculture and Animal Science* 22:319–328.

———. 1966. Studies of Macropodidae in Queensland. 4. Social organization of the grey kangaroo (*Macropus giganteus*). *Queensland Journal of Agriculture and Animal Science* 23:317–322.

Kirsch, J. 1984. First occurrence of an eastern male grey kangaroo cross with western female grey kangaroos. *Australian Mammal Society Bulletin* 8(2):133.

Kirsch, J. A. W. and W. E. Poole. 1967. Serological evidence for speciation in the grey kangaroo, *Macropus giganteus* Shaw 1790 (Marsupialia: Macropodidae). *Nature* 215:1097–1098.

———. 1972. Taxonomy and distribution of the grey kangaroos, *Macropus giganteus* Shaw and *Macropus fuliginosus* (Desmarest), and their subspecies (Marsupialia: Macropodidae). *Australian Journal of Zoology* 20:315–339.

Kleiber, M. 1961. *The Fire of Life.* New York: Wiley.

Krebs, C. J. 1989. *Ecological Methodology.* New York: Harper and Row.

Laake, J. L., S. T. Buckland, D. R. Anderson, and K. P. Burnham. 1993. *DISTANCE User's Guide, V2.0.* Fort Collins: Colorado Cooperative Fish and Wildlife Research Unit, Colorado State University.

Lancaster, R. J. 1949. Estimation of digestibility of grazed pastures from faeces nitrogen. *Nature* 163:330–331.

Lee, A. K. and A. Cockburn. 1985. *Evolutionary Ecology of Marsupials.* Cambridge: Cambridge University Press.

Lee, A. K. and S. J. Ward. 1989. Life histories of macropodoid marsupials. In G. Grigg, J. Jarman, and I. Hume, eds., *Kangaroos, Wallabies, and Rat-kangaroos,* pp. 105–115. Chipping Norton, Australia: Surrey Beatty.

Leopold, A. S. and T. O. Wolfe. 1970. Food habits of nesting wedge-tailed eagles *Aquila audax,* in southeastern Australia. *CSIRO Wildlife Research* 15:1–17.

Leslie, D. M., Jr. and E. E. Starkey. 1985. Fecal indices to dietary quality in cervids in old-growth forests. *Journal of Wildlife Management* 49: 142–146.

———. 1987. Fecal indices to diet quality: a reply. *Journal of Wildlife Management* 51:321–325.

Low, B. 1978. Environmental uncertainty and the parental strategies of marsupials and placentals. *American Naturalist* 112:197–213.

———. 1979. The predictability of rain and the foraging patterns of the red kangaroo (*Megaleia rufa*) in central Australia. *Journal of Arid Environments* 2:61–76.

Low, B. S., E. Birk, C. Lendon, and W. A. Low. 1973. Community organization by cattle and kangaroos in mulga near Alice Springs, N.T. *Tropical Grasslands* 7:149–156.

Main, A. R. 1987. Evolution and radiation of the terrestrial fauna. In G. R. Dyne and D. W. Walton, eds., *Fauna of Australia: Vol. 1A, General Articles,* pp. 136–155. Canberra: Australian Government Publishing Service.

Main, M. B., F. W. Weckerly, and V. C. Bleich. 1996. Sexual segregation in ungulates: directions for research. *Journal of Mammalogy* 77:449–461.

Marsack, P. and G. Campbell. 1990. Feeding behavior and diet of dingoes in the Nullarbor region, Western Australia. *Australian Wildlife Research* 17:349–357.

Martin, C. J. 1902. Thermal adjustments and respiratory exchanges in monotremes and marsupials. *Philosophical Transactions of the Royal Society of London,* B. 195:1–37.

Martin, P. S. and R. G. Klein, eds. 1984. *Quaternary Extinctions: A Prehistoric Revolution.* Tucson: University of Arizona Press.

Massey, B. N., F. W. Weckerly, C. E. Vaughn, and D. R. McCullough. 1994. Correlations between fecal nitrogen and diet composition in free-ranging black-tailed deer. *Southwestern Naturalist* 39:165–170.

Maynes, G. 1976. Growth of the parma wallaby *Macropus parma* Waterhouse. *Australian Journal of Zoology* 24:217–236.

———. 1989. Zoogeography in the Macropodoidea. In G. Grigg, J. Jarman, and I. Hume, eds., *Kangaroos, Wallabies, and Rat-kangaroos,* pp. 47–66. Chipping Norton, Australia: Surrey Beatty.

McCabe, R. E. and T. R. McCabe. 1984. Of slings and arrows: an historical retrospective. In L. K. Halls, ed., *White-tailed Deer: Ecology and Management,* pp. 19–72. Harrisburg, Penn.: Stackpole Books.

McCarron, H. C. K. and T. J. Dawson. 1989. Thermal relations of Macropodoidea in hot environments. In G. Grigg, J. Jarman, and I. Hume, eds., *Kangaroos, Wallabies, and Rat-kangaroos,* pp. 255–263. Chipping Norton, Australia: Surrey Beatty.

McCullough, D. R. 1969. The tule elk: its history, behavior, and ecology. *University of California Publications in Zoology* 88:1–209.

———. 1979. *The George Reserve Deer Herd: Population Ecology of a K-Selected Species.* Ann Arbor: University of Michigan Press.

———. 1982. Evaluation of night spotlighting as a deer study technique. *Journal of Wildlife Management* 46:963–973.

———. 1985. Variables influencing food habits of white-tailed deer on the George Reserve. *Journal of Mammalogy* 66:682–692.

———. 1987. The theory and management of *Odocoileus* populations. In C. M. Wemmer, ed., *Biology and Management of the Cervidae,* pp. 535–549. (Research Symposia of the National Zoological Park). Washington, D.C., and London: Smithsonian Institution Press.

———. 1992. Concepts of large herbivore population dynamics. In D. R. McCullough and R. H. Barrett, eds., *Wildlife 2001: Populations,* pp. 967–984. London: Elsevier Science.

———. 1997. Breeding by female fawns in black-tailed deer. *Wildlife Society Bulletin* 25:296–297.

McCullough, D. R., and D. H. Hirth. 1988. Evaluation of the Petersen-Lincoln estimator for a white-tailed deer population. *Journal of Wildlife Management* 52:534–544.

McCullough, D. R., D. H. Hirth, and S. J. Newhouse. 1989. Resource partitioning between sexes in white-tailed deer. *Journal of Wildlife Management* 53:277–283.

McCullough, D. R. and D. E. Ullrey. 1985. *Chemical Composition and Gross Energy of Deer Forage Plants on the George Reserve, Michigan.* East Lansing: Michigan Agricultural Experiment Station Research Report 465.

McCullough, Y. 1980. Niche separation of seven North American ungulates on the National Bison Range, Montana. Ph.D. diss., University of Michigan, Ann Arbor.

McIntosh, D. L. 1966. The digestibility of two roughages and the rates of passage of their residues by the red kangaroo, *Megaleia rufa* (Desmarest) and the Merino sheep. *CSIRO Wildlife Research* 11:125–135.

McLeod, S. R. 1997. Is the concept of carrying capacity useful in variable environments? *Oikos* 79:529–542.

McNab, B. K. 1963. Bioenergetics and the determination of home range size. *American Naturalist* 97:133–140.

McShea, W. J., H. B. Underwood, and J. H. Rappole, eds. 1997. *The Science of Overabundance. Deer Ecology and Population Management.* Washington, D. C. and London, U.K.: Smithsonian Institution Press.

Merchant, J. C. 1989. Lactation in macropodoid marsupials. In G. Grigg, J. Jarman, and I. Hume, eds., *Kangaroos, Wallabies, and Rat-kangaroos,* pp. 355–366. Chipping Norton, Australia: Surrey Beatty.

Merrilees, D. 1984. Comings and goings of Late Quaternary mammals in extreme south-western Australia. In P. S. Martin, and R. G. Klein, eds., *Quaternary Extinctions: A Prehistoric Revolution,* pp. 629–638. Tucson: University of Arizona Press.

Millar, J. S. 1977. Adaptive features of mammalian reproduction. *Evolution* 31:370–386.

Miller, G. H., J. W. Magee, B. J. Johnson, M. L. Fogel, N. A. Spooner, M. T. McCulloch, and L. K. Ayliffe. 1999. Pleistocene extinction of *Genyornis newtoni:* Human impact on Australian megafauna. *Science* 283:205–208.

Morisita, M. 1959. Measuring of interspecific association and similarity between communities. *Memoirs of the Faculty of Sciences, Kyushu University, Series E, Biology* 3:65–80.

Moss, G. L. and D. B. Croft. 1999. Body condition of the red kangaroo (*Macropus rufus*) in arid Australia: the effect of environmental condition, sex and reproduction. *Australian Journal of Ecology* 24:97–109.

Mould, E. D. and C. T. Robbins. 1981. Nitrogen metabolism in elk. *Journal of Wildlife Management* 45:323–334.

Mubanga, G., J. L. Holecheck, R. Valdez, and S. D. Schemnitz. 1985. Relationships between diet and fecal nutritive quality in mule deer. *Southwestern Naturalist* 30:573–578.

Murray, P. 1984. Extinctions down under: a bestiary of extinct Late Pleistocene monotremes and marsupials. In P. S. Martin and R. G. Klein, eds., *Quaternary Extinctions: A Prehistoric Revolution,* pp. 600–627. Tucson: University of Arizona Press.

———. 1991. The Pleistocene megafauna of Australia. In P. Vickers-Rich, J. M. Monaghan, R. F. Baird, and T. H. Rich, eds., *Vertebrate Paleontology of Australasia,* pp. 1071–1164. Melbourne: Monash University.

Myers, P. 1978. Sexual dimorphism in size of vespertilionid bats. *American Naturalist* 112:701–711.

Mykytowycz, R. 1964. Coccidea in wild populations of the red kangaroo, *Megaleia rufa* (Desmarest), and the grey kangaroo, *Macropus canguru* (Muller). *Parasitology* 54:105–115.

Nelson, D. W. and L. E. Sommers. 1972. A simple digestion procedure of estimation of total nitrogen in soils and sediments. *Journal of Environmental Quality* 1:423–425.

———. 1973. Determination of total nitrogen in plant material. *Agronomy Journal* 65:109–112.

———. 1980. Total nitrogen analysis of soil and plant tissues. *Journal of the Association of Official Analytical Chemists* 63:770–778.

Nelson, M. E. and L. D. Mech. 1981. Deer social organization and wolf predation in northeastern Minnesota. *Wildlife Monographs* 77:1–53.

Newsome, A. E. 1964. Anoestrus in the red kangaroo *Megaleia rufa* (Desmarest). *Australian Journal of Zoology* 12:9–17.

———. 1965. The abundance of red kangaroos, *Megaleia rufa* (Desmarest), in central Australia. *Australian Journal of Zoology* 13:269–287.

———. 1971. The ecology of red kangaroos. *Australian Zoology* 16:32–50.

———. 1973. Cellular degeneration in the testis of red kangaroos during hot weather and drought in central Australia. *Journal of Reproduction and Fertility Supplement* 19:191–201.

———. 1975. An ecological comparison of the two arid zone kangaroos of Australia and their anomalous prosperity since the introduction of ruminant stock to their environment. *Quarterly Review of Biology* 50:389–424.

———. 1977a. The red kangaroo—an example of biological indicators of environmental change. In H. Messel and S. T. Butler, eds., *Australian Animals and their Environment,* pp. 25–48. Sydney: Shakespeare Head Press.

———. 1977b. Imbalance in the sex-ratio and age-structure of the red kangaroo in central Australia. In B. Stonehouse and D. Gilmore, eds., *The Biology of Marsupials,* pp. 221–233. Baltimore: University Park Press.

———. 1980. Differences in the diets of male and female red kangaroos in central Australia. *African Journal of Ecology* 18:27–31.

Newsome, A. E., J. C. Merchant, B. L. Bolton, and M. L. Dudzinski. 1977. Sexual dimorphism in molar progression and eruption in the agile wallaby. *Australian Wildlife Research* 4:1–5.

Nicholls, E. M. and K. G. Rienits. 1971. Tryptophan derivatives and pigment in the hair of some Australian marsupials. *International Journal of Biochemistry* 2:593–603.

Nicholls, N. 1991. The El Niño–Southern Oscillation and Australian vegetation. *Vegetatio* 91:23–36.

Norbury, G. L., G. M. Coulson, and B. L. Walters. 1988. Aspects of the demography of the western grey kangaroo, *Macropus fuliginosus melanops,* in semiarid north-west Victoria. *Australian Wildlife Research* 15:257–266.

Norbury, G. L., D. C. Norbury, and A. J. Oliver. 1994. Facultative behaviour in unpredictable environments: mobility of red kangaroos in arid Western Australia. *Journal of Animal Ecology* 63:410–418.

Taylor, W. P., ed. 1956. *The Deer of North America: The White-tailed, Mule and Black-tailed Deer, Genus Odocoileus.* Harrisburg, Penn.: Stackpole Books.

Templeton, A. R. 1981. Mechanisms of speciation: a population genetic approach. *Annual Review of Ecology and Systematics* 12:23–48.

Thomas, J. W. and D. E. Toweill. 1982. *Elk of North America: Ecology and Management.* Harrisburg, Penn.: Stackpole Books.

Thomson, P. C. 1992. The behavioural ecology of dingoes in north-western Australia. III. Hunting and feeding behaviour, and diet. *Wildlife Research* 19:531–541.

Tunbridge, D. 1988. *Flinders Ranges Dreaming.* Canberra: Aboriginal Studies Press.

———. 1991. *The Story of the Flinders Ranges Mammals.* Kenthurst, Australia: Kangaroo Press.

Tyndale-Biscoe, C. H. 1989. The adaptiveness of reproductive processes. In G. Grigg, J. Jarman, and I. Hume, eds., *Kangaroos, Wallabies, and Rat-kangaroos,* pp. 277–285. Chipping Norton, Australia: Surrey Beatty.

Tyndale-Biscoe, C. H. and M. B. Renfree. 1987. *Reproductive Physiology of Marsupials.* Cambridge: Cambridge University Press.

van Oorschot, R. A. H. and D. W. Cooper. 1989. Twinning in the genus *Macropus* especially *M. eugenii* (Marsupialia: Macropodidae). *Australian Mammalogy* 12:83–84.

Wallmo, O. C. 1981. Mule and black-tailed deer distribution and habitats. In O. C. Wallmo, ed., *Mule and Black-tailed Deer of North America,* pp. 1–25. Lincoln: University of Nebraska Press.

Walton, D. W. and B. J. Richardson, eds. 1989. *Fauna of Australia: Volume 1B, Mammalia.* Canberra: Australian Government Publishing Service.

Ward, J. V. and J. A. Stanford. 1983. The intermediate disturbance hypothesis: an explanation for biotic diversity patterns in lotic ecosystems. In T. D. Fontaine and S. M. Bartell, eds., *Dynamics of Lotic Ecosystems,* pp. 347–356. Ann Arbor, Mich.: Ann Arbor Science.

Wasson, R. J. 1976. Holocene aeolian landforms in the Belarabon area, southwest of Cobar, New South Wales. *Journal and Proceedings of the Royal Society of New South Wales* 109:91–110.

Watson, D. M. and T. J. Dawson. 1993. The effects of age, sex, reproductive status, and temporal factors on the time-use of free-ranging red kangaroos (*Macropus rufus*) in western New South Wales. *Wildlife Research* 20:785–801.

Weckerly, F. W. 1998. Sexual-size dimorphism: influence of mass and mating systems in the most dimorphic mammals. *Journal of Mammalogy* 79: 33–52.

Wehausen, J. D. 1980. Sierra Nevada bighorn sheep: history and population ecology. Ph.D. diss., University of Michigan, Ann Arbor.

Wellard, G. 1987. The effect of weather on soil moisture and plant growth in the arid zone. In G. Caughley, N. Shepherd, and J. Short, eds., *Kangaroos: Their Ecology and Management in the Sheep Rangelands of Australia,* pp. 35–49. Cambridge: Cambridge University Press.

Wells, R. G., D. R. Horton, and P. Rogers. 1982. *Thylacoleo carnifex* Owen (Thylacoleonidae, Marsupialia): marsupial carnivore? In M. Archer, ed., *Carnivorous Marsupials Volume 2,* pp. 573–576. Sydney: Royal Zoological Society of New South Wales.

Whitehouse, S. J. O. 1977. The diet of the dingo in Western Australia. *Australian Wildlife Research* 4:145–150.

Wilkinson, L., M. A. Hill, P. Howe, and S. Miceli. 1992. *SYSTAT* (4 vol.). Evanston, Ill.: SYSTAT Inc.

Wilson, G. R. 1975. Age structures of populations of kangaroos (Macropodidae) taken by professional shooters in New South Wales. *Australian Wildlife Research* 2:1–9.

Windsor, D. E. and A. I. Dagg. 1971. The gaits of the Macropodidae (Marsupialia). *Journal of Zoology* 163:165–175.

Wrangham, R. W. and D. I. Rubenstein. 1986. Social evolution of birds and mammals. In D. I. Rubenstein and R. W. Wrangham, eds., *Ecological Aspects of Social Evolution,* pp. 452–470, Princeton, N.J.: Princeton University Press.

Wright, S. M. 1993. Observations of the behaviour of male eastern grey kangaroos when attacked by dingoes. *Wildlife Research* 20:845–849.

Young, S. P. 1956. The deer, the Indians and the American pioneers. In W. P. Taylor, ed., *The Deer of North America: The White-tailed, Mule and Black-tailed Deer, Genus Odocoileus,* pp. 1–27. Harrisburg, Penn.: Stackpole Books.

Index

Aborigines, 12, 16, 19, 24, 46, 234, 256
   and fire, 41
Acacia aneura, 37
Acacia spp. (mulga), 35
activity, kangaroo
   and cold, 145
   daily, 125–133
   and environmental state, 133–135
   and feeding, 144, 148–149
   and heat, 145–147
   1nighttime, 139
   seasonal, 121–124
   sex and species differences, 149–151
   and weather, 73–74, 135–144
adaptive kernel (home range), 105–106
age ratio, 84–86, 91
   and reproduction, 85–86
aggregations
   feeding concentrations, 211–212
   and predator avoidance, 167
aggressive behavior
   fighting, 179–182
   grass-pulling, 182–183
   threats, 178

alarm behavior, 59–60, 155–156, 173–
      178, 262–263
   and observer distance, 177–178
   species differences, 173–178
   and time of day, 174–175
   and wind, 140, 173
Alces americana, 260
ANOVA (analysis of variance), 74, 77,
      79
antilopine kangaroo, 3, 5
Apophyllum anomalum, 39
Aquila audax, 45
   See also wedge-tailed eagle
area requirement constant (ARC), 106–
      109
Aristida behriana, 39
Australia
   climate, 2
   geologic history, 3, 4, 255–256
   rainfall, 19, 29–30
   topography, 17

Bailey's formula, 53–54
bats, 2

behavior
  direct observation, 74–75
  and weather, 60–61
  See also methods (ad hoc)
bias
  capture, 87–89
  detection, 222–223
  resighting, 98
  See also mark-recapture population
    estimate
bimble box, 14 fig. 1.4a, 19, 37, 75,
    104–105, 210, 222, 223, 224, 225,
    227
biogeography, 255, 261
black-tailed deer, 16, 267
  See also Odocoileus h. hemionus
Blue Mountain, 24
Blue Mountains, 17, 19
body posture, 169–170
  aggressive, 178
  feeding, 218
body size
  and home range, 109
  and K-selection, 265–266
Bogan River, 19
Bovidae, 3
Brachychiton populneus, 39
Brassica tournefortii, 39
breeding
  correlation with local conditions,
    195–197
  seasons, 190
  timing, 193–199
brushtail possum, 3

CALHOME (program), 69
Callitris columellaris, 37
Canberra, ix, xi, xii
Canis familiaris dingo, 5
  See also dingo
Canis latrans, 11
Canis lupus, 11

Capra hirca, 42
  See also goat, feral
carrying capacity, 93, 246, 255
Carthamus lanatus, 41
Casuarina cristata, 38
cat, feral, 46, 234
Centaurea melitensis, 39
"centripetality," 98
Cervidae, 3
Cervus elaphus, 11
Cirsium vulgare, 39
Citrullus lanatus, 41
climate change, 255–256
  and kangaroo distribution, 257–259
Cobar, 18, 19, 25, 26, 30
coevolution, 256
community structuring, 254
competition, 1, 6, 242, 248, 251–255,
    258–259, 269–270
Connachaetes taurinus, 167
coyote, 11
CSIRO Wildlife and Rangelands Re-
    search Division, ix, 35
Cucumis myriocarpus, 41
cuscus, 3
cypress-pine, 37–38, 203

Danthonia caespitosa, 39
Darling River, 17, 19
diet preferences, 201
digestive anatomy, 200
dingo, x, 5, 12, 16, 140, 234–235
  predation on kangaroos, 235
dispersal, 98, 230
DISTANCE (program), 54–55, 89
distribution, and climate, 86, 149, 258
Dromaius novaehollandiae, 44, 237
  See also emu
drought, 30–31, 32 fig. 2.8, 33, 97,
    203, 238, 258, 268, 269
  and reproductive synchrony, 193–
    195

population effects, 80, 81
population response, 91–92, 95
Durokoppin Nature Reserve, 117

eartags, 53
eastern grey kangaroo, ix, x, 3, 4, 5, 11,
    13 fig. 1.2, 42, 219 fig. 9.9b
    feeding sites, 210
    habitat and distribution, 12
    habitat preference, 223
    See also Macropus giganteus
echidna, 42
Echium plantagineum, 41
Eimeria purchasi, 217
El Niño–Southern Oscillation/La Niña
    events, 80
elk, 11
embryonic diapause, 198–199
Emex australis, 39
emu, 44, 237 fig. 11.1, 238, 269
    competition with kangaroos, 242, 248
    group size, 238
    habitat selection, 238–239
    niche overlap, 242–244
    population size, 239–241
    reproductive failure, 238
environmental states, 49
Eptesicus vulturnus, 42
Equus burchelli, 167
Eragrostis parviflora, 41
Erodium crinitum, 39
Eucalyptus dumosa, 37
E. gracilis, 37
E. intertexta, 37
E. microcarpa, 37
E. morrisi, 37
E. populnea, 19, 37
E. socialis, 37
Eucalyptus spp., 36
euro, x, 3, 5, 9 fig. 1.1, 11, 42, 93, 269
    habitat and distribution, 14
    See also Macropus robustus

European rabbit, x, 39, 42, 43 fig. 2.13,
    234, 237, 269
    as competitors with kangaroos, 244–
    245, 246–247
European red fox, x, 45 fig. 2.15, 234

fecal collection, 78
fecal nitrogen
    analytic method, 77–79
    seasonal pattern, 216–217
    species comparison, 216–217
feces and feed conditions, 214–215
feeding aggregations, 154–155
feeding ecology, species differences,
    218–220
feeding site
    microsites, 213
    selection and seasonal changes, 208–
    210
    species comparison, 204–205
Felis catus, 46
    See also cat, feral
fire, 39, 41, 203
fitness, 250
flehmen, 184–185
flooding, 28 fig. 2.6, 39
fog, 52
forage preferences, 205–206
Fowler's Gap, 33, 99, 143, 153, 201,
    202–203, 246, 252, 256

Gastrolobium, 200
Gazella granti, 167
G. thomsoni, 167
Geijera parviflora, 38
"ghosts of competition past" model,
    254, 256
Gnaphalium luteoalbum, 37
goat, feral, 42, 43 fig. 2.13, 203, 237,
    245
    as competitors, 247–248
    See also Capra hirca

Gondwanaland, 235

grassland, 39, 40 fig. 2.12, 93, 105, 223, 224, 225, 227–229

graziers, 6, 59, 97, 248

and fire, 41

Great Dividing Range, 17

green feed

burned areas, 207–208

growth, 33, 35

seasonality, 201–203

temperature correlation, 33, 35

grey box, 37, 222, 223, 224, 225, 227

Griffith, 25

grooming, 171–172

group composition

sex and species differences, 156–157

group size, 153

and habitat, 166

and population density, 153–154, 166

habitat

and burn status, 58–59

and home range movements, 100–101

habitat use

diel patterns, 59

seasonality, 224

shrub density, 227–229

species differences, 223–224, 231–232

time of day, 224

tree density, 225–227

harmonic mean, 105–106

Hattah-Kulkyne National Park, 93–94, 100, 117, 154

Helichrysum bracteatum, 41

Heliotropium europaeum, 41

herbaceous cover, 51

Heterodendrum oleifolium, 39

Hillston, 18, 19

home range, 96–97, 99

adaptive kernel, 105–106

and energy requirements, 118

establishment, 115–116

extensions, 113–114

minimum convex polygon 105, 108–109

shift to new, 114–116

size and sex and species differences, 101–109, 117

and social behavior, 118

Hordeum leporinum, 39, 208, 211, 212

hunting, xiii, 2, 12, 16, 234

hybridization, 164–165

insects, response to, 172–173

juveniles, 57, 156

K-selection, 267–269

kangaroo

abundance, 5–6

behavior, 6–7

comparison with deer, 10–11

diet, 2, 14–15

dispersal patterns, 8

distribution, 5, 9 fig. 1.1, 11

female kinship, 7

home range, 8

lactation, 2

locomotion, 2, 170–171, 263

mortality, 30–31

nomadism, 7

predation, 5

reproduction, 1, 2

social organization, 7

species abundance at Yathong, 153–154

taxonomy, 3, 4

temperature regulation, 2

and ungulate models, 270–271

water requirements, 2

Keginni Creek, 24

Kinchega National Park, ix, x, 31, 33, 93, 98, 99, 119, 120, 143, 153, 197–198
king brown snake, 45
KROS (kangaroo radio observation station), 70, 71, 72 fig. 3.3, 73

Lachlan River, 17, 19
Lactuca serriola, 41
Lake Cargelligo, 17–18, 19
line transect population estimate, 49, 54–55
livestock, 269
locomotion, 170–171, 263
longevity, 91

Macropodidae, 3
Macropods, evolution, 3
Macropus, 4
Macropus antilopinus, 3, 5
M. *bernardus,* 3, 5
M. *fuliginosus,* 3
    See also western grey kangaroo
M. *giganteus,* 3
    See also eastern grey kangaroo
M. *parryi,* 55
M. *robustus,* 3, 5
M. *rufogriseus,* 7
M. *rufus,* 3
See also red kangaroo
M. *titan,* 4
Maireana pyramidata, 41
male dominance behavior, 193
mallee, 25 fig. 2.4, 36, 40 fig. 2.12, 51, 58, 75, 93, 116
    use by kangaroos, 92, 173, 219, 222–225, 232, 268
mark-recapture population estimate, 49, 52–53
Marrubium vulgare, 41
marsupial lion, 5, 12
McPAAL (program), 69

Medicago polymorpha, 39
Menindee, ix, 31, 98, 100
Merrimerawa Ridge, 24, 93, 148
methods, xii
    ad hoc, 47
    standardization, 47
    systematic, 47
    systematic survey, 49
microbial fermentation, 200
minimum convex polygon, 105, 108–109
mixed groups, 153
mixed-sex groups, 158–159
mixed-species groups, 166–167
mobs, 97, 213, 153, 154
monitor lizard, 5
moose, 266
mortality, 30, 31, 67
    and arrested development, 186
    drought-caused, 80, 203
    juvenile, 156
    male-biased, 84
mother-young interactions, 190
    pouch emergence, 189
Mount Hope, 18, 197
mountain lion, 11
movements
    long-range, 118
    and rainfall, 119–120
    short-range, 110–113
    short-term, 117–118
mule deer, x, 9 fig. 1.1, 10, 11, 16, 267, 270
    See also Odocoileus hemionus
mulga, 35, 37
    use by kangaroos, 92–93, 223
Mus musculus, 46
myxomatosis, 46, 244

New South Wales, ix, x, 18 fig 2.1
New South Wales Park and Wildlife Service, 21, 64, 244, 245

niche, 3, 5, 8, 10, 16, 221, 229

niche breadth (*FT* values), 61, 62–63, 229–230

niche overlap (measure *C*), 61–62, 231
with emus, 242–244
sex comparison, 232, 254
species comparison, 232, 254

niche relationships, 62–63

niche separation, 213
evolution of, 256–258, 270

niche spacing, 254–255

Nicotiana glauca, 41

N. suaveolens, 41

nomadism, 97–98, 99, 100–101, 119–120

nursing, 189

Nymagee, 25

Odocoileus, 10, 16
distribution, 9 fig. 1.1

Odocoileus hemionus, 9 fig. 1.1, 10
See also mule deer

O. h. columbianus, 16

O. h. hemionus, 16

O. h. sitkensis, 237

O. virginianus, 9 fig. 1.1, 10
See also white-tailed deer

Oryctolagus cuniculus, 42
See also European rabbit

Osphranter (subgenus), 4

Oxylobium, 200

Palorchestes, 4

Petersen-Lincoln method, 86

Phalangeridae, 3

pig, feral, 44, 237, 245
See also Sus scrofa

pine associations, 39, 40 fig. 2.12, 75, 225

pine-box, 37, 44 fig. 2.14, 116, 219
fig. 9.9, 223–225

play-fighting, 190

poaching, 144, 246

Populus tremuloides, 37

population density, 87, 89–90, 92–95

population estimate
line transect, 89–91
mark-recapture, 80–89

post-partum estrus, 190

pouch young, 57–58, 64, 186–189

predation, 11–12, 16, 144, 245–246
defense, 261–262, 263
escape response, 235

Proboscidea louisiana, 41

productivity, population response to, 94–95

Pseudechis australis, 45

Pseudostsuga menziesii, 267

Puma concolor, 11

r-selection, 265–266
and habitat, 266
and kangaroos, 267–269

radio telemetry, xii, 52, 63, 66–68, 99, 121, 221–222
activity monitoring, 70–74
and habitat use, 69–70

rainfall, 32, 100
population response, 93, 94, 95

red kangaroo, 3, 9 fig. 1.1, 11, 14
fig. 1.4, 42, 202 fig. 9.1, 219
fig. 9.9a
cold stress, 86
habitat and distribution, 12, 14
habitat preference, 223
See also Macropus rufus

red-necked wallaby, 7

regurgitation, 201

reproduction, 186
timing at Yathong, 187–189

reproductive success, 185

resource partitioning, 229
See also niche

Rhodanthe corymbiflora, 41

Rumex brownii, 41

Salix babylonica, 38
Salsola kali, 41
Scarlett O'Hara, 66, 69, 73, 104, 112,
      114–116, 155, 157, 236
Sequoia sempervirens, 267
sex and age classification, 56–57
sex ratio, 81–84, 90–91
   and habitat use, 82–84
   and season, 81–82
sexual behavior
   consort pairs, 183
   copulation, 184–185
   and male dominance relationships,
      185–186
   tending bond, 184, 192–193, 264
sexual dimorphism, 15 fig. 1.5, 16, 62,
      91, 249–250, 251, 264–265
   and fecal nitrogen, 217
sexual selection, 250
Shearers' Quarters, x, 23 fig. 2.3, 31, 32
      fig. 2.8, 50, 60–61, 73, 136, 212
sheep grazing, x, 39
sheep stations, x, 19, 20–21
"shingle-back," 237–238
shrub density, 51, 225–226, 229
shrubs as browse, 203–204, 220, 247
sibling species, 3, 5, 246
Sitka deer, 267
social groups
   mixed sex, 158–159
   mixed species, 163–164
   and reproduction, 166
   seasonal changes, 159–162
sodium fluoroacetate, 200, 219–220
Solanum nigrum, 41
Sonchus oleraceus, 41
spear grass, 41
   See also Stipa variabilis
Stipa variabilis, 39, 42, 203, 205, 210,
      212, 213, 214, 218, 220
stronghold and overlap model, 257
stunning, 63–64, 65 fig. 3.2, 66–67,
      101, 263

Sus scrofa, 44, 238
   See also pig, feral
Sydney, vii, xi, 17
"Symphyomyrtus" woodland, 35
systematic surveys, 50–52

Tachyglossus aculeatus, 42
Tasmanian wolf, 11–12
temperature, 52
thylacine wolf, 5, 234, 235
Thylacinus, 5
Thylacoleo, 5
Tibooburra, 100, 119
Trachydosaurus rugosus, 237
transilience, 256
tree density, 51, 224–225
Triodia (grasslands), 4
tule elk, 140, 262

ungulates, 140, 270
   behavior, 168, 184, 191, 264
   breeding seasons, 198, 250–251, 264
   carrying capacity, 246
   competition, 250
   mixed species groups, 166–167
   niche relationships, 62
   predator avoidance, 236, 262, 262
   sex ratio, 84
University Research Expedition Pro-
      gram (UREP), xi, 76

Varanidae, 5
vegetation types, 58
Verbascum thapsus, 41
vicariance, 255
Vulpes vulpes, 45
   See also European red fox

Wallabia bicolor, 263
Wallaby Creek, 143–144, 168–169
wallaroo, 3, 5, 9 fig. 1.1, 11, 113
water requirements, 147–148
wedge-tailed eagle, 45 fig. 2.15, 233–234

western grey kangaroo, ix, x, 3, 4, 5, 9
   fig. 1.1, 11, 13 fig. 1.3, 42
   habitat and distribution, 12
   habitat preference 223–224
   See also Macropus fuliginosus
whiptailed wallaby, 55
white-tailed deer, 9 fig. 1.1, 10, 11,
   266–267, 270
   habitat and distribution, 12
   See also Odocoileus virginianus
wilga-belah, 38 fig. 2.11, 40 fig. 2.12,
   105, 222–225
Willandra Billabong, 19
Willandra National Park, ix, x
wind, 52, 136, 140–143, 173
wolf, 11
woodland, 93, 95, 223, 231–232

Xanthium spinosum, 41

Yathong Homestead, 19, 21, 23, 27–29,
   116

Yathong Nature Reserve, ix, x–xi, 8,
   17, 18 fig. 2.1, 19, 22 fig. 2.2
   climate and rainfall, x–xi, 24–26, 33
   drought, 31, 33
   fire, 41–42
   geology, 24
   history, 19–24
   mammals of, 42–46
   predators, 44–46
temperature, 26–27, 29
   topography, xi
   vegetation, 35, 36 fig. 2.10, 37–41
   vegetation monitoring
      photo points, 75–76
      step point surveys, 76, 77
   weather, 136–137

young-at-foot, 56–58, 84, 156, 186–
   189
   and predation, 223–224

GPSR Authorized Representative: Easy Access System Europe, Mustamäe tee 50, 10621 Tallinn, Estonia, gpsr.requests@easproject.com